Knowledge and Space

Klaus Tschira Symposia

Volume 7

Series editor
Peter Meusburger, Department of Geography, Heidelberg University, Germany

Knowledge and Space

This book series entitled "Knowledge and Space" is dedicated to topics dealing with the production, dissemination, spatial distribution, and application of knowledge. Recent work on the spatial dimension of knowledge, education, and science; learning organizations; and creative milieus has underlined the importance of spatial disparities and local contexts in the creation, legitimation, diffusion, and application of new knowledge. These studies have shown that spatial disparities in knowledge and creativity are not short-term transitional events but rather a fundamental structural element of society and the economy.

The volumes in the series on Knowledge and Space cover a broad range of topics relevant to all disciplines in the humanities and social sciences focusing on knowledge, intellectual capital, and human capital: clashes of knowledge; milieus of creativity; geographies of science; cultural memories; knowledge and the economy; learning organizations; knowledge and power; ethnic and cultural dimensions of knowledge; knowledge and action; and the spatial mobility of knowledge. These topics are analyzed and discussed by scholars from a range of disciplines, schools of thought, and academic cultures.

Knowledge and Space is the outcome of an agreement concluded by the Klaus Tschira Foundation and Springer in 2006

More information about this series at http://www.springer.com/series/7568

Peter Meusburger • Derek Gregory
Laura Suarsana
Editors

Geographies of Knowledge and Power

 Springer

Editors
Peter Meusburger
Department of Geography
Heidelberg University
Heidelberg, Germany

Derek Gregory
Department of Geography
The University of British Columbia
Vancouver, BC, Canada

Laura Suarsana
Department of Geography
Heidelberg University
Heidelberg, Germany

Peter Wall Institute for Advanced Studies
    University Centre
University of British Columbia
Vancouver, BC, Canada

ISSN 1877-9220
Knowledge and Space
ISBN 978-94-017-9959-1   ISBN 978-94-017-9960-7   (eBook)
DOI 10.1007/978-94-017-9960-7

Library of Congress Control Number: 2015943675

Springer Dordrecht Heidelberg New York London

Printed on acid-free paper

Springer Science+Business Media B.V. Dordrecht is part of Springer Science+Business Media (www.springer.com)

Acknowledgments

The editors thank the Klaus Tschira Stiftung for funding the symposia and book series on Knowledge and Space. The staff of the Klaus Tschira Stiftung and Studio Villa Bosch always contribute a great deal to the success of the symposia.

Together with all the authors in this volume, we are especially grateful to David Antal, James Bell, Patricia Callow, and Ginger A. Diekmann for their tireless dedication to quality as technical editors of the chapters and as translators for some of them. Volker Schniepp at the Department of Geography at Heidelberg University has been an enormous support in getting the figures and maps to meet the high standards of publication. We also thank the students of Heidelberg University's Department of Geography who helped organize the 7th symposium and prepare this publication, especially Amadeus Barth, Julia Brasche, Helen Dorn, Claudia Kämper, Laura Krauß, Melanie Kudermann, Inga Labuhn, Julia Lekander, Martina Ries, and Tina Thiele.

Contents

Contributors

John Agnew Department of Geography, University of California, Los Angeles, CA, USA

Trevor J. Barnes Department of Geography, University of British Columbia, Vancouver, BC, Canada

Dale F. Eickelman Department of Anthropology, Dartmouth College, Hanover, NH, USA

Derek Gregory Department of Geography, The University of British Columbia, Vancouver, BC, Canada

Peter Wall Institute for Advanced Studies University Centre, University of British Columbia, Vancouver, BC, Canada

Róbert Győri Department of Social and Economic Geography, Eötvös Loránd University, Budapest, Hungary

Ferenc Gyuris Department of Regional Science, Eötvös Loránd University, Budapest, Hungary

Robert Jewett Department of Scientific Theology, University of Heidelberg, Heidelberg, Germany

Sarah de Leeuw Northern Medical Program, University of Northern British Columbia, Faculty of Medicine, UBC, Prince George, BC, Canada

Stefan M. Maul Department of Languages and Cultures of the Near East – Assyriology, Heidelberg University, Heidelberg, Germany

Peter Meusburger Department of Geography, Heidelberg University, Heidelberg, Germany

Anssi Paasi Department of Geography, University of Oulu, Linnanmaa, Finland

Richard Peet Graduate School of Geography, Clark University, Worcester, MA, USA

Jo Reichertz Institute for Communication Science, University of Duisburg-Essen, Essen, Germany

Nico Stehr Cultural Studies, Institute for Political and Social Sciences, Zeppelin University, Friedrichshafen, Germany

Laura Suarsana Department of Geography, Heidelberg University, Heidelberg, Germany

Jürgen Wilke Department of Communication, Mainz University, Mainz, Germany

Graeme Wynn Department of Geography, University of British Columbia, Vancouver, BC, Canada

Power, Knowledge, and Space: A Geographical Introduction

1

Derek Gregory, Peter Meusburger, and Laura Suarsana

> *Knowledge and power are integrated with one another, and there is no point in dreaming of a time when knowledge will cease to depend on power*
>
> Foucault (1980, p. 52).

Long before they became interested in the ideas of Gramsci and Foucault, geographers had developed interest in the rule and coordination of social systems in space, in power relations within and between territories, in the social construction and symbolic meaning of centers and peripheries and in the hierarchy of settlement systems. All of these were entangled with questions of power and knowledge. It was Foucault (1980, pp. 63–77) who emphasized that the term *region* has the same etymological root as the Latin verb *regere*, to rule, from which we derive the English regulate, regime, and regiment, and he did so in an interview with geographers that followed directly from his own investigations in the relations between power and knowledge. Originally, then, region meant a space that was ruled, organized, coordinated, controlled, and influenced by a constellation of power that presided over a social system (Berthoin Antal, Meusburger, & Suarsana, 2014, p. 2). Today that sense is often conveyed by the term territory, and we now have a genealogy of

D. Gregory (✉)
Department of Geography, The University of British Columbia,
1984 West Mall, Vancouver, BC V6T 1Z2, Canada

Peter Wall Institute for Advanced Studies University Centre,
University of British Columbia, 6331 Crescent Road, Vancouver, BC V6T 1Z2, Canada
e-mail: derek.gregory@geog.ubc.ca

P. Meusburger • L. Suarsana
Department of Geography, Heidelberg University, Berliner Straße 48,
69120 Heidelberg, Germany
e-mail: peter.meusburger@geog.uni-heidelberg.de; laura.suarsana@geog.uni-heidelberg.de

© Springer Netherlands 2015
P. Meusburger et al. (eds.), *Geographies of Knowledge and Power*,
Knowledge and Space 7, DOI 10.1007/978-94-017-9960-7_1

its European origins (Elden, 2013). Raffestin (1980, pp. 44–56, 1984, 1986a, 1986b) developed the related concept of territoriality[1] as an intrinsically and constitutively geographical theory of power, "moving beyond the traditional politico-geographical focus on state actors and state territory" (Klauser, 2014, p. 27). "Space becomes territory within any social relation of communication" (Raffestin, 1980, p. 133; translation by Klauser, 2012, p. 111; see also Raffestin, 1977). He focuses on the "codes, ideas and semiotic systems that mediate the territorialisation, de- and re-territorialisation of space" (Klauser, 2014, p. 26). As communication is an essential basis for the functioning of organizations and the glue that holds the spatially distributed parts of a goal-oriented social system together, each new invention of storing and transmitting information has changed the relationships between power, knowledge, and space. The invention of scripts, of papyrus and paper, of printing machine, telegraph, telephone, wireless radio transmission, computer, satellite communication, and the most recent surveillance technologies have influenced the ways in which those in power can generate, store, evaluate and transmit information; the distance over which rulers or headquarters of organizations can give orders and execute control; and the easiness by which powerful people can manipulate public opinion and strengthen the social coherence, convictions, and identities of their supporters. These inventions have influenced the form of rule (e.g., ambulant rule in illiterate societies, bureaucratic rule in residencies after society has reached a certain level of literacy), the speed of information transmission, the spatial division of labor, the scope of surveillance, and the optimal locations for exercising power. These innovations also transformed the space of resistance for those who would challenge established systems of power and authority.

A question of special interest is the spatial concentration of knowledge and power, or to put it differently: Why do higher echelons of power inevitably tend to spatial concentration? Why are the places where powerful decision-makers and highly skilled experts work so densely clustered in a few centers? Why do many leading representatives of knowledge themselves seek proximity to political or economic power? To avoid misunderstandings, we should be clear that the definition of the term *center* in this context is organizational, not topographic. Centers and peripheries are always socially constructed. They appear at various scales, ranging from a primary group of persons to a global perspective of world cities. The center of a social system is the place from which the rest of the social system is ruled, guided, and coordinated. The center is a point of reference and orientation, it provides perspectives and worldviews on how "the other" should be seen; it lends meaning and direction to the social system's actions and is the anchor that provides security and identity (Turner, 1979, p. 33). By elucidating the concept of Orientalism, Said (1978), Gregory (1998), and others disclosed some of the ways in which power, knowledge, and geography were brought together under the worldview of Eurocentrism. In the Eurocentric vision of the world, Europe "was constituted as the locus from which sovereign meaning was to be endowed" (Gregory, 1998, p. 20).

---

[1] "[Territoriality] reverses the usual geographical approach. Its starting point lies not anymore in the analysis of space but in social actors' instruments and codes which are leaving marks and indications in territory" (Raffestin, 1986a, p. 94; translation from French by Klauser, 2012, p. 114).

Important decision makers and highly skilled experts in various domains seem to profit from relational and spatial proximity to each other for functional and symbolic reasons. Power and knowledge depend on each other and incorporate each other; both have enabling and innovative effects. Knowledge consolidates power, and power attracts and sometimes legitimates knowledge. Throughout history, those who enjoyed privileged access to defining (sometimes divining) and distributing knowledge—priests, ideologues—have depended on support from those executing political power, but the converse is also true. After all, the interpretation of oracles and dreams, the critical exegesis of the Bible, and the interpretation of the works of Marx, Lenin, and Mao can lead to very different results in different places. Usually, the conveyors of orientation knowledge[2] are able to maintain their power, privileges, and monopolies on the interpretation of texts, events, and situations only as long as they enjoy the confidence and protection of their political masters.

Academics and professionals may seem more independent of (political) power, but they, too, are greatly attracted to its centers. Neither the scholars of the Middle Ages and the Renaissance nor the natural scientists of the nineteenth century nor the pundits at the think tanks of today shied away from allying themselves with power to some degree in order to promote their scientific ideas, ensure their livelihoods, or assume a position of influence (Kintzinger, 2003, p. 191). Proximity to one of the power centers can clearly facilitate academic activities, scientific processes, and projects that would have no chance were it not for the support from the powers that be.

Relational proximity to the center of a domain offers priority access to resources, privileges, and crucial information unavailable to those at the margin (see Ibarra & Andrews, 1993). Being at or near the center of a domain or being supported by powerful people also has psychological significance because it denotes importance, reputation, competence, and trustworthiness.[3] Centrality in a social system is the spatial manifestation of authority and prestige. Relational proximity to the power center of a domain increases the chances that experts and scholars will receive public attention and be able to influence key decision-makers.

Edward Said in his penultimate Reith Lecture[4] called upon public intellectuals—he meant critical intellectuals—to speak truth to power, but there is no doubt that many experts and intellectuals seek to prove their usefulness to power; they strive for recognition and influence; and quite often they deliberately serve the interests of political power (see Barnes's Chap. 9 in this volume). Centers attract experts like magnets. Important decision-makers, professionals, artists, and intellectuals striving for prestige, influence, or success sooner or later find their place in one of the centers of their professional domain or create a new center. Even those who have taken up Said's challenge often play their oppositional role to greatest effect in established centers of power.

---

[2] Some authors, e.g., Scheler (1926), call it salvation knowledge.

[3] This may be one of the reasons why fraudulent researchers appear to always find highly respected senior scientists to coauthor their articles (Stroebe, Postmes, & Spears, 2012, pp. 672, 678).

[4] Retrieved on 1 October 2014 from http://www.independent.co.uk/life-style/the-reith-lectures-speaking-truth-to-power-in-his-penultimate-reith-lecture-edward-said-considers-the-basic-question-for-the-intellectual-how-does-one-speak-the-truth-this-is-an-edited-text-of-last-nights-radio-4-broadcast-1486359.html

The reason for that is more than functional proximity; a center also has great symbolic and psychological meaning and can confer significance and even legitimation through its centrality. It is associated with social attributes such as power, authority, dominance, prestige, control, attractiveness, and influence. According to Eliade (1969) and Turner (1979, pp. 9–10), the center was identical in many religions to a sacred place where the gods or ancestors dwelt or revealed themselves. The temple was the point of communication between heaven and earth or between the priests and the supernatural beings or forebears. The Assyrians called their center Babylon, which means "gate of the gods." World cities today are no longer the gates to divine power, but we acknowledge what Simmel (1900/2004) once called the magical power of money to bridge distances, and they thus remain seats of the most powerful political and financial institutions. According to the "Matthew effect,"[5] advantage begets advantage: Power accumulates in centers, where different forms of power tend to attract each other (see Popitz, 1992, p. 36).[6]

These sorts of inquiries were reinvigorated by critical engagements with Foucault (1969, 1979, 1984, 1994, 2007a, 2007b) and Gramsci (1971) (for details see Elden & Crampton, 2007; Klauser, 2013, 2014; Peet, 2007; Peet's Chap. 13 in this book; Raffestin, 2007; Raffestin & Butler, 2012). They have confirmed that "space, knowledge and power are necessarily related" (Elden & Crampton, 2007, p. 9). For Foucault (1984), "space is fundamental in any exercise of power" (p. 252). It is "a vital part of the battle for control and surveillance of individuals, but it is a battle and not a question of domination" (Elden & Crampton, 2007, p. 2). Elden (2007), Elden and Crampton (2007), Fall (2007), Hannah (1997, 2000, 2007), Huxley (2007), Klauser (2013), Legg (2007), Philo (1992), Raffestin (2007), Thrift (2007), and many others have shown the importance of geography to Foucault's work—as he himself conceded towards the end of that interview with Hérodote—and how his work in turn provides all manner of insights into the tense triangle between power, knowledge and space (see Foucault, 1980).

In this context, *space* is usually understood as relative or relational space, which is to say that space is conceptualized as a product of interrelations and interactions. Relative space is never a closed system; it is always "in a process of becoming, always being made" (Massey, 1999, p. 28).[7] With his book *The Production of Space,* Henri Lefebvre (1991) initiated an understanding of relative space that is intrinsically connected with social action (for details see Gregory, 1994; Belina,

---

[5] The expression, first used by Merton (1968), stems from the Gospel of Matthew 25:29: "For unto every one that hath shall be given, and he shall have abundance: but from him that hath not shall be taken even that which he hath" (King James Version). In modern parlance, the rich get richer, the poor get poorer.

[6] The role of networking platforms, the intrigues of lobbyists of big money and influential media, and the buzz of power plays in today's world centers of political power are described in a witty, entertaining, and troubling way by the political correspondent Leibovich (2013).

[7] Theoretical concepts of place and space have already been addressed extensively in the volumes 1–6 of this series and hundreds of other publications. Nevertheless, as this book addresses various disciplines it seems necessary to repeat the gist of these discussions in this introduction in order to prevent misunderstandings among those readers who still adhere to traditional concepts of (absolute) space or overrate the role of distance.

2013; Klauser, 2014; Schmid, 2005). In his understanding, space is not only a social product but also a producer and regulator of social action (Lefebvre, 1991, p. 358). For him spatial order is both the cause and effect of social ordering (p. 143). Space is the "locus, object and tool of surveillance" (Klauser, 2014, p. 7) and represents both a structural principle of the social dimension and a matrix of social relations and processes. It serves as a medium of perception and social communication and can perform an important function in representing differences of power and status. In religious liturgy, political staging (e.g., court etiquette of princes and kings, state receptions held by governments, party conferences, and military parades), diplomatic negotiations, and many other settings, one can make the role and relative status of people visible through their positioning in space and through the very performance of space itself, a veritable choreography of bodies, practices and flows.

But there are other terms that are also vital in analyzing the spatiality of power and knowledge, and these two require a multidimensional understanding. Place, for example, denotes a location characterized by specific configurations, facilities, and resources that enable or impede certain actions; it signifies a position in a hierarchy or network and as such is always porous, open to the world and thus in constant motion; and it operates as a "discursively constructed setting" (Feld & Basso, 1996, p. 5) that has a symbolic and emotional meaning, provides an identity, and communicates a complex history of events, cultural memories, and attachments (Canter, 1977; Manzo, 2005; Rowles, 2008; Scannel & Gifford, 2010). Places can be studied from a wide variety of philosophical perspectives. They are "known, imagined, yearned for, held, remembered, voiced, lived, contested, and struggled over . . . and metaphorically tied to identities" (Feld & Basso, 1996, p. 11). People are rooted in and emotionally attached to places. As Gregory (1998), Jöns (2003), Latour (1987), Livingstone (1995, 2003), Meusburger and Schuch (2012), and others have shown, places can function as centers of calculation, truth spots, and sites of knowledge generation, information control, and power execution. We know this not least from our histories of geographical exploration in the nineteenth century:

> What seemed plausible in the lecture hall of the Royal Geographical Society in London, for example, might well become a half-truth on the ground. As imaginative geographies were circulated from field to library and back again, they passed through different sites of knowledge production that were connected in ramifying networks of exchange, and they inevitably interpenetrated, confounded and reworked one another. (Gregory, 1998, p. 21)

What would count *as* knowledge depended on the status of its collectors and producers; the supposedly 'modest witnesses' of nineteenth-century science were white, propertied men. Their exploits almost always relied on local informants but these indigenous knowledges were often disqualified from the status of objective truth and had to be brought home in contraband, covert form. What this means, in short, is that the very spaces and places of science, as Livingstone (1995, 2003) calls them, and the complex circuits through which its knowledges travel, enter intimately into the form, fate and fortune of knowledge. We put this in the present tense because it would be idle to pretend that this process of conditioning, appropriating, filtering and reformulating is confined to the past.

One of the objectives of this volume is to enrich the conversation on knowledge, power, and space by bringing the voices of various disciplines, different theoretical concepts, and different scales of analysis together. The relations between knowledge and power are not as simple as the much cited slogan "knowledge is power" might suggest. It would be naïve to assume that knowledge and power are *directly* related. Knowledge does not automatically lead to power, and power does not necessarily command, attract, or produce knowledge. It depends on the scale of analysis (individual agents, organizations, territories), the time-frame of a study, the context of action, the category of knowledge, and many other criteria whether such a relationship can be detected or not and how strong it is. Many powerful people have little knowledge—Pred (2007) once castigated what he called their "situated ignorance", and many politicians reject 'evidence-based policy' for what they see as the courage of their convictions—and many prominent scholars have little power. In most cases knowledge can only be transformed into power via the support of organizations and institutions. Not all categories of knowledge have an equally close relationship to power. Therefore, *Peter Meusburger* presents in the first chapter an overview about the complex interrelations between knowledge and power, and elaborates on the questions of how power and knowledge can be conceptualized and how the relations between them can be explained. The fact that many definitions of knowledge contain the term *capacity* suggests that knowledge or professional skills may be left unused (see Stehr, 1992, p. 114) or may go unapplied because of political pressure, restrictive local milieus, or lack of resources. Political power frequently ignores expertise that contradicts its self-portrayal and goals.[8] Therefore, the intensity, indeed the very existence, of a relationship between knowledge and power depends on various preconditions.

Discussing the semantic difference between factual knowledge and orientation knowledge, Meusburger analyzes the functions that these two categories of knowledge have in the acquisition and stabilization of power. He discusses also the methods by which the powers that be have tried again and again to affect or control the creation and dissemination of knowledge and information, and to manipulate public attention. This question is a central issue of the book and therefore is taken up from various points of view in the chapters of Agnew, Eickelman, Gyuris and Győri, Jewett, de Leeuw, Paasi, Peet, Stehr, and Wilke.

Another topic addressed in various chapters of this book is the credibility of experts. How can we know whether somebody is an expert or not. How is expertise defined and by whom? How have knowledge claims and the definition of expertise changed in the course of history? How do we know when we are being informed and how do we know when we are being manipulated (Wilkin, 1997, p. 12)? Expertise today is much more contested than it was in earlier periods, and our knowledge society faces an undermining of the epistemological monopoly held by gatekeepers of the scientific disciplines. *Nico Stehr* worries in his

---

[8]The Russian poet Alexander Sergejewitsch Pushkin said: "The illusion which exalts us is dearer to us than ten thousand truths". (retrieved on 1 July 2014 from http://www.quoteid.com/Aleksandr_ Pushkin.html)

contribution that the gap between the expert skills of powerful agents and the knowledge of laypersons has dramatically and irreversibly widened recently. He notes that large segments of modern society do not know enough to participate intelligently in policy discourse and democratic decision-making in the emerging expert society so that rapid advances in scientific knowledge might evolve into a burden on democracy. His answer to this problem is that the social sciences and the humanities have generated two models for dealing with scientific knowledge claims. The *model of instrumentality* argues that science speaks to society and does so not only with considerable authority but also with significant success, whereas society has little if any opportunity to talk back. The alternative approach is the *capacity model.* According to this model the social sciences and the humanities operate as meaning producers or "mind makers". It stresses that the agents who employ social science knowledge are *active* agents who transform, re-issue, and otherwise redesign social science knowledge. Stehr argues that social science knowledge is an intellectual resource that is open and complex and thus can be molded in the course of "travel" from the social science community to society. This model therefore accepts that people may critically engage social science knowledge using local knowledge resources and thus make social science accountable to the public.

Important decision-makers constantly need situation analyses, forecasts of risks and chances, projections, and prognoses to help them come to grips with the uncertainties of the future and to anticipate future developments. This is a main reason why powerful institutions need the support of science and reconnaissance, an issue that is addressed in the chapters of Gregory, Maul, and Meusburger. *Derek Gregory* takes up the interesting question of how and why we perceive different kinds of reality and how different perceptions and experiences may influence analysis of situations, decision making, and the execution of power. He focuses on the contrast between what Clausewitz (1780–1831) called more generally *paper war* and *real war.* According to Gregory this contrast bedevils all conflicts, but the lack of what today would be called geospatial intelligence proved to be catastrophic in World War I. Gregory compares the British experience in East Africa and on the Western Front in 1914–1918. In contrast to the East African campaign where a soldier named Gabriel despaired of inaccurate and useless maps, war in Belgium and France was fought with increasingly sophisticated, highly detailed geospatial intelligence. Gregory describes a combination of mapping and sketching, aerial reconnaissance and sound-ranging that transformed the battlefield into a highly regulated, quasi-mathematical space: the abstract space of a military Reason whose material instruments were aircraft, artillery, and machine-guns. The author contrasts this sophisticated cartography and its intrinsically optical-visual logic with the muddy, mutilated, and shell-torn *slimescapes* in which the infantry were immersed month after month. He calls the radically different knowledges that the war-weary soldiers improvised as a matter of sheer survival a *corpography*: a way of apprehending the battle space through the body as an acutely physical field in which the senses of sound, smell, and touch were increasingly privileged in the construction of a profoundly haptic or somatic geography.

Stefan Maul describes how in ancient Babylonia and Assyria divinatory procedures attained the status of a "science of telling the future." The future prospects of a plan were regularly determined from the color and shape of the liver of a sheep that had been slaughtered for this very purpose. As certain features on the surface of the liver were interpreted as favorable or unfavorable signs, it was assumed that the appearance of a sheep's liver correlated with future events. The signs of the liver were also correlated with equivalent astral signs. The goal was to identify the laws that governed the world, in order to utilize them in the realm of politics. Such a procedure to tell the future stands in opposition to modern science and may be ridiculed from today's point of view. But Maul argues that it would be unwise to dismiss ancient Near Eastern divination as mere superstition or aberration. He explains why this procedure had a high reputation and was such an effective form of political decision-making. By giving shape to the future, creating space for negotiation, and helping to build consensus, it was a decisive means of reaching political goals. The Mesopotamians saw the mastery of such divinatory procedures as a decisive reason for the lasting cultural and geopolitical success of Babylonia and Assyria.

The question, who has the power to interpret the past, evaluate truth claims, and construct collective identities is taken up in the chapters of Eickelman, Gyuris and Győri, and Meusburger. *Dale F. Eickelman* notes in his chapter that there is no singular public Islam, but rather a multiplicity of identities and practices based on competing readings of the past. In the Muslim-majority world, the role of religion in society and community life never receded, though it did change and develop in ways often underemphasized by Western observers. The authority of conventional religious scholars remains strong in the modern Islamic world. Traditional Islamic scholars regard themselves alone as capable of interpreting the Qur'an through their expertise in the science of scriptural hermeneutics. However, modern Muslims increasingly are disinclined to allow conventionally trained religious scholars the final word in interpreting such vital questions as "What is Islam?" "How is it important to my life?" and "How do I interpret the past?" Some fundamentalists attempt to imitate the life of the Prophet Muhammad literally. Other Muslims emphasize the necessity of interpreting the Qur'an as if it were revealed in the present and in interpreting the life and sayings of the Prophet metaphorically and not literally, engaging critical reason. The latter seek open discussion of issues related to the "common good" (*al-maslaha al-'amma*), an essentially contested concept that is at the core of public life in Muslim-majority countries. Mass higher education in Islamic countries and the increasing availability of new media had the consequence that religious authority has become more fragmented, which encourages increasingly open debates over topics such as innovation and the common good and blurs the formerly sharp distinctions between the religious and the secular. The problem of defining what knowledge is valued and how it relates to faith and authority is increasingly a subject of intense debate in Muslim societies. Like concepts of good governance, duty, and social justice, innovation in Islamic thought and practice is impossible to define once and for all. People can justify why they hold one interpretation over others, and authorities can attempt to block public debate, but the "proper" meaning of an essentially contested concept cannot by definition be settled once and for all.

Robert Jewett presents another prominent example of the linkages between religion, ideology, and political power. He describes how millennialism influenced early American colonists in New England who had a distinctive sense of mission to redeem the entire world. The colonists saw New England as the Protestant realm that fulfilled the ideal of the heavenly 1,000-year kingdom descending to earth after the battle of Armageddon (book of Revelation 20). The colonies outside New England were mostly non-millennial until the middle of the eighteenth century, but by the eve of the American Revolution, the ideology of millennialism was clearly developed and around 1840 it was popularized by priests, politicians, and media and closely linked with generating a national American identity. The concept of Manifest Destiny (one nation under God) rested on the premise that the United States was the "holy nation" referred to in Exodus 19:6 and that it had a divine mission. The idea of being a chosen people and a nation with the destiny to fulfill God's will on earth and to advance freedom, democracy, and peace around the world remains a characteristic feature of American civil religion and determined expansionist American politics from the outset. Through American history down to the present, victory against God's alleged enemies ("savage nations" will submit to the rule of the saints) has assumed a high priority, whoever those enemies happen to be. The idea of battling against the forces of the evil, and the division of the world into good and bad people became a crucial component of American politics, visible in the French and Indian War, the American Revolution, World Wars I and II, the Cold War of 1945–90, the Vietnam War, the Gulf Wars, and the current war on terrorism.

The power of words is also discussed by *Graeme Wynn*, who analyses the impacts of two books, namely George Perkins Marsh's *Man and Nature*, published in 1864, and *Silent Spring* by Rachel Carsons, published in 1962. Long celebrated as a key work in the study of human-environment relations, as offering a radical new interpretation of society's capacity to alter nature, and as the fountainhead of the conservation movement, *Man and Nature* holds a revered place in the pantheon of geographical and environmental writing. It was certainly influential and expanded the understanding on human-environment relations. But according to Wynn few have asked the questions: Were its arguments unprecedented? Where did they come from? Where did they go? How did they work? Were they framed in particularly novel and/ or compelling ways? What facilitated their dissemination? How did they gain purchase? When and where were they challenged (if at all)? When and where were they most influential? Who supported Marsh's ideas? Who advanced his reputation? Many of these questions speak directly to the relations between knowledge and power. After exploring these questions the chapter turns to a brief comparison of the production and reception of *Man and Nature* with the writing of and reaction to that other great environmental classic, *Silent Spring*, published almost exactly a century after Marsh's book, to ponder the effects of context and contingency in shaping an author's capacity to speak truth to power, and to offer an assessment of the ways in which the words of Marsh and Carson worked in the world.

Meusburger argues that the current crisis of expertise may partly be due to the fact that history of science knows a number of examples where scholars, scientists, and intellectuals have allowed themselves to be exploited by political power, have curried favor with totalitarian systems, or have supported those in power by manipulating facts and faking documents. An extreme form of complicity of scholars with political power is presented by *Trevor Barnes*. He shows in great detail how academics may be tempted to serve the interests of a totalitarian system if they see a chance to carry into effect their theories and expertise in the real world and thus gain reputation. He discusses the role of the geographer Walter Christaller whose task in National Socialist Germany was to plan the newly annexed territory of western Poland in conformance with his central place theory. Although Christaller was not a "desk killer" in the same sense as Adolf Eichmann, he participated at least as a bureaucrat in the "Generalplan Ost" and clearly played a crucial role in a project that aimed at transforming Western Poland into German land and resulted in the forced expulsion and death of many people. Barnes is especially interested in how Christaller, who was fearful of the National Socialists before the war began, and who became a communist after the war came to an end, could be a National Socialist during the war, and why he voluntarily and enthusiastically put his knowledge at the disposal of the National Socialist regime. Christaller certainly did not want to become part of the National Socialist war machine, but he could not resist the temptation of power-knowledge. He sought academic credibility and political relevance; he wanted to show the importance of his theory and was convinced that his ideas were capable of remaking the world.

A similar ingratiation to political power can be stated with some writers and intellectuals. One example of the long list of intellectuals and writers who were prone to embracing totalitarian systems is Lion Feuchtwanger (1884–1958)[9] who was an uncritical propagandist of Stalin's reign of terror. He praised the constitution of Stalin's Soviet Union as the world's most democratic constitution and defended Stalin's political show trials and mass murders (for details see Sternburg, 2014). One of the protagonists in his 1939 novel *Exil* [Exile], states: "Writing makes sense only . . . if one is allied with power" (Feuchtwanger, 1983, p. 105) and "A truth without power is no truth" (p. 99).

Ferenc Gyuris and *Róbert Győri* illuminate the other side of the coin, in other words how a totalitarian system can suppress, change, and control a scientific discipline to fit the regime's ideology. After the violent establishment of the Communist system in Hungary, human geography was stigmatized as *reactionary*, *bourgeois*, and without *practical utilization*. It was demolished and replaced by a new, Marxist-Leninist economic geography based on Soviet principles. State research institutes, the Hungarian Academy of Sciences, and the Central Planning Office became the most important "centers of calculation" where Soviet-type "big science" was established, whereas universities lost much of their former importance. Many geographers were pensioned off or exiled from academia. Others were driven to the periphery or forced to compromise with the system. Important positions in academic geography were given to politically loyal *newcomers*, some of them

---

[9] Feuchtwanger was one of the most published German authors during the years of the Weimar Republic (1919–1933). As an opponent of National Socialism, he had to leave Germany.

having not even studied geography. Geography as a discipline was expected to contribute to the *construction of socialism* and to participate in the propagation of Communist goals. Thus, Sovietization thoroughly reshaped Hungarian geography and changed its social, political, and economic roles as a field of science.

Other case studies of how academic disciplines may be shaped by hegemonic institutions are presented by Agnew and Paasi. *John Agnew* is interested in how the local or national does become the global. He discusses various ways in which the geography of knowledge can be related to world politics or what he is calling the geopolitics of knowledge. He focuses on the specific case of how *international relations*—a theory largely consisting of U.S.-originated academic ideas about the nature of statehood and the world economy—came to dominate much academic thinking about world politics outside the United States. He argues that the dominant ways in which intellectuals and political elites around the world have come to think about world politics are not the result of either an open "search" for the best perspective or theory or a reflection of an essentially "local" perspective. It is rather the influence of American schools and authors that in turn reflect the current geopolitical order. He describes the founding of the academic discipline *international relations* in the early postwar United States, its travels around the world as a function of American hegemony, and the story of two alternatives: the English School, to illustrate the limits of pluralism, and the rise of an international relations theory with "Chinese characteristics," to show how an alternative with hegemonic potential can begin to emerge.

A very important and seldom discussed type of hegemonic pressure on academic practice, generation of scientific knowledge, and style of publication is discussed by *Anssi Paasi.* In his chapter he challenges so-called Anglophonic hegemony in the publication industry and criticizes the neglect of prominent research in other world languages. Having a look at the list of references in Anglophone journals, it can be argued that human geography—especially as far as the exchange and directions of flows of geographical ideas across national borders are concerned—was more international prior to the 1960s, with researchers willing to use foreign languages in their research work to closely examine what was taking place in academic geography in other linguistic contexts.

The often naïve and one-sidedly mechanical measurement of research output and citations, the neglect of different research cultures, the neoliberalization of university life, and the fierce struggle over students and research money have created an evaluation industry that has dramatically influenced how relevant knowledge is defined, produced, and controlled, and how it has also changed the forms of publishing. Paasi scrutinizes both the roots and increasing importance of the English language in the international academic market and notes how this has become a particular challenge for social scientists operating outside the English-speaking world. The authorities responsible for academic governance and current neoliberal competition in many countries have raised claims that research should be published in English and in the "best" journals, which more often than not means journals that have been classified by one North American company, Thomson Reuters, which has a monopoly in compiling the Web of Science (WoS) citation data and in the choice of the journals that are represented in these data. What is understood as international science may often just be standard national science as practiced in the core region, but it nonetheless tends to shape the criteria for excellence in the peripheries as well. Very problematic is the fact that scholars in humanities and social sciences are compelled by some journal editors and reviewers to adapt to agendas reflecting certain Anglophone research themes and theoretical orientations as well as to restrict their lists of references to English even if the research frontier is represented by other languages, and research traditions. Nobody will question the role of English as *lingua franca* in the natural sciences. But humanities and social sciences have different research cultures in which languages and books play a more important role than in the hard sciences. This situation may entail two provoking questions: Have the groundbreaking publications of Baudrillard, Bourdieu, Deleuze, Foucault, Gadamer, Habermas, Heidegger, Jaspers, Lefebvre, Luhmann, Marx, Virilio, or Max Weber been parochial as long as they were not translated into English or irrelevant as far as they were published as books and not in international journals? Or should it rather be interpreted as an intellectual deficit of monolingual authors and editors if they were not able to read the works of these outstanding scholars in the original language? According to Paasi, internationalization in academia is not leading to a borderless world in which academic ideas flow and interact, but to the rise of new, uneven spatial patterns and new hegemonic centers.

Paasi's chapter triggers some further (provocative) research questions not dealt with in this book, for example, about the impact of power, competition, and reward systems on the quality of scientific research and the frequency of frauds in science. To what extent do neoliberal pressure and competition related to impact factors, research money, or other questionable indicators influence the scientific quality, creativity, and originality of research? To what extent can the steep increase of academic fraud since the 1990s (see Fang & Casadevall, 2011; Steen, 2011; Stroebe, Postmes, & Spears 2012)[10] be related to the current evaluation and reviewing "industry" that overestimates unreliable impact factors?[11] Is there empirical evidence that the standardization and homogenization of scientific practice and the neoliberalization of universities have lowered the level of scientific creativity and originality of young scholars? How can it be explained that some of the most astonishing frauds in the history of science have been published in some of the most prestigious international journals with the highest impact factors (e.g., in the biosciences: *New England Journal of Medicine, Science, Cell, Nature, Journal of Experimental Medicine, Lancet*), and that there is "a surprisingly robust correlation between the journal retraction index and its impact factor" (Fang & Casadevall, 2011, p. 3856).

---

[10] Using the PubMed database, Steen (2011) studied 742 papers (distributed among 404 journals) for which retraction notices could be obtained. The number of papers retracted for fraud increased more than sevenfold in the 6 years from 2004 to 2009 (p. 251). Similar results were presented by Bhutta and Crane (2014).

[11] In scientometry and many disciplines, it is widely acknowledged that impact factor is a flawed measure of scientific quality and importance (Bloch & Walter, 2001; Casadevall & Fang, 2014; Fang & Casadevall, 2011; Fersht, 2009; Hansson, 1995; Seglen, 1997a, 1997b; Smith, 2008).

Why are journal impact factor and individual article citation rate so poorly correlated (for details see Casadevall & Fang, 2014)?

Richard Peet studies relations between knowledge and power on a global macroscale. He argues that the financial crisis in the period from 2007 to 2010 forces a reconsideration of institutional and discourse analysis of power complexes in the tradition of Antonio Gramsci and Michel Foucault and a movement "back" toward structural analysis in the tradition of Karl Marx. He uses the term *institution* in the Foucauldian sense of a *community of experts*, an elite group of highly connected individuals controlling an area of knowledge and expertise. This community of experts shares the same ideas and ideals, takes the same things for granted, and is self-policing. In his view the essence of power is the creation of hegemony in places that control belief systems and interpretative frameworks in which social life occurs. He argues that the power centers of the modern capitalist world do not produce commodities in the form of physical objects; they mainly create the ideas that order and control the production of objects. In the Marxist tradition, then, power takes the form of persuasive ideologies, circulating through dominant clusters of highly interconnected institutions. He distinguishes between economic, ideological, and political power centers and argues that in finance capitalism, economic centers of power predominate over the others, using the political centers to marshal collective power on their behalf, ideological centers to manufacture the myths that legitimate their actions, and media to manipulate people.

Jürgen Wilke devotes his chapter to media control and censorship, and evaluates the state of press freedom and media control on a global scale. He shows that measures to control the spread of publications are as old as the invention of Gutenberg's printing machine. Censorship and confiscation of already printed books and pamphlets as well as regulations regarding the settlement of printers (in Europe originally they were only allowed in certain cities) and other measures prevented the production of printed work undesired by those in power. However, in states where the enforcement of censorship depended on local authorities, media control did not have the same impact as in states where media control was centralized. Wilke explains why England became the "motherland" of press freedom at the end of seventeenth century, and why in continental Europe there was a succession of occasional signs of progress and new setbacks, depending on the respective understanding and exercise of power.

At the end of the nineteenth century, pre-censorship had generally been abolished in the European countries. In the twentieth century however, media control achieved an unprecedented level of totality. In order to re-educate their population according to their ideology, the totalitarian systems in the Soviet Union, Germany, and Italy practiced a monopolization of information to indoctrinate their population. The control of broadcasting and film censorship were easier than censorship of printed material, because the new media could be controlled centrally by effective means. After 1945, the Allied powers exerted strict control over the media in their occupation zones. Whereas the Western Allies gradually established a pluralist press system in their occupation zones, the Soviet Union was able to extend its political governance to Eastern and Central Europe, where a multi-faceted control and propaganda apparatus was effective until the end of the 1980s. Surveys conducted by several non-governmental organizations (e.g., *Reporters Without Borders*) show that at the beginning of the twenty-first century there are still great interstate disparities of press freedom and media control. The fact that freedom of opinion and press freedom is declared by law does not speak for the total absence of media control. In many cases, this guarantee only exists on paper, but is restricted in real life (see also Chap. 2 of Meusburger).

Inversely to the geography of power, there exists also a geography of resistance (Keith & Pile, 2013; Sharp, 2000; Staeheli, 1994). Following Foucault (1982), who understood power as a "mode of action upon the actions of others" (p. 790), those in power will always meet some degree of resistance. If some people were not able to resist enterprises of intellectual domination, there would be no progress or freedom in human society (Wilkin, 1997, p. 4). *Sarah De Leeuw* aims in her chapter to contribute to broader discussions in geography about power, knowledge, and the colonial education of indigenous peoples, and theories of resistance. She examines historical and contemporary education systems designed with indigenous peoples in mind and underlines the need to reinvigorate social justice considerations within research of human geography. Taking as her starting point intimate colonial geographies lived by First Nations peoples in northern British Columbia, Canada, De Leeuw argues that theories of resistance do not allow for adequate theorizing of the ways in which Indigenous subjects navigate powerful forces, especially educational ones, that are intent on assimilating and de-indigenizing them. Schools, classrooms, and the curricula taught within them are conceptualized in this contribution as tense political sites where conflicting modes of knowledge clash and where, ultimately, indigenous children grapple with (as opposed to simply resist) expressions of (neo)colonial power.

In the last chapter of the book, *Jo Reichertz* explores the sources of power in communication and the impact of communication processes on power. Without any doubt, communication can move people to do things that they otherwise would rather not do. The article investigates why communication has that power and what the sources of that power are—namely beyond violence and authority. In alignment with the sociological, communication science, and speech-act literature, the thesis is developed that the power of communication emerges when the communication partners have developed a relationship of respect. In such a relationship communication possesses the power to strengthen identity or to damage it. The power of communication therefore rests on a relationship of respect and the identity-making ability of communication.

Recent conflicts in eastern Ukraine and the Middle East recall the old saying that truth is always the first victim of war; and they promote again the old research question which institutions or centers of power distort, control, and manipulate information in which way and why some areas or ethnic groups show more "intellectual self-defence" (Chomsky, 1987, pp. 610–631) or are better able to resist hegemonic ideologies and propaganda than others. Political and cultural elites employ various methods in trying to influence the access to information, construct identities and cultural memories, and affect the use and spatial range of ideological vocabulary (e.g., God's own country, axis of evil), which produces thematic places (areas) in the sense of Lossau and Flitner (2005) and Mattissek (2007).

The goal of the Klaus Tschira Symposia and the series "Knowledge and Space" is to create new spaces where theoretical concepts, methods, and issues of various disciplines dealing with knowledge and space can intensively be disputed. We editors hope that the co-presence of different and even contradictory approaches and provocative questions in one book will encourage readers to cross disciplinary borders and perhaps to challenge some beloved research traditions and paradigms. We are very grateful to the Klaus Tschira Foundation for providing the Studio of the Villa Bosch and the "venture capital" for this enterprise.

References

Belina, B. (2013). *Raum* [Space]. Münster, Germany: Westfälisches Dampfboot.

Berthoin Antal, A., Meusburger, P., & Suarsana, L. (2014). The importance of knowledge environments and spatial relations for organizational learning: An introduction. In A. Berthoin Antal, P. Meusburger, & L. Suarsana (Eds.), *Learning organizations. Extending the field* (Knowledge and space, Vol. 6, pp. 1–16). Dordrecht, The Netherlands: Springer. doi:10.1007/978-94-007-7220-5_1.

Bhutta, Z. A., & Crane, J. (2014). Should research fraud be a crime? *BMJ: British Medical Journal, 349*, g4532. doi:10.1136/bmj.g4532.

Bloch, S., & Walter, G. (2001). The impact factor: Time for change. *The Australian and New Zealand Journal of Psychiatry, 35*, 563–568. doi:10.1046/j.1440-1614.2001.00918.x.

Canter, D. (1977). *The psychology of places*. London: Architectural Press.

Casadevall, A., & Fang, F. C. (2014). Causes for the persistence of impact factor mania. *mBio, 5*(2), e00064–14. doi:10.1128/mBio.00064-14.

Chomsky, N. (1987). *Turning the tide*. Montreal, Canada: Black Rose Books.

Elden, S. (2007). Governmentality, calculation, territory. *Environment and Planning D: Society and Space, 25*, 562–580. doi:10.1068/d428t.

Elden, S. (2013). *The birth of territory*. Chicago: The University of Chicago Press.

Elden, S., & Crampton, J. W. (2007). Introduction—Space, knowledge and power: Foucault and geography. In J. W. Crampton & S. Elden (Eds.), *Space, knowledge and power: Foucault and geography* (pp. 1–18). Aldershot, UK: Ashgate.

Eliade, M. (1969). *Images and symbols: Studies in religious symbolism* (P. Mairet, Trans.). New York: Sheed & Ward.

Fall, J. J. (2007). Catalysts and converts: Sparking interest for Foucault among Francophone geographers. In J. W. Crampton & S. Elden (Eds.), *Space, knowledge and power: Foucault and geography* (pp. 107–128). Aldershot, UK: Ashgate.

Fang, F. C., & Casadevall, A. (2011). Retracted science and the retraction index. *Infection and Immunity, 79*, 3855–3859. doi:10.1128/IAI.05661-11.

Feld, S., & Basso, K. H. (1996). Introduction. In S. Feld & K. H. Basso (Eds.), *Senses of place* (pp. 3–11). Santa Fe, NM: School of American Research Press.

Fersht, A. (2009). The most influential journals: Impact factor and Eigenfactor. *Proceedings of the National Academy of Science of the United States of America, 106*, 6883–6884. doi:10.1073/pnas.0903307106.

Feuchtwanger, L. (1983). *Exil* [Exile]. Frankfurt am Main, Germany: S. Fischer. (Original work published 1939)

Foucault, M. (1969). *L'archéologie du savoir* [The archaeology of knowledge]. Paris: Gallimard.

Foucault, M. (1979). *Discipline and punish: The birth of the prison*. New York: Vintage Books.

Foucault, M. (1980). *Power/knowledge: Selected interviews & other writings, 1972–1977* (C. Gordon, Ed.). New York: Pantheon.

Foucault, M. (1982). The subject and power. *Critical Inquiry, 8*, 777–795.

Foucault, M. (1984). Space, knowledge and power. In P. Rabinow (Ed.), *The Foucault reader* (pp. 239–256). Harmondsworth, UK: Penguin.

Foucault, M. (1994). *The order of things: An archaeology of the human sciences*. New York: Vintage Books.

Foucault, M. (2007a). The language of space. In J. W. Crampton & S. Elden (Eds.), *Space, knowledge and power: Foucault and geography* (pp. 163–167). Aldershot, UK: Ashgate.

Foucault, M. (2007b). Questions on geography. In J. W. Crampton & S. Elden (Eds.), *Space, knowledge and power: Foucault and geography* (pp. 173–182). Aldershot, UK: Ashgate.

Gramsci, A. (1971). *Selection from the prison notebooks*. London: Lawrence and Wishart.

Gregory, D. (1994). *Geographical imaginations*. Oxford, UK: Basil Blackwell.

Gregory, D. (1998). Power, knowledge and geography. In *Explorations in critical human geography* (Hettner-lecture, Vol. 1, pp. 9–40). Heidelberg, Germany: Department of Geography, Heidelberg University.

Hannah, M. (1997). Space and the structuring of disciplinary power: An interpretive review. *Geografiska Annaler: Series B, Human Geography, 79*, 171–180. http://www.jstor.org/stable/490655

Hannah, M. (2000). *Governmentality and the mastery of territory in nineteenth-century America*. Cambridge, UK: Cambridge University Press.

Hannah, M. (2007). Formations of "Foucault" in Anglo-American geography: An archaeological sketch. In J. W. Crampton & S. Elden (Eds.), *Space, knowledge and power: Foucault and geography* (pp. 83–105). Aldershot, UK: Ashgate.

Hansson, S. (1995). Impact factor as a misleading tool in evaluation of medical journals. *Lancet, 346*(8979), 906. doi:10.1016/S0140-6736(95)92749-2.

Huxley, M. (2007). Geographies of governmentality. In J. W. Crampton & S. Elden (Eds.), *Space, knowledge and power: Foucault and geography* (pp. 185–204). Aldershot, UK: Ashgate.

Ibarra, H., & Andrews, S. B. (1993). Power, social influence, and sense making: Effects of network centrality and proximity on employee perceptions. *Administrative Science Quarterly, 38*, 277–303.

Jöns, H. (2003). *Grenzüberschreitende Mobilität und Kooperation in den Wissenschaften. Deutschlandaufenthalte US-amerikanischer Humboldt-Forschungspreisträger aus einer erweiterten Akteursnetzwerkperspektive* [Cross-boundary mobility and cooperation in the sciences: U.S. Humboldt Research Award winners in Germany from an expanded actor-network perspective] (Heidelberger Geographische Arbeiten, Vol. 116). Heidelberg, Germany: Heidelberg University, Selbstverlag des Geographischen Instituts.

Keith, M., & Pile, S. (Eds.). (2013). *Geographies of resistance*. New York: Routledge. (First published in 1997)

Kintzinger, M. (2003). *Wissen wird Macht. Bildung im Mittelalter* [Knowledge becomes power: Education in the Middle Ages]. Darmstadt, Germany: Jan Thorbecke.

Klauser, F. R. (2012). Thinking through territoriality: Introducing Claude Raffestin to Anglophone socio-spatial theory. *Environment and Planning D: Society and Space, 30*, 106–120. doi:10.1068/d20711.

Klauser, F. R. (2013). Re-visiting Michel Foucault: Towards a political geography of mediation. *Geographica Helvetica, 68*, 95–104.

Klauser, F. R. (2014). Introduction. Foundations for a political geography of surveillance. In F. Klauser, *Governing the everyday in the information age: Towards a political geography of surveillance* (pp. 2–46). Unpublished habilitation thesis, University of Berne, Berne, Switzerland.

Latour, B. (1987). *Science in action*. Milton Keynes, UK: Open University Press.

Lefebvre, H. (1991). *The production of space* (Donald Nicholson-Smith, Trans.). Oxford, UK: Blackwell. (French original 1974)

Legg, S. (2007). Beyond the European province: Foucault and postcolonialism. In J. W. Crampton & S. Elden (Eds.), *Space, knowledge and power: Foucault and geography* (pp. 265–289). Aldershot, UK: Ashgate.

Leibovich, M. (2013). *This town. Two parties and a funeral - plus plenty of valet parking! - in America's gilded capital*. New York: Blue Rider Press.

Livingstone, D. N. (1995). The spaces of knowledge: Contributions towards a historical geography of science. *Environment and Planning D: Society and Space, 13*, 5–34. doi:10.1068/d130005.

Livingstone, D. N. (2003). *Putting science in its place: Geographies of scientific knowledge*. Chicago: University of Chicago Press.

Lossau, J., & Flitner, M. (2005). Ortsbesichtigung. Eine Einleitung [Visiting the scene: An introduction]. In M. Flitner & J. Lossau (Eds.), *Themenorte* (pp. 7–23). Münster, Germany: Lit.

Manzo, L. C. (2005). For better or worse: Exploring multiple dimensions of place meaning. *Journal of Environmental Psychology, 25*, 67–86. doi:10.1016/j.jenvp.2005.01.002.

Massey, D. (1999). Philosophy and politics of spatiality: Some considerations. In D. Massey (Ed.), *Power-geometries and the politics of space-time* (Hettner-lecture, Vol. 2, pp. 27–42). Heidelberg, Germany: Heidelberg University, Department of Geography.

Mattissek, A. (2007). Diskursanalyse in der Humangeographie. State of the art [Discourse analysis in human geography: State of the art]. *Geographische Zeitschrift, 95*, 37–55.

Merton, R. K. (1968). The Matthew effect in science. *Science, 159*(3810), 56–63. doi:10.1126/science.159.3810.56.

Meusburger, P., & Schuch, T. (Eds.). (2012). *Wissenschaftsatlas of Heidelberg University. Spatio-temporal relations of academic knowledge production*. Knittlingen, Germany: Bibliotheca Palatina.

Peet, R. (2007). *Geography of power*. London: Zed Books.

Philo, C. (1992). Foucault's geography. *Environment and Planning D: Society and Space, 10*, 137–161. doi:10.1068/d100137.

Popitz, H. (1992). *Phänomene der Macht* [Phenomena of power] (2nd enlarged ed.). Tübingen, Germany: Mohr Siebeck.

Pred, A. (2007). Situated ignorance and state terrorism. Silences, W.M.D., collective amnesia, and the manufacture of fear. In D. Gregory & A. Pred (Eds.), *Violent geographies. Fear, terror, and political violence* (pp. 363–384). New York: Routledge.

Raffestin, C. (1977). Paysage et territorialité [Landscape and territoriality]. *Cahiers de Géographie du Québec, 21*(53/54), 123–134.

Raffestin, C. (1980). *Pour une géographie du pouvoir* [For a geography of power]. Paris: Litec.

Raffestin, C. (1984). Territoriality: A reflection of the discrepancies between the organization of space and individual liberty. *International Political Science Review, 5*, 139–146. doi:10.1177/019251218400500205.

Raffestin, C. (1986a). Territorialité: Concept ou paradigme de la géographie sociale? [Territoriality: A concept or paradigm in social geography ?]. *Geographica Helvetica, 41*, 91–96.

Raffestin, C. (1986b). Elements for a theory of the frontier. *Diogenes, 34*(134), 1–18.

Raffestin, C. (2007). Could Foucault have revolutionized geography? In J. W. Crampton & S. Elden (Eds.), *Space, knowledge and power: Foucault and geography* (pp. 129–137). Aldershot, UK: Ashgate.

Raffestin, C., & Butler, S. (2012). Space, territory, and territoriality. *Environment and Planning D: Space and Society, 30*, 121–141. doi:10.1068/d21311.

Rowles, G. D. (2008). The meaning of place. In E. B. Crepeau, E. S. Cohn, & B. A. Boyt Schell (Eds.), *Willard and Spackman's occupational therapy* (11th ed., pp. 80–89). Philadelphia: Wolters Kluwer/Lippincott Williams & Wilkins.

Said, E. W. (1978). *Orientalism*. New York: Vintage.

Scannel, L., & Gifford, R. (2010). Defining place attachment: A tripartite organizing framework. *Journal of Environmental Psychology, 30*, 1–10. doi:10.1016/j.jenvp.2009.09.006.

Scheler, M. (1926). *Die Wissensformen und die Gesellschaft. Probleme einer Soziologie des Wissens* [The forms of knowledge and society: Problems of a sociology of knowledge]. Leipzig, Germany: Der Neue Geist Verlag.

Schmid, C. (2005). *Stadt, Raum und Gesellschaft. Henri Lefebvre und die Theorie der Produktion des Raumes* [City, space and society. Henri Lefebvre and the theory of the production of space] (Sozialgeographische Bibliothek, Vol. 1). Stuttgart, Germany: Franz Steiner.

Seglen, P. O. (1997a). Why the impact factor of journals should not be used for evaluating research. *British Medical Journal*, *314*(7079), 498–502. doi:http://dx.doi.org/10.1136/bmj.314.7079.497.

Seglen, P. O. (1997b). Citations and journal impact factors: Questionable indicators of research quality. *Allergy, 52*, 1050–1056. doi:10.1111/j.1398-9995.1997.tb00175.x.

Sharp, J. P. (Ed.). (2000). *Entanglements of power: Geographies of domination/resistance* (Critical geographies, Vol. 5). London: Psychology Press.

Simmel, G. (2004). *The Philosophy of money* (3rd enlarged ed.) (T. Bottomore & D. Frisby, Trans.) (D. Frisby, Ed.). New York: Routledge. (Original work published 1900)

Smith, R. (2008). Beware the tyranny of impact factors. *The Journal of Bone and Joint Surgery, 90*, 125–126. doi:10.1302/0301-620X.90B2.20258.

Staeheli, L. A. (1994). Empowering political struggle: Spaces and scales of resistance. *Political Geography, 13*, 387–391. doi:10.1016/0962-6298(94)90046-9.

Steen, R. G. (2011). Retractions in the scientific literature: Is the incidence of research fraud increasing? *Journal of Medical Ethics, 37*, 249–253. doi:10.1136/jme.2010.040923.

Stehr, N. (1992). Experts, counselors and advisers. In N. Stehr & V. R. Ericson (Eds.), *The culture and power of knowledge: Inquiries into contemporary societies* (pp. 107–155). Berlin, Germany: Walter de Gruyter.

Sternburg, von W. (2014). *Lion Feuchtwanger. Die Biographie*. [Lion Feuchtwanger. The biography.] (Extended edition). Berlin: Aufbau Verlag.

Stroebe, W., Postmes, T., & Spears, R. (2012). Scientific misconduct and the myth of self-correction in science. *Perspectives on Psychological Science, 7*, 670–688. doi:10.1177/1745691612460687.

Thrift, N. J. (2007). Overcome by space: Reworking Foucault. In J. W. Crampton & S. Elden (Eds.), *Space, knowledge and power: Foucault and geography* (pp. 53–58). Aldershot, UK: Ashgate.

Turner, H. W. (1979). *From temple to meeting house: The phenomenology and theology of places of worship*. The Hague, The Netherlands: Mouton.

Wilkin, P. (1997). *Noam Chomsky: On power, knowledge and human nature*. New York: St. Martin's Press.

Relations Between Knowledge and Power: An Overview of Research Questions and Concepts

Peter Meusburger

No Power Without Knowledge, No Knowledge Without Power[1]

Since the dawn of civilization, rulers have been convinced that they need forecasts to help them come to grips with the uncertainties of the future and that they need a lead in knowledge[2] and forethought[3] to acquire and exercise power. Depending on the culture and historical period, political and military rulers preparing to make a vital decision have first consulted oracles, dream interpreters, astrologers, augurs,[4] haruspices,[5] priests, shamans, prophets, and other "sages" contending that they have prophetic abilities, contact with the gods or ancestors, or uncommonly great knowledge (see Barton, 1994; Mann, 1986; Maul, 1994, 2003, 2013; and Chap. 5 by Maul in this volume).

Many rulers have endeavored to consolidate or widen their power and their epistemological advantage by setting up centers of knowledge (e.g., academies and universities) or monopolizing divinatory expertise, that is, the "knowledge and techniques of looking into the future" (see Chap. 5 by Maul in this volume). Persian

---

[1] I borrow this phrase from Kammler (2008, p. 305). It means that the exercise of power uses and generates knowledge and, conversely, that knowledge coincides with certain effects of power.

[2] This chapter's general references to knowledge are solely to categories and agents of knowledge that are capable of enhancing or jeopardizing power. Of course, there are categories of knowledge that have little or nothing to do with power. The Aristotelian concepts of *episteme*, *techne*, and *doxa* also apply in this context, but their highly dissimilar use by various authors today (see Löbl, 1997, 2003) could lead to misunderstandings.

[3] According to Russell (1958), civilization is "a manner of life due to the combination of knowledge and forethought" (p. 159).

[4] Augurs were concerned with interpreting the movements and cries of birds (Barton, 1994, p. 33).

[5] The haruspices interpreted omens and the entrails of sacrificial animals (Barton, 1994, p. 34).

P. Meusburger (✉)
Department of Geography, Heidelberg University,
Berliner Straße 48, 69120 Heidelberg, Germany
e-mail: peter.meusburger@geog.uni-heidelberg.de

© Springer Netherlands 2015
P. Meusburger et al. (eds.), *Geographies of Knowledge and Power*,
Knowledge and Space 7, DOI 10.1007/978-94-017-9960-7_2

King Cambyses II (558–522 B.C.) had wise men and priests brought to Babylon from Egypt, Chaldea, Assyria, Persia, Judea, Syria, Asia Minor, and other lands he had conquered (see Brunés, 1967, p. 237), probably with the idea of concentrating all available knowledge in his power center. Caliph Harun al-Raschid (A.D. 786–809) and his son al-Ma'mun established the "House of Knowledge" at their seat of government in Baghdad, where Greek, Indian, and Persian tracts were translated into Arabic (Ahmed, 1988, p. 333). In past centuries several universities were founded by European sovereigns in their respective cities of residence, primarily to further their ambitions in the power politics of their day. The highly educated, scientifically minded, and multilingual Frederick II (1194–1250)[6] was supposedly the first European sovereign to pursue his own policy on science and knowledge generation independently of the church. In 1224 he founded a university of administration, the University of Naples, to secure his claim to power and to centralize and stabilize his government (Kintzinger, 2003, pp. 116–119).

> [Frederick II] surrounded himself with outstanding scholars, took part in their scientific endeavors, had them engage in debates at his court, and supported them however he could without restricting their work. In addition to theoretical learnedness, Frederick was also always concerned with its practical application. He regulated the vocational training of physicians and introduced scientific examinations and prescribed curricula. He wanted all the sciences to be taught at the University of Naples. The main objective, though, was to train lawyers for the kingdom's government and administration. This promotion of the sciences by a medieval prince was unique in its way, but it, too, had its shortcomings. Interested Sicilian subjects were forbidden to study abroad, and the knowledge gained in Naples was not allowed to be used elsewhere. (pp. 116–117)[7]

Because knowledge is "a part and an instrument of legitimate authority (*Herrschaft*) and social order" (Kintzinger, 2003, p. 33) and power "is a basic principle of modern society's development and integration" (Kneer, 2012, p. 267), the powers that be must continually try to attract exceptional exponents of knowledge to their goals, to incorporate those persons into consensual networks, and to prevent the formation of rival coalitions that could threaten their hold on power (see Popitz, 1992, pp. 201–211). A power center whose goals fail to win sufficient backing from scientists, engineers, intellectuals, journalists, artists, and experts from various other domains will eventually lose out to other aspiring power centers. Power is not stable; it must be attained, consolidated, exhibited, and legitimated again and again.[8]

---

[6] Frederick II became King of Sicily (1198), King of the Germans (1211–1212), and Holy Roman Emperor (1220–1250).

[7] Unless otherwise specified, the English translations of quotations in this chapter are my own in collaboration with D. Antal.

[8] To Max Weber legitimate authority was institutionalized power (see also Popitz, 1992, p. 232) and was the indispensable sociological category as opposed to power because it was, as he stated, objectively and verifiably linked to effects rooted in order (Maurer, 2012, p. 361). Many authors use the terms *power* and *authority* synonymously. Others discriminate between them: "Whereas power is thought of as something mobile, dynamic, and malleable, authority [Foucault] is conceived of as something stable, irreversible, rigid. . . . In relations involving authority the mobility and dynamics observable in power relations are thus more or less completely expunged. Authority is thereby *reified, rigidified* power" (Kneer, 2012, p. 279).

But scholars were not called to the courts of kings, dukes, and princes only to advise them on their decisions, advocate their hegemonic interests, and thereby guarantee the viability and self-preservation of the given political system. Their presence in the centers of power or "centers of calculation" (Latour, 1987, pp. 215–257) also served to legitimate the decisions of rulers, reinforce the status of those in power, and meet their need for economic, political, and ideological appearances (see also Göhler, 1997; Kintzinger, 2003, p. 33).

With Kintzinger (2003) in mind, one can thus say that the power of authority has always tended to take advantage of the power of knowledge (p. 191). The only things to have changed over the centuries are the kind of knowledge that those in power demand and the relations between knowledge and power. After the rise of the natural sciences in the sixteenth century (see Taylor, Hoyler, & Evans, 2010), the importance of knowledge for political power became ever more apparent. Realizing its significance, Humanists such as Philip Melanchthon (1497–1560) saw the founding of schools and universities as a path to political power and economic wealth (see Meusburger, 2013, p. 22). As Francis Bacon (1561–1626) wrote a few decades after Melanchthon, "The roads to human power and to human knowledge lie close together and are nearly the same" (Bacon, 1620/1863, Aphorisms, IV [Book 2]). And "human knowledge and human power meet in one" (Aphorisms, III [Book 1]; see also Röttgers, 1980, p. 595). Not entirely agreeing with Bacon's idea that knowledge is power, Gottfried W. Leibniz (1646–1716) countered by stating, "Although each science extends power over external things, it has another use, namely, the culmination of the spirit"[9] (quoted in Meier-Oeser, 2004, p. 909). Other authors, too, stress that knowledge is "a good in itself, or a means of creating a broad and humane outlook on life in general [and not] merely an ingredient in technical skill" (Russell, 1958, p. 35).

But Leibniz and Russell seem to have overlooked that the knowledge central to power is not only about technical superiority, military prowess, and natural sciences but also about cultural knowledge, cultivation of the whole person (*Bildung*), a certain mindset, and moral position. Hence, if key decision-makers in a social system arrive at a refinement or culmination of the spirit (*Vervollkommnung des Geistes*), then this achievement can contribute to reducing the errors and misconceptions in their situational analyses, problem-solving, and decision making. It can raise the likelihood that these decision-makers will recognize the long-term unintended consequences of their actions or help them recognize those impacts earlier than would otherwise be the case. It can also mean that their erudition, ethics, and mindset may bring them to desist from certain actions that uneducated or ideologically fixated people in power would blithely execute (see also Russell, 1958, p. 41). The decoupling of science, cultivation of the whole person, and morality in the wake of modern science's development has triggered multiple crises and disasters (see Mittelstraß, 1982, pp. 103–107). Environmental catastrophes and armed conflict show what can ensue when technical knowledge is used without moral grounding.

---

[9] *Quanquam . . . omnis scientia potentiam in externa quoque augeat . . . , est tamen alius ejus usus . . . , ipsa scilicet perfectio mentis.*

Granted, scientific insights and innovative engineering have honed the efficiency of transport, industrial production, surveying, communications, surveillance technology, and military armament. In the nineteenth century, science became "a weapon in constellations of competition" (Schimank, 1992, p. 218). It is also true that new technical knowledge has assisted in perfecting the long-term exercise of power in many areas (Popitz, 1992, pp. 179–180). Decision-makers should nevertheless resist the temptation to distinguish useful (utilitarian) knowledge (usually meaning that of the natural sciences) from unuseful (nonutilitarian) knowledge. The assessment of whether or not particular knowledge is useful can change very quickly. If important political, economic, or military decision-makers lack wisdom, education, knowledge about foreign cultures, empathy, and experienced-based intuition, then their system will ultimately profit little from technological superiority.

In the nineteenth century, politicians or institutions holding political power came by two additional instruments to influence the generation and diffusion of knowledge. One was the introduction of compulsory schooling, which shifted the control of formal education from the church to the state and turned the system of state education into a kind of "disciplinary apparatus" in the sense meant by Foucault (1979, 2007; see also Speth, 1997). The second source of influence on the production and dissemination of knowledge was the process of nation-building, which was advanced by the spread of literacy and mass media; the construction of museums, monuments, and other places of memory; and national rituals and "heroic" historiography. Newly emerging nation-states aspired not only to political, administrative, and military sovereignty over a certain territory but also to a homogeneity of culture, memory, and identity. By controlling educational systems and cultural institutions (e.g., museums, memorials, national exhibitions, and media), the state or other power elites managed to manufacture national consent, shape firm convictions and interpretations of the world, promote an official language, and construct an ideological hegemony or domination over other ethnic, religious, or societal groups (see Gramsci, 1971; Gregory, 1998; Herman & Chomsky, 1988; Meusburger, 2011; Meusburger, Heffernan, & Wunder, 2011; Simonds, 1989; Tanner, 1999). "Hegemony, in Gramsci's writings, refers to non-violent forms of control exercised through the whole range of dominant cultural institutions and social practices, from schooling, museums, and political parties to religious practice, architectural forms, and the mass media" (Mitchell, 1990, p. 553).

In the course of history, the essence of domination, the ways in which power has been exercised, the forms in which it has been asserted and stabilized, and thus also the relations between knowledge and power have repeatedly changed, of course (see Imbusch, 2012a; Mann, 1986; Popitz, 1992).

> The early modern period marked the first time that the ruler's power was restricted by a contract between the sovereign and the people, and the process of secularization raised matters regarding the legitimacy of dominion. Sovereignty was thus no longer something naturally bequeathed or divinely willed; it henceforth appeared as something of human origin and, hence, as something historically changeable. This shift was prepared by the philosophy of the Enlightenment and by rational natural law. (Imbusch, 2012a, pp. 21–22)

In modern times access to positions of power has had to be justified in different ways than in earlier periods. Since the early nineteenth century, the relations between knowledge and power have become progressively institutionalized and formalized, with power coming to be exercised more and more through organizational structures, rules, and stipulated procedures (see Popitz, 1992, p. 234). These developments have stemmed partly from a number of social megatrends, including the rising rates of literacy, the advent of compulsory education, the bureaucratization of government administration[10] and major organizations, the rise of meritocracy, the professionalization of many occupations, the ever greater reliance on scientific methods and theory in production processes and the overall economy (see Meusburger, 2013), new communications technologies, new modes of governmentality and surveillance, and the democratization of political systems.

These developments have resulted to some extent in a depersonalization of power relations and in the emergence of abstract power structures in which various positions are vested with different responsibilities, decision-making authority, prerogatives, and privileges. Access to these positions, at least in meritocratic societies, has been regulated increasingly by proof of qualification, educational degrees, examinations, screenings, and other selection procedures. With the regulatory system now being "largely impersonal and objective (i.e., without reference to specific persons and social relations)" (Maurer, 2012, p. 364), many positions and acts of exercising power are less visible than they once were. According to Mutschler (2005, p. 259), it is essential that power becomes invisible if it is to be stabilized successfully. However, Münkler (1995) underscores that power enjoys both visible and invisible elements or characteristics and capacities (p. 213; see also Gordon, 2002; Mutschler; Rehberg, 2005; Tanner, 2005). There are circumstances in which power is supposed to be as invisible as possible (e.g., censure, torture, interception of emails, and the falsification of data) and those in which it is ostentatious (e.g., court etiquette, military parades, and press conferences). "Rendering [power] completely invisible divests it of its formative impact" (Münkler, 1995, p. 213). Baum and Kron (2012, pp. 345–346, 353) argue that *liquid modernity* is characterized by an even more successful (more perfidious) concealment of structures and relations of politics (*Herrschaft*) than is *solid modernity* (see also Bauman, 2000; Bauman & Haugaard, 2008). In solid modernity power relations were more visible than in liquid modernity.

In summary, the power of the spirit has played an ever greater role in the exercise of power. In the words of Gustav L. Radbruch (1878–1949), the renowned Heidelberg philosopher of law and Reich Minister of Justice under the Weimar Republic, "Power is spirit: In the end, all power is power over souls. . . . All power rests on the willing or unwilling recognition of those subject to it" (Radbruch, 1993, p. 311). As Napoleon reportedly said after his abortive invasion of Russia, "Do you know what amazes me most in this world? It is the powerlessness of material force. There are only two things in the world, the sword and the spirit. In the long term, it

---

[10]"Bureaucratic administration means fundamentally the exercise of control on the basis of knowledge" (Weber, 1922/1964, p. 339).

is always the spirit that will triumph over the sword" (as cited in Radbruch, 1993, p. 311; see also p. 156).

Of course, education, training, new technologies, and learning processes in the widest sense were immensely important for economic and social development in past centuries, too. Yet many observers share the view that, since the 1960s, a knowledge society has arisen in which knowledge, research, qualifications, and inventions have higher value than ever before (Bell, 1973; Drucker, 1969; Meusburger, 1998; Richta, 1969, 1977; Rueschemeyer, 1986; Stehr & Ericson, 1992).

This chapter addresses several questions: What interrelations are there between knowledge and power? Can a meaningful semantic difference exist between factual knowledge and orientation knowledge? What functions do factual knowledge and orientation knowledge have in the acquisition and stabilization of power? By which means and why do leaders of orientation knowledge make moral judgments on the Self and the Other? What is the function of myths, legends, collective memories, cultural traditions, and collective identities? Why are propaganda, persuasion, disinformation, censorship, and manipulation of information central features of politics and hegemonic practices? Why do many key purveyors of factual knowledge and orientation knowledge seek proximity to power? And with what methods do the powers that be try to affect the creation and dissemination of knowledge?

Factual Knowledge and Orientation Knowledge: Differences Between Logos and Mythos

What is truth, reality, and objective knowledge? What are the differences between opinion, belief, faith, and knowledge? What categories of knowledge should be discerned? Philosophical questions of this kind have been discussed by a multitude of authors going back as far as Plato (for details see Abel, 2008; Stegmaier, 2008; Stenmark, 2008; Welker, 2008; Wieland, 1982). There is no need to repeat their discussions in this chapter. However, the long-standing efforts in philosophy to tell *logos* from *mythos*, knowledge from faith, and rationality from irrationality are seminal for any research focusing on the relations between knowledge and power. In this chapter part of the old dichotomy between *logos* and *mythos* is represented by the categories called *factual knowledge* and *orientation knowledge*.

Factual Knowledge

Factual knowledge can be thought of as subsuming a wide range of facets: the sum of what has been perceived, discovered, or learned by means of a methodically well-regulated procedure bound to justification, truth, and verification (Abel, 2008, p. 12); empirically verifiable findings; professional skills; expertise required for causal analysis and scientific explanation; practical experience allowing for a degree of predictability; and so-called technoknowledge, which helps solve problems of a technical or scientific nature. The term factual knowledge is thus widely equivalent to what Mittelstraß (1982) calls *Verfügungswissen* (pp. 16, 62, 103; 2001, pp. 75–76; 2010, p. 22). This category of knowledge is needed in order to achieve a realistic description and analysis of a given situation, to master complexity, to cope with competition, and to manage risks under uncertainty.

To avoid misunderstandings, several categories of factual knowledge should be differentiated. Factual knowledge can be regarded as widely shared, canonized knowledge that is generated by experts and taken as true on the basis of the prevailing state of the art in research. This kind of factual knowledge, according to Felder (2013, p. 14), is divisible into (a) indisputable matters (e.g., $4 \times 5 = 20$; the distance between A and B is 12,678 miles; the sum of the angles in a triangle equals $180°$) and (b) contestable matters provable as true or false only through lengthy empirical examination (e.g., humans influence the climate, viruses can trigger cancer).

Factual knowledge can be distinguished further according to the level of abstraction or generalization by which it is represented. Abstraction and generalization are needed to reduce the information overload, to have principles and laws at one's disposal, and to focus on those categories of information that are most relevant for certain decisions. In different problem-solving situations, decision-makers have to rely on information gathered and represented at different levels of abstraction and generalization. The crucial point is how to choose the adequate level. A map in the scale of 1:200,000 has a higher degree of generalization than a map in the scale 1:10,000, which shows much more detail but may be useless in certain decision-making situations because of its information overload. Gregory (Chap. 4 in this volume) and Leed (1981) demonstrate the gap between abstract factual knowledge and factual knowledge gained by personal experience. In World War I, generals using maps or aerial photographs for their decisions had a different kind of factual knowledge about the battlefield than did the infantry crawling through the mud of the trenches.

> Trench war is an environment that can never be known abstractly or from the outside. Onlookers could never understand a reality that must be crawled through and lived in. This life, in turn, equips the inhabitant with a knowledge that is difficult to generalize or explain. (Leed, 1981, p. 79)

There is also the distinction between abilities that a person acquires subjectively for the most part through repetitive activities over an extended period—long learning processes, for example—and knowledge of facts (experiential knowledge), which can be imparted socially without the individual personally having to submit to the particular experience and without having to go through years of learning (for details see Schütz & Luckmann, 1973). The first variant of this differentiation is exemplified by the craft trades, the ability to play the violin, and the physical and mental performance of an experienced mountain climber. The second variant is illustrated well by a child's socially acquired knowledge not to touch a hot stove. In this second, socially imparted kind of knowledge, intersubjective recognition of knowledge in the sense of communicative constructivism (Christmann, 2013; Keller, 2013; Keller, Knoblauch, & Reichertz, 2013; Knoblauch, 2013a, 2013b) plays a greater role than in the first, subjective category (e.g., the physical performance of the mountain climber). There are situations in which actors do not depend on whether their competencies or factual knowledge are accepted by others. In some competitive situations, an intersubjective acceptance of new factual knowledge is not desired at all; secret factual knowledge can mean competitive advantage.

Orientation Knowledge

The multifaceted nature and diverse use of the term *orientation* calls for an explanation of its inception. Stegmaier (2008) minutely describes why the concepts of orientation and getting oriented have drawn increasing attention since the nineteenth century, how various philosophers (e.g., Kant, Herder, Fichte, Schopenhauer, Schleiermacher, Buber, Heidegger, Cassirer, and Mittelstraß) have treated the concepts of orientation and orientation knowledge, which categories and definitions they have used, and why the concept of orientation can avoid paradoxes that can confound logic.

The term *orientation knowledge* was created by Kant (1786/1996)[11] and was later specified and popularized by Mittelstraß (1982, pp. 16–20, 50–51, 82, 103). In this chapter I use it[12] generally to refer to revealed knowledge (*Heilswissen*: the salvation knowledge of *religion* and *ideology*), subjective—objectively unjustified—knowledge (myths and legends), spirituality, cultural traditions, and experiences of transcendence.

Orientation knowledge stands in contrast to reality, empirically verifiable facts, and scientific knowledge that is gained incrementally in controlled fashion. As Kant (1786/1996) wrote, "All believing is a holding true which is subjectively sufficient, but *consciously* regarded as objectively insufficient; thus it is contrasted with *knowing*"[13] (p. 13; for details see Stegmaier, 1992, p. 298). Schleiermacher (1814–1815, 1833/1988) conceived of orienting oneself as the "supplement of all real knowledge not attained by way of science" (p. 9).[14] Fichte (1845–1846, p. 195), too, saw orientation as a supplement of real knowledge (Stegmaier, 2008, pp. 103–110). Likewise, the young Martin Buber (1913/1965) juxtaposed the concept of *Orientierungswissen*

---

[11] This work, first published in October 1786 in the *Berlinische Monatsschrift* (pp. 304–330), became the "most significant document in the critical philosophy of orientation" (Stegmaier, 2008, p. 79). It is where Kant introduced the term *orientation* for "the moral and practical use of reason" (Stegmaier, 1992, p. 298).

[12] In this context the term means religious, ideological, or cultural orientation knowledge, that is, an orientation to values rather than an orientation in space or to facts. Unfortunately, the broad, common use of the word *orientation* can lead to misunderstandings. I retain the term orientation knowledge because the alternatives—*redemption knowledge, salvation knowledge, revealed knowledge, spiritual knowledge, religious knowledge, religiosity,* and *invisible religion* (Luckmann, 1967)—are too narrow.

[13] *Aller Glaube ist nun ein subjektiv zureichendes, objektiv aber mit Bewußtsein unzureichendes Fürwahrhalten; also wird er dem Wissen entgegengesetzt.* Retrieved October 7, 2014, from tenth paragraph at http://www.zeno.org/Philosophie/M/Kant,+Immanuel/Was+hei%C3%9Ft%3A+sich +im+Denken+orientieren

[14] To Schleiermacher, all knowledge formation was orientation: "Accordingly, all knowledge needs orientation, and no knowledge comes about without it" (Stegmaier, 2008, p. 107).

and reality (for details see Stegmaier, 2008, p. 126). As interpreted by Stegmaier (2008, p. 134), Heidegger linked the concept of orientation with worldview, which guides the life of the individual. Cassirer (1907/1922) classified orientation under "mythical thinking" (p. 619). Scheler (1926), a pioneer of the sociology of knowledge, identified three forms of knowledge:

- *Leistungs- und Herrschaftswissen:* instrumental knowledge and power/knowledge to accomplish practical goals. This category is more or less equivalent to factual knowledge.
- *Bildungswissen*: formative, or self-formative, knowledge to shape the individual's personality
- *Erlösungswissen* and *Heilswissen:* the redemption knowledge and salvation knowledge offered by religions, ideologies, knowledge of aims, and worldviews.[15] This category is equivalent to orientation knowledge.

Orientation knowledge, occasionally also called *symbolic knowledge*,[16] consists chiefly of belief systems, values, cultural traditions, worldviews, ideologies, religions, moral positions, mindsets, action-guiding norms (*handlungsleitende Normen*), and reflection about the ethical conduct of one's life (*Reflexion über die Ethik der Lebensführung*). In other words, it encompasses overall perspectives from which one sees and interprets the world (for details see Mittelstraß, 1982, 2001, 2010; Stegmaier, 2008; Tanner, 1999; see Fig. 2.1).

Orientation knowledge lays a basis for making moral valuations; providing actors and societal systems with a moral compass, ideologies, goals, values, a cultural memory, and a collective identity; strengthening the motivation and internal cohesion of societal systems; and offering rituals to their members and meeting their spiritual needs. "The major mechanisms of power have [always] been accompanied by ideological productions" (Foucault, 1980, p. 102). The same is true both for great cultural achievements and inimical developments. Domination (imperialism, colonialism), for example, endures only if supported by an intellectual discourse or by ideologies and worldviews (Baum & Kron, 2012, p. 344). As Russell (1958) put it:

> Whenever the few have acquired power over the many, they have been assisted by some superstition which dominated the many. Ancient Egyptian priests discovered how to predict eclipses, which were still viewed with terror by the populace; in this way they were able to extort gifts and power which they could not otherwise have obtained. (p. 78)

---

[15] To Scheler (1926), salvation knowledge, the only noninstrumental variety of knowledge, had the highest value. This view was a notable misunderstanding, however, for religions and ideologies are by no means noninstrumental from the perspective of the person wielding power. On the contrary, they can be among its foremost sources. For further discussion of this concept, see Meusburger (2008, especially pp. 58, 71, 73; 2011, pp. 54–57).

[16] In past publications I, too, have used the term symbolic knowledge (Meusburger, 2005, 2007b). It can lead to confusion, however, because some authors take it to mean knowledge about the meaning of symbols.

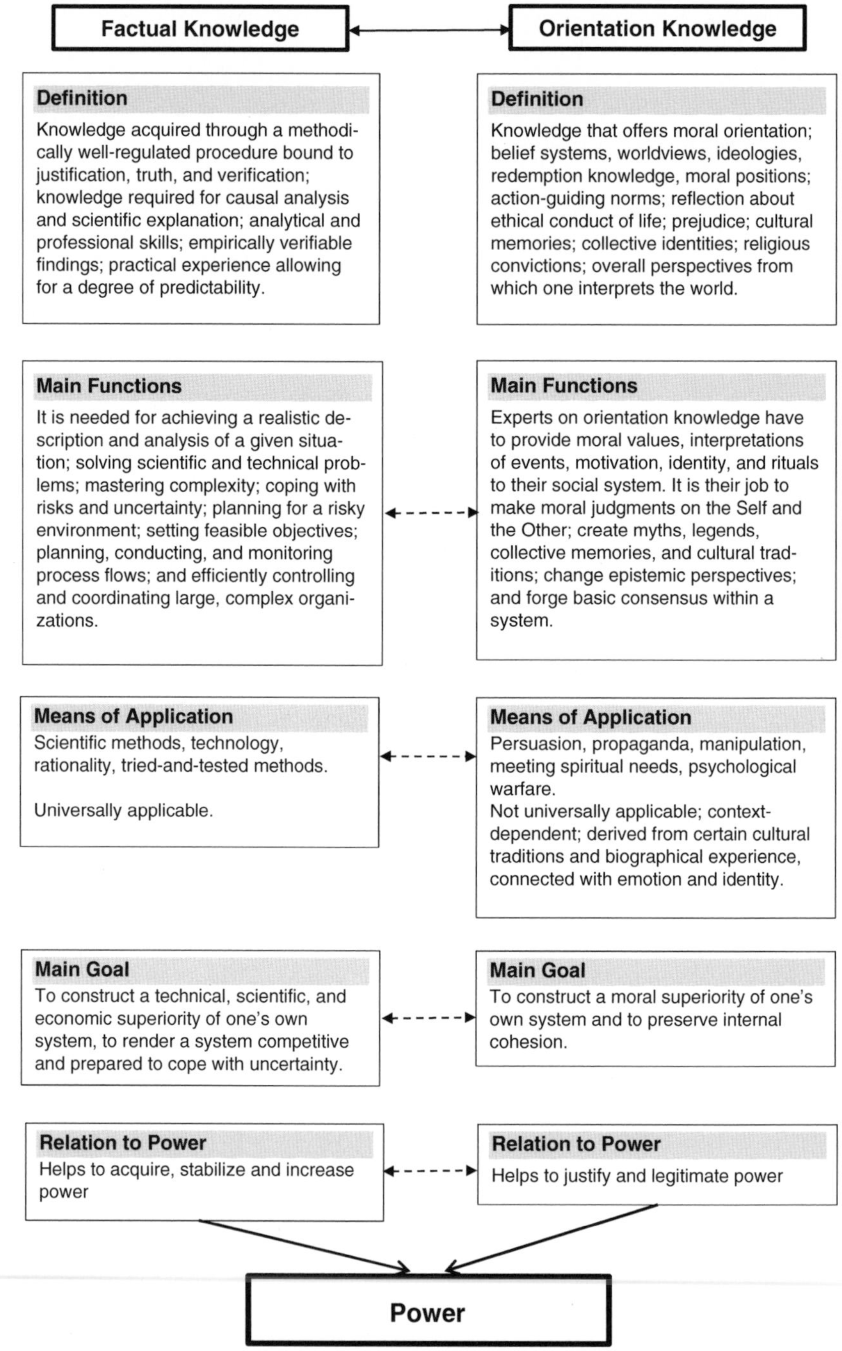

| Factual Knowledge | ⟷ | Orientation Knowledge |

Definition

Knowledge acquired through a methodically well-regulated procedure bound to justification, truth, and verification; knowledge required for causal analysis and scientific explanation; analytical and professional skills; empirically verifiable findings; practical experience allowing for a degree of predictability.

Definition

Knowledge that offers moral orientation; belief systems, worldviews, ideologies, redemption knowledge, moral positions; action-guiding norms; reflection about ethical conduct of life; prejudice; cultural memories; collective identities; religious convictions; overall perspectives from which one interprets the world.

Main Functions

It is needed for achieving a realistic description and analysis of a given situation; solving scientific and technical problems; mastering complexity; coping with risks and uncertainty; planning for a risky environment; setting feasible objectives; planning, conducting, and monitoring process flows; and efficiently controlling and coordinating large, complex organizations.

Main Functions

Experts on orientation knowledge have to provide moral values, interpretations of events, motivation, identity, and rituals to their social system. It is their job to make moral judgments on the Self and the Other; create myths, legends, collective memories, and cultural traditions; change epistemic perspectives; and forge basic consensus within a system.

Means of Application

Scientific methods, technology, rationality, tried-and-tested methods.

Universally applicable.

Means of Application

Persuasion, propaganda, manipulation, meeting spiritual needs, psychological warfare.
Not universally applicable; context-dependent; derived from certain cultural traditions and biographical experience, connected with emotion and identity.

Main Goal

To construct a technical, scientific, and economic superiority of one's own system, to render a system competitive and prepared to cope with uncertainty.

Main Goal

To construct a moral superiority of one's own system and to preserve internal cohesion.

Relation to Power
Helps to acquire, stabilize and increase power

Relation to Power
Helps to justify and legitimate power

Power

Fig. 2.1 Functions of factual knowledge and orientation knowledge in the acquisition and retention of power (Design and copyright by the author)

Today one is not likely to speak so much of superstition as of worldview, ideology, or religion, but in principle Russell's statement applies to the present as well.

Conceptions and Definitions of Power and Their Relationship to Knowledge

How Can Power Be Conceptualized and Defined?

Conceptions and definitions of power are as manifold and diverse as those of knowledge. One categorization distinguishes between actor-based and system-based conceptions of power. An early proponent of the actor-based view was Weber (1922/1978), who defined power (*Macht*) as actor-specific resources used out of self-interest or as influence *despite* resistance. Power "is the probability that one actor within a social relationship will be in a position to carry out his own will despite resistance, regardless of the basis on which this probability rests" (p. 53).

Parsons (1967) offered a system-related approach. He defined power as "the capacity of a social system to mobilize resources to realize collective goals" (p. 193). To Arendt (1970) "power is never held by an individual; it is possessed by a group and exists only as long as the group remains intact" (p. 44). In other words, power is "the human ability not just to act but to act in concert" (Avelino & Rotmans, 2009, p. 547; Gordon, 2002, p. 133). Systems-related approaches of power tie into concepts such as control and coordination, discourses, decision-making within organizations, and the creation and diffusion of knowledge within and between social systems. "By organizing and arranging their social relations, people simultaneously distribute power" (Imbusch, 2012b, p. 191).

A second categorization differentiates between instrumental, structuralist, and discursive interpretations of power.

> Instrumental perspectives view power as actor-specific resources used in the pursuit of self-interests, referring to Weber's definition. In contrast, structuralist perspectives on power stress that material structures and institutional processes *predetermine* the behavioral options of decision-makers. In addition, discursive perspectives on power emphasize the dominance of ideas, frames, norms, discourses, perspectives, beliefs, and so on. Within 'discursive' interpretations there are those that emphasize the *structural* nature of discourse (such as Foucault) and those that emphasize the *agent-based* nature of discourse (such as Habermas). In some debates 'power and structural constraint are theorized as *opposite ends* of a continuous spectrum', in which power is directly related to agency (Haugaard, 2002, p. 38, italics added). In contrast, Foucault has analyzed power as an inherently *non-subjective* phenomenon that it is exercised *by* structures and *through* actors, contending that individuals are not the subjects, but rather the *vehicles* of power (Foucault, 1980, p. 101). (Avelino & Rotmans, 2009, p. 546)

A third categorization differentiates between innovative, constitutive, transformative, and systemic power (Avelino & Rotmans, 2009; Borch, 2005). *Innovative power* is defined as "the capacity of actors to create or discover new resources" (Avelino & Rotmans, 2009, p. 552). *Constitutive power* is the ability to distribute resources. It is related to institutions and structures that promote social order by shaping and stabilizing the distribution of resources (p. 552). *Transformative power* is defined as "the ability to transform the distribution of resources, . . . by redistributing resources and/or by replacing old resources with new resources. This involves the development of new structures and new institutions" (p. 553). *Systemic power* is defined as

> the combined capacity of actors to mobilize resources for the survival of a societal system, i.e., a particular continent, region, nation, sector, industry or business (depending on the chosen level of analysis). The extent to which actors are able to mobilize resources for the survival of a system defines the level of 'systemic power' exercised by those actors within that system. (p. 553)

An especially fruitful categorization is Pitkin's (1972) distinction between *power over* and *power to*[17] (see also Göhler, 2011, pp. 225–234). Power over means that an actor has power over other individuals. To put it differently, that person is in a position to follow through on his or her own intentions vis-à-vis those of other people. He or she is able to restrict the range of choices and actions of others. Power over can thus be formulated only within the framework of a social relation.

> Analysis of power relations described as *power over* requires at least one of the participants to be in a position to exercise more power than its addressees can in the power relation. In this case power is a given; it must already exist before it can be exercised. (Göhler, 2011, p. 229)

By contrast, power to does not refer to social relations to other persons. It means an individual ability to exercise power, a power to act, an ability or capacity to move something or reach a goal irrespective of what other people think about it. Of course, power to is also an ability to resist (Göhler, 2011, p. 229). "From the perspective of *power to*, autonomy is construed; from the perspective of *power over*, options for action are restricted" (p. 226).[18]

Yet another categorization differentiates between transitive and intransitive power (Berthold, 1997; Göhler, 2011; Speth & Buchstein, 1997). "Power referenced to the external world is transitive power, that is, power that transmits one's will to others and exerts influence on them. Power referenced to one's own group is intransitive power" (Göhler, p. 236).

In the administrative sphere, power exists mostly as official authority, or *Amtsgewalt* (*potestas*: rule, force, strength, ability, or control), vested with competencies and mandates (see Kobusch & Oeing-Hanhoff, 1980). As the term suggests, such authority is linked not to persons but rather to offices or positions in organizations. It is granted to actors only for a clearly defined period and often only within a specified territory. In Europe official authority has been gaining significance since about

---

[17] "One may have *power over* another or others, and that sort of power is indeed relational . . . But he may have power to do or accomplish something all by himself, and that power is not relational at all; it may involve other people if what he has power to do is a social or political action, but it need not" (Pitkin, 1972, p. 277).

[18] Drawing on Allen (1999), Göhler (2011) speaks also of *power with* (p. 234), which is understood to mean an ability not just to take action together but to stand shoulder to shoulder in the process.

the twelfth century, when Roman law was reintroduced and the foundations for a state administration were laid by the growing literacy of officials. Jurists had an important part in the organization of administrative power. The resurrection of Roman law in the twelfth century "had in effect a technical and constitutive role to play in the establishment of the authoritarian, administrative, and, in the final analysis, absolute power of the monarchy" (Foucault, 1980, p. 94).

How Can Relations Between Knowledge and Power Be Conceptualized and Explained?

The close relationship between knowledge and power is evident by the very fact that knowledge and power have the same etymological roots. The term *power* derives from the Latin word *potere* (to be able). The Latin noun *potentia* denotes an ability, capacity, or aptitude to affect outcomes, to make something possible. It can therefore be translated both as knowledge and power (see also Avelino & Rotmans, 2009, p. 550; Moldaschl & Stehr, 2010, p. 9; Schönrich, 2005, p. 383).

But knowledge is not just an instrument of power; it is more than something that serves or helps attain it. Several authors assert that there is an internal relation between power and knowledge. Tanner (2005, p. 5) states that power and wisdom are already linked in the Old Testament (e.g., Job 36).[19] Barton (1994) argues "that power cannot be divorced from any communication that presents itself as the truth" (p. 20). Nietzsche (1968), Foucault (1979), and other authors not only equate power with violence, coercion, and repression but also see a productive dynamic in it: "Power has innovative, power-enhancing effects[.] . . . Power releases energies, creates, invents, generates" (Kneer, 2012, p. 269; see also Bublitz, 2008, p. 274). "The exercise of power uses and generates knowledge, and, conversely, knowledge coincides with certain effects of power. In short, no power without knowledge and no knowledge without power" (Kammler, 2008, p. 305). Foucault (1980) coined the double word "power/knowledge" (*pouvoir-savoir*) to show that power and knowledge incorporate each other. To win a measure of insight into the connection between knowledge and power, consider some of Foucault's important statements on the subject:

> . . . the exercise of power itself creates and causes to emerge new objects of knowledge and accumulates new bodies of information. (p. 51)

> The exercise of power perpetually creates knowledge and, conversely, knowledge constantly induces effects of power. (p. 52)

> Knowledge and power are integrated with one another, and there is no point in dreaming of a time when knowledge will cease to depend on power. (p. 52)

> It is not possible for power to be exercised without knowledge, it is impossible for knowledge not to engender power. (p. 52)

---

[19] For details see http://www.biblegateway.com/keyword/?search=wisdom&version=KJV&searchtype=all

Many authors call attention to the dynamic interrelation of knowledge and power. One of them is Kneer (2012):

> The classical type of power also had a close tie with knowledge, but in modern times there has been a peculiar, reciprocal increase of power and knowledge: Incessant surveillance and control of individuals is bringing forth systematic knowledge, and, conversely, this knowledge serves the continuing increase in power. (p. 273)

Avelino and Rotmans (2009) argue that "knowledge is a meta-condition to meet the four conditions of power (access, strategies, skills and willingness); and . . . that creating or communicating knowledge is also a form of power exercise in itself" (p. 558). The context in which power is exercised relates to both an actor's position (function) within a social system and the place where an action occurs.

To affect other persons and their goals, values, and actions, actors wielding power—regardless of their personal abilities and knowledge—need specific discretionary authority, resources, institutional support, and ways to engage spontaneously in face-to-face contact with other influential and highly qualified actors. These essentials, however, are not available everywhere; they are tied to specific positions within a certain organization and to particular places and milieus. Foucault (1980) noted that power can be exercised especially through operations and interaction within organizations and networks: "Power is employed and exercised through a net-like organization" (p. 98). An academic, high official, chief executive officer, or politician who gives up all institutional affiliations upon retirement may still retain a degree of influence by dint of personal charisma or may occasionally be consulted. But the moment all formal authority, means of power, and resources are relinquished, this person also loses the ability to overcome resistance to his or her goals.

The position a person has in an organization is not the only factor determining what that individual can achieve with his or her abilities. The local potential for spontaneous high-level contact, the knowledge milieu, and the prestige of the place at which an actor discharges most of his or her functions has a bearing as well. This is one of the reasons why academics, top managers, journalists, and politicians, for instance, can be more effective in some places or milieus than in others. When it comes to exercising power, certain places and spatial contexts have always been more important than others.

Power is exercised not only through actions (requests, demands, commands, or attendant gestures) appropriate "for changing another actor's system of convictions and preferences. . . . The very presence of a powerful actor or the presentation of power-coded signs can be an act of exercising power" (Schönrich, 2005, p. 384). What is lacking in official authority or resources can sometimes be made up in prestige (*auctoritas*). For whoever possesses that quality can indirectly exert considerably influence and, hence, power.[20]

In many societal and economic fields, technical competence, domain-specific knowledge, experience, occupational success, and personal integrity are prerequisites

---

[20] *Potestas* and *auctoritas* were differentiated as two forms of power by Cicero's time (106–43 B.C.) in the Latin-speaking realm (Kobusch & Oeing-Hanhoff, 1980, p. 586).

for building prestige and authority. The art, or science, of persuasion is another foundation of power for exerting influence and moving things (Cialdini, 2008; Kobusch & Oeing-Hanhoff, 1980, p. 586; Tanner, 2005, p. 11).

Asymmetry of Power Relations

Seeking and exercising power is always about creating, preserving, or diminishing asymmetries between the actor (social system) who has it and the actor subjugated to it. One may distinguish between stable and dynamic asymmetries. In some social relations such as those between parent and small child, teacher and student, or jailer and prisoner, the asymmetries in the distribution of power are clear from the outset. In other spheres they stem from differences in resources and privileges; levels of training, qualification, and information; or the results of an occupational selection process, economic competition, armed struggle, or political conflict. That is, this second category of asymmetries in the distribution of power can be changed again. Creating dynamic asymmetries of this nature is about attaining and at least temporarily keeping an edge in knowledge, information, organizational abilities, technologies, and resources and about exerting influence on the production and spread of knowledge.

Such asymmetries are realizable in many ways. When it comes to factual knowledge, some of the possibilities are heavy investment in education, research, and development; immigration of highly qualified actors; development of superior technologies and weapons; research secrecy (see Lappo & Poljan, 1997, 2007; Westwick, 2000); communications espionage (e.g., the scandal currently engulfing the U.S. National Security Agency); betrayal; the encrypting or decrypting of secret information; censorship (Boyer, 2003; Burt, 1998; Malý, 2005; Post, 1998); bans on research; and the plundering of patents in conquered countries (Gimbel, 1990; Harmssen, 1951; Lasby, 1971). As far as orientation knowledge is concerned, such asymmetries arise mostly when standards of definitions or the moral or legal norms applied to oneself differ from those applied to others (Elias & Scotson, 1994; Imbusch, 2012b, p. 185), as when otherness is put down and demonized (the axis of evil) and one's own world and experience is morally glorified.

Most asymmetries of knowledge and power have a spatial dimension and can be studied in various spatial scales. They are expressed by spatial inequalities of various kinds, appear in the hierarchy of central places, and in many domains influence the attractiveness of places, the distribution of resources, and the migration of people.

Can Factual Knowledge Be Clearly Differentiated from Orientation Knowledge?

Before detailed examination of the various functions that factual knowledge and orientation knowledge do have in the acquisition and exercise of power, it is necessary to discuss whether these two kinds of knowledge can be differentiated clearly. The answer to this question depends on the level of analysis (individual person or goal-oriented social system) and on the type of problem that has to be solved. A person's cognitive processes and actions are based on both factual and orientation knowledge and on emotions, intuitive insights, and automatized (subconscious) routines. However, in some problem-solving situations the individual needs factual knowledge first. In other situations orientation knowledge plays the dominant role. Factual knowledge is acquired and applied in the everyday life-world.[21] It must also prove itself there. Factual knowledge and orientation knowledge complement and influence each other. Orientation knowledge can adversely affect the perception and acceptance of available factual knowledge. By the same token, newly won factual knowledge can modify existing orientation knowledge (e.g., prejudice) whether or not anyone is aware of it. At the level of the individual, it is thus analytically difficult and sometimes not even purposeful to distinguish clearly between factual knowledge and orientation knowledge.

Yet in terms of social systems characterized by a high division of labor, complex structures, and the will to keep them viable, it makes complete sense to distinguish between factual knowledge and orientation knowledge, if only for practical reasons. At that level of aggregation, the two kinds of knowledge serve different purposes. Specialists in generating and imparting orientation knowledge (e.g., priests, mullahs, rabbis, ideologues, propagandists, and spin doctors) have other tasks and roles within a system, need other kinds of occupational competence, and therefore undergo training different from that of people who generate and impart factual knowledge (e.g., engineers, scientists, or medical doctors).

In summary, there are decision-making situations in which one must definitely separate factual knowledge from orientation knowledge because complex sociotechnical systems would otherwise cease to work and would no longer reach their objectives. Without appropriate factual knowledge, it would be impossible to manufacture an airplane, carry out chemical analysis, launch a satellite into geocentric orbit, program software, conduct a research project, or even build a sturdy house. However, every social system requires a body of orientation knowledge in order to define its goals and preserve its internal cohesion, motivation, cultural identity, and collective memory. And the powers that be use orientation knowledge to mobilize their followers, create collective identities, and consolidate power. Once drawn, though, the line between factual knowledge and orientation knowledge is not engraved in stone. It is contingent on culture and time, as shown by the following account of development in the sciences (see also Hanegraaff, 2008; Stenmark, 2008; Welker, 2008).

Unlike the case in the Arabic cultural space, science in medieval Christian Europe served primarily moral, ethical, and theological objectives. Meier-Oeser (2004,

---

[21] Schütz and Luckmann (1973, pp. 22–34) distinguish between various provinces of reality, namely, those of the everyday life-world, fantasy worlds, and the dream world. "The life-world is something to be mastered according to my particular interests. I project my own plans into the life-world, and it resists the realization of my goals, in terms of which some things become feasible for me and others do not" (p. 15). Only in the everyday life-world do materiality and physicality operate and technologies and competition play a role. Only there can a lead in knowledge develop into economic or political significance.

p. 903) points out that in the Middle Ages *scientia* initially meant something like *doctrina* (a principle or body of principles presented for acceptance or belief) and *disciplina* (a branch of knowledge or teaching). Knowledge of the natural sciences tended to be seen as less important. Medieval science was subordinated to theology. The highest truth was revealed religious truth. "Purely logical reasoning and the testimony of the organs of the senses have only a subsidiary role, and only in so far as they do not contradict the truth of the revealed Scripture" (Sorokin, 1985, p. 229). The notion that rationality (*ratio*) should be emancipated from faith (*fides*) and that scientific thinking should be liberated from the confines of ecclesiastical control was not proclaimed until the twelfth century, the period of academic awakening in the French cathedral schools (Kintzinger, 2003, pp. 142–143).

What is recognized as factual knowledge at time A can be defined as orientation knowledge at time B and vice versa. What is defined as superstition, ceremony, or ritual in European society, shaped as it is by rationalist thought and the credo of individuality, can be regarded by South Sea Islanders as factual knowledge. Gunter Senft's chapter in volume 8 of the series on Knowledge and Space beautifully shows that the knowledge of how to make a traditional canoe on the Trobriand Islands consists not only in the way one chooses and then works a tree trunk but also in the message that a canoe can be made only if one knows the traditional rites associated with each step in the work. Without these rites, the diverse steps in the work cannot be executed. If the rites are forgotten, then it becomes impossible to continue making traditional canoes.

Moreover, it is important with orientation knowledge to tell the external from the internal perspective. What is superstition, faith, or ideology to the external observer can be seen as objective knowledge by the members of a faith community or the disciples of an ideology, for they are more or less convinced that their religion or worldview is true or correct. Adherents of creationism do not doubt that they possess solid knowledge. Many new religious movements (New Age, kabbalah, esotericism) claim that their beliefs are scientifically proven (Belyaev, 2008; Lewis & Hammer, 2011; Zeller, 2011). Many Marxists are convinced that Marx discovered scientific laws of history and that Marxism is an objective science based on facts (for details see Gyuris, 2014, pp. 115–116).

The hard thing for a social system is to find the balance between these two kinds of knowledge and to know which of them should have precedence in which situations, depending on the challenge and problem at hand. In many situations calling for a decision, stressing orientation knowledge more than factual knowledge dulls perception, complicates sober situational analysis, limits self-critical insight and receptiveness to information contrary to favored stances, and even allows inadequately qualified decision-makers into a social system's positions of responsibility, where they eventually harm their own system.

In the course of history there have been repeated attempts—especially by totalitarian systems—to place higher value on orientation knowledge than on factual knowledge. As Russell (1958) put it, "Revolts against reason . . . are a recurrent phenomenon in history" (p. 88). Győri and Gyuris (Chap. 10 in this volume) describe an example of such ideologically driven aberrations. Intent on emulating policies of the Soviet

Union, senior decision-makers in communist Hungary's planned economy of the late 1940s pursued the cultivation of subtropical plants such as cotton, a move that the area's climatic conditions naturally doomed to failure. One of the most remarkable historical blunders due to overemphasizing orientation knowledge and neglecting factual knowledge was The Great Leap Forward in the People's Republic of China (1958–1961). This campaign was ordered by Mao Zedong with the goal of rapidly transforming the country from an agrarian economy into an industrialized society with the help of unskilled people. The Great Leap Forward ended in economic disaster and tens of millions of excess deaths. Most totalitarian states have failed in the long run because they attached more importance to their ideology than to the analysis of empirically verifiable facts and ultimately believed in their own propaganda. They managed to remain in power so long only because they were able to control the spread of information almost completely and because their monopoly on propaganda enabled them to mold much of the population's orientation knowledge effectively over a long period.

Orientation knowledge differs from factual knowledge in many other ways, too. First, orientation knowledge can be reactivated with relative ease even after long phases of repression or censorship following socially controllable learning and information processes. But factual knowledge, once it is proven wrong, only very seldom makes a comeback. Second, tried-and-tested factual knowledge is universally applicable, whereas orientation knowledge has developed within certain cultural traditions and biographical experience. "Key human experiences are condensed and interpreted in orientation knowledge. It remains linked to communities, cultural contexts, and particular institutions that make it possible to cluster, deepen, and abidingly pursue communication about contentious matters" (Tanner, 1999, p. 233).

Tried-and-tested factual knowledge is compatible with many different world-views, but various categories of orientation knowledge are mutually exclusive for the most part. The adherents of any religion or worldview can use scientific findings (mathematics, chemistry) and technologies (airplanes, computers, weapons) and can benefit from a spread of literacy, a scientific study, or specific qualifications (foreign languages). But it is difficult to imagine someone being both a practicing Moslem and practicing Catholic at the same time or supporting both a communist and a conservative party in the same election campaign.

There are thus exciting, yet little researched, questions to explore. Do the relation and distance between factual knowledge and orientation knowledge change from one era, culture, and ideology to the next? If so, how much? What situations demand a clear demarcation between factual knowledge and orientation knowledge so that a social system remains viable? And in what situations is it unnecessary or even impossible to separate factual knowledge from orientation knowledge because they are too closely intertwined? Telling them apart is surely easier if one goes further and breaks down factual knowledge into natural, experience-based, descriptive, and interpretive sciences, for the divergence between interpretive sciences and orientation knowledge is considerably smaller than it is between natural sciences and orientation knowledge.

Factual Knowledge and Power

Functions of Factual Knowledge in Acquiring and Retaining Power

Whenever those in power want to accomplish their stated goals, secure or expand their dominion and resources over long periods, protect or widen their technological or economic lead, or make their worldview prevail, the social system they control (e.g., an organization, institution, company, army, or state) must continually solve problems and weather crises, competition, transformation, and conflicts. Social systems operating in a contested or highly unstable, dynamic environment can ensure their long-term existence or finally succeed against a rival only if they have sufficient factual knowledge, competence, and absorptive capacity and can avoid making too many poor decisions during recurrent situational analyses and problem-solving.[22] Flawed situational analyses, ideologically incurred lack of self-criticism, and poor judgment owing to inadequate knowledge and information waste resources, lead to political and military defeats, undercut the system's competitiveness, undermine the leadership's authority, and weaken the cohesion of the social system in question. Whoever shares Foucault's (1979, 1990) view that changes and discontinuities are an important trademark of society and describes "social relations as confrontation, as the interaction of operative forces, as continuous overt and covert violence, as war, and as subjugation, but particularly as struggle" (Kneer, 2012, p. 267) will almost inevitably have to address the role of factual knowledge in coping with uncertainty.

Factual knowledge and the capacity for reflection are needed partly for carrying out situational analysis that is as realistic as possible; setting feasible objectives; solving technical and scientific problems; determining the efficient use of energy and resources; planning, conducting, and monitoring process flows; and efficiently controlling and coordinating large, complex organizations. To be successful, outlast competition, or reap business profits, though, one does not need knowledge per se but rather a knowledge-related edge over rivals. Such an advantage in factual knowledge can consist in technological head starts, inventions, or scientific findings. Or it may lie in superior absorptive and analytical capacity and in creativity or intuition that facilitate a social system's detection of possible problems or opportunities and risks of new developments earlier than its competitors do (for details see Meusburger, 2013, p. 17–18).

An operation's success or failure and the longevity of a goal-oriented social system thus heavily depend on how something is perceived, experienced, represented, analyzed, and interpreted in the many iterative steps of the decision-making

---

[22] The fact that this chapter focuses on competitive societies confronted with an uncertain environment does not mean that possible achievements of collaboration, friendship, or altruism are underestimated. Knowledge and power gained from collaboration and partnership may be even more significant for addressing certain issues. The question is in which scale (family, firm, state, global institutions) and under which preconditions such noncompetitive environments will be feasible.

process; how a social system deals with the knowledge and contradictions of its actors and with its own fragility; and how the knowledge that is thereby obtained influences activities. At root lies Descartes' (1637/2001) old issue of how to distinguish the true from the false (for details see Ricœur, 2005). "How do we know when we are being informed and how do we know when we are being manipulated?" (Wilkin, 1997, p. 12). At each link in a chain of perceptions, analyses, and decisions, mistakes can be made, resources wasted, or advantages gained over competitors. The timely perception of a problem and the apprehension and description of a situation depend primarily on the prior knowledge, capacity for reflection, cognitive abilities, and personal experience of the actors involved. These skills decide whether and how available information is perceived, analyzed, and evaluated by them and whether it enters and broadens their body of knowledge.

The cornerstone of a social system's doom already lies in place if the facts and contexts important for a decision are not sufficiently well known[23]; if problems and developments are not perceived in time; if the information and knowledge needed for a situation's analysis are absent or too abstract; if unqualified actors occupy positions of decision-making responsibility; if one's resources and abilities are overestimated and those of the rivals are underestimated; if the opportunities and risks of a technological, economic, or political development are misjudged; that is, if the social constructs of the important decision-makers are too removed from an intelligible situation or perceivable material reality. This need for clarity is one of the reasons why espionage, deception, and camouflage play such an important role in modern warfare.

History abounds with examples illustrating how highly qualified decision-makers in politics, business, science, and the military or leads in research, technology, productivity, and secret-service intelligence eventually fosters growth in political, military, and economic power and how this edge is lost through technical incompetence, wrong perception, misjudgment, or insufficient adaptability of important decision-makers. Every social system makes mistakes but can partially recover from them by committing additional resources, investing in relevant learning, or exploiting the mistakes of competitors. But a social system suffering from an accrual of poor decisions with onerous consequences winds up squandering resources, creating dependence, eroding reputation, and weakening competitiveness. To survive for long in a dynamic, highly uncertain external world, a social system must be capable of learning and adapting, must have high-ranking contacts to other important systems, and must be able to recognize (anticipate) new developments, risks, and opportunities early. For these reasons those in power need the skills of experts, consultants, and scientists to analyze situations, set achievable goals, seek

---

[23] The importance of a lead in information was already underlined by the Chinese military general and philosopher Sun Tzu (544-496 B.C.) in his work *The Art of War*: "The general who wins a battle makes many calculations in his temple ere the battle is fought. The general who loses the battle makes but a few calculations beforehand. Thus do many calculations lead to victory and few calculations to defeat." Retrieved November 24, 2013, from http://www.military-quotes.com/Sun-Tzu.htm. Some modern scholars believe that *The Art of War* contains not only the thoughts of its original author but also commentary and clarifications by later military theorists.

alternatives, find solutions to problems, manage major organizations efficiently, build technological leads, and sustain competitiveness. It also pays to keep in mind that the competence, knowledge, and information from which a system gleans a competitive advantage are always rare (see Meusburger, 2013).

In principle, competition between different power centers is about building at least a temporary lead over others in knowledge, technology, productivity, and information that can contribute to political, military, economic, or scientific superiority in a given situation. It need not entail momentous innovations such as the invention of the steam engine, the telephone, the airplane, or the computer. History shows that even small technical changes in a chariot (the Egyptians invented the yoke saddle for their chariot horses in 1500 B.C.), a bow (Hungary, ninth century A.D.), or an equestrian saddle (the Mongol invasion in the thirteenth century A.D.) had great historical import because they gave the corresponding armies a significant military advantage and changed the balance of power for a certain period. In later centuries it was maps, navigational instruments, secret cosmographies, new ship designs, weapons, encrypting machines, missiles, and nuclear bombs that each affected the military, political, and economic power relations for a time.

It is true that scientific disciplines have provided new tools and technologies to improve production, communication, transport, energy use, and space exploration. Modern science has spawned new materials, reshaped industry, changed the management of firms, created new weapons, and has thereby altered the planning and execution of military operations. Most important, modern science "has been decisive in the reproduction of elites and their cultural capital; and it has been central in offering new ideals and social goals, new ways of thinking about the world, nature, and society alike" (Pestre, 2003, p. 247).

The Search for Absolute Truth or Getting on in the Life World?

If the concerns of coping with life, staying competitive, retaining power, or understanding the evolution of social systems are the main interest, there is no need to recount the vast philosophical literature about what truth is, whether there is absolute or objective truth, or whether an objective reality can exist beyond the human perceptual world and language (for an overview see Anacker, 2004; Arndt, 2004; Chomsky, 1987; Gehring, 2004; Hardy & Meier-Oeser, 2004; Knebel, 2004; Knoblauch, 2013a, 2013b; Onnasch, 2004; Pulte, 2004a, 2004b; Zachhuber, 2004). Humans have a basic need for "objective clarity" (Felder, 2013, p. 20), "objective reality" (Berger & Luckmann, 1966, p. 65–146), "recipe knowledge" (pp. 57, 83; Schütz & Luckmann, 1973, pp. 225–226) or reliable knowledge rooted in experience and experiments. They seek assurance that their perception of the world they inhabit is as real as possible and that they can accurately assess the opportunities and risks of what they do.

But aside from solving technical problems (e.g., designing an airplane, building a safe bridge), which calls for proven, experimentally tested and absolutely reliable knowledge,[24] many situations requiring a decision or successful management of uncertainty do not depend on apprehending an objective reality or possessing an absolute truth. There is often not even the time to accumulate factual knowledge about all the circumstances a situation entails. In those cases, decisions just have to rest on the information and experience one has (see Stegmaier, 2008, p. 14). It is frequently possible to find good, albeit not unshakable, reasons that a particular statement about the independent world is more likely to be true than a competing statement (Gadenne, 1999). "As a consequence of evolutionary development, our mental world, which our brain pieces together with the help of our senses, is so good at replicating the real world, at least some of its key attributes, that we can operate in it successfully" (Penzlin, 2002, p. 73).

Studying the connection between factual knowledge and power is not about grasping and describing a reality that exists independently of human sense perceptions but rather about realistically judging a situation, finding one's way in the world, solving practical problems, performing specific tasks, and coping with unforeseeable challenges. Humans, with their limited cognitive abilities, will never be able to grasp reality in its totality. They can only try to approach fragments of reality asymptotically. If actors or social systems are to survive in an extremely competitive, volatile environment, they must be able to adequately size up the constraints of the external world in which they want to reach a specified goal, to ascertain their resources and possibilities, and to draw the proper conclusions from those considerations. In such situations the materiality of the environment and the corporality of acting individuals have a special importance.

The position taken in this section appreciates the concepts of communicative constructivism (for details see Christmann, 2013; Keller, 2013; Keller et al., 2013; Knoblauch, 2013a, 2013b) but takes exception to the arbitrariness of radical constructivism, which casts the world as nothing but a construct of the brain and recognizes only the existence of subjective truths. The statements of radical constructivism are banal if one does not simultaneously ask why actors with different experiences and disciplinary qualifications arrive at different social constructions and which impacts realistic and unrealistic constructs can have for the actors or the system to which they belong. A social construct's quality, or "verisimilitude" (Pulte, 2004a), depends mostly on knowledge and abilities resting on earlier experiences and learning processes that enable the actor to glean patterns from clues and incomplete information (Liebenberg, 1990), to analyze and interpret those patterns, and to come to the conclusions that are correct or helpful for the attainment of a particular objective.

When analyzing a situation, solving a problem, accomplishing a goal, or reducing uncertainty, a person can distinguish realistic or appropriate, useful, and adequate social constructs (e.g., situational analyses, circumstantial judgments, and market

---

[24] Even well-founded knowledge is not always the "truth" but rather knowledge acquired according to the prevailing rules by means of approved measurement methods. In other words, it can turn out differently depending on what measurement methods are applied (see Cicourel, 1974). Views may differ on the rules and the measurement methods to be used. The submission of results, however, makes it possible to judge how reliable a given knowledge was and whether it was justifiably relied on.

analyses) from unrealistic, inadequate, or loss-incurring ones. When results have become available, it is possible to say, at least with hindsight, that particular statements have described a given situation or the perceivable world more accurately than others, that certain social constructs have weathered competition or conflict situations better than others, and that "the subjective criteria of truth . . . must not stray arbitrarily far from their objective grounds" (Puster, 1999, p. 99). As aptly (and ironically) stated by the evolutionary biologist George G. Simpson (1963), "the monkey who did not have a realistic perception of the tree branch he jumped for was soon a dead monkey—and therefore did not become one of our ancestors" (p. 84). Unrealistic social constructs based on deficient knowledge and lack of experience and information are among the most salient causes of failure of actors, organizations, and states.

Orientation Knowledge and Power

What Functions Does Orientation Knowledge Have for a Social System?

In addition to a basic need for clarity, human beings have a basic need for moral and cultural orientation, especially when looking for meaning or, as political animals (*zoon politikon*), when making decisions shaped by interwoven interests (Felder, 2013, p. 20). Imparting orientation knowledge is not about the search for a scientifically verifiable truth and not about objectively provable facts. Nor is it about exact and objective descriptions or situational analyses that are as realistic as possible. Specialists in generating and imparting orientation knowledge have the task of communicating goals; interpreting events[25]; giving values, motivation, identity, and legitimation to their social system; and forging basic consensus within a system. It is their job to make moral judgments on the Self and the Other and to change epistemic perspectives. They are expected to manufacture myths, legends, collective memories, cultural traditions, and identities; make them prevail over rival stocks of knowledge; and hand them down to the next generation. Religions, ideologies, cultural memories, and worldviews are the binding agents of the social system and are the lever that is used in attempts to morally denigrate, discredit, or demonize opposing systems. Concepts of orientation knowledge are not purely intellectual devices; Eurocentrism was not just an idea. "The accounts drawn up under its sign had acutely material consequences" (Gregory, 1998, p. 14).

It goes without saying that there exist numerous historical and contemporary examples of how orientation knowledge (e.g., world religions) has contributed to mutual understanding, societal harmony, peaceful coexistence, personal and collective identity, empowerment, and great cultural achievements. However, orientation

---

[25] For Nietzsche, all will to power is interpretation. The stronger person determines the moral standards and the criteria of truth and defines the worldview. The weaker person is subjected to outside perspectives and values (Speth, 1997, pp. 274, 277).

knowledge has also been misused by institutions of power and applied in order to persuade, to manipulate, to pursue psychological warfare, and to feign moral superiority. The rest of this chapter focuses on such abuse because examples thereof disclose with striking clarity the main mechanisms of how orientation knowledge conduces to political power in competitive situations.

Orientation knowledge is often used to legitimate power. Every kind of power needs justification and therefore strives for legitimation (Popitz, 1992, pp. 17, 66). The legitimation of power can occur in very different ways, depending on the era, the authority structures, and the culture involved. However, many attempts to legitimate power are assertions that the center of power (e.g., tribal chief, government, party boss, corporate management, army high command, or university president) knows more or is wiser or more competent than the rest of the organization, can judge better than others what is right or promising, and has sources of information that others do not. This point is especially conspicuous in totalitarian states or fundamentalist theocracies, where there is just one truth and one correct interpretation and where protest is punished. In totalitarian systems (e.g., Stalinism, Maoism, and National Socialism), the center—usually the mass leader—claims to be infallible and can never admit error. "The assumption of infallibility, [however], is based not so much on superior intelligence as on the correct interpretation of the essentially reliable forces in history or nature, forces which neither defeat nor ruin can prove wrong because they are bound to assert themselves in the long run" (Arendt, 1951, p. 339).

Depending on the era, exponents of orientation knowledge have claimed to receive messages from the gods or from ancestors; to have the ability to interpret signs from gods, dreams, and oracles; to be inspired by the Holy Spirit; to possess sacred books, to own secret knowledge or divine wisdom revealed from generation to generation only to a small elite of insiders (Dan, 2007; Halbertal, 2007), to stand for the will of God on earth, or to be better positioned than others to interpret holy scriptures (the Bible, the Koran) or the publications of Marx, Lenin, and Mao. It has been argued in Christian Europe and other cultures that the rulers have been granted their power by God or the gods (see Fig. 2.2). Emperors and kings in Christian Europe ruled "by the grace of God" (Tanner, 2005, p. 4).[26] Their power was thus doubly protected—by popular obedience from below and by their function as a servant of God from above (Röttgers, 1980, p. 592).

The manifestation of such needs for legitimation lay also in the fact that the emperors of the Holy Roman Empire were crowned by the pope, an act staged to express the divine will sanctioning their rule. These contentions by the communicators of meaning that they are able to appeal directly to God or mediate between God and the people have long been represented pictorially. In Christianity the dove has stood as the symbol of the Holy Spirit, as evidence of a direct link to God. In Fig. 2.3, for example, the thoughts that Saint Gregory is to write down are received

---

[26] This understanding of power comes from Paul's letter to the Romans, "Let every person be subject to the governing authorities; for there is no authority except from God, and those authorities that exist have been instituted by God" (Romans 13:1, New Revised Standard Version).

Fig. 2.2 The apotheosis of Holy Roman Emperor Otto III (A.D. 980–1002). This representative example of medieval dynastic iconography shows Otto surrounded by an aureole, which is otherwise confined to depictions of Christ. In keeping with the medieval concept of rule, the image expresses the idea that Otto, by virtue of his imperial coronation, has himself become Christ, the Annointed One. Otto's status is confirmed by the Hand of God that appears in the blue nimbus above across. The Hand is crowning the emperor, who spreads his arms in the pose of crucifixion (Source: Liuthar Evangelar. Copyright: Domkapitel Aachen. Photograph: Pit Siebigs. Reprinted with permission)

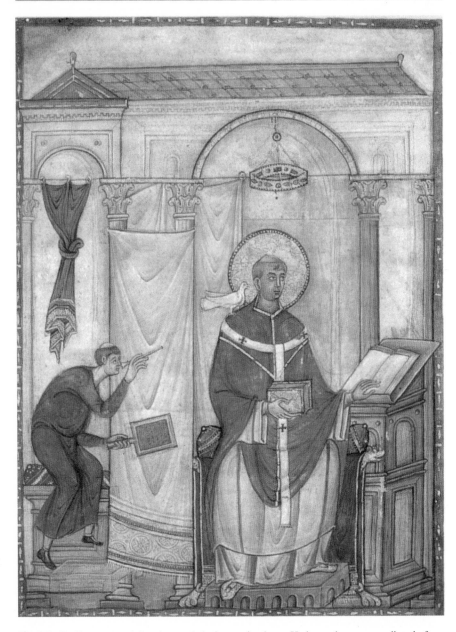

Fig. 2.3 St. Gregory receiving the words he is to write down. He hears the message directly from the Holy Spirit, which is symbolized by the dove sitting on his shoulder (Source: Meister des Registrum Gregorii. (Trier, Stadtbibliothek, Hs. 171/1626). Reprinted with permission)

by him through the dove sitting on his shoulder, that is, directly through the Holy Spirit. For centuries, therefore, the assertion of having a lead in knowledge has been one of the most effective ways to legitimate power.

Another way to legitimate power is to produce narratives maintaining that a dynasty is of divine origin (e.g., the Japanese imperial family) or that the roots of the ruling family extend far back in history. Some rulers in Islam used to attribute their legitimacy to their descent from Mohammed. Reza Pahlavi, the Shah of Persia, traced the ancestry of his dominion back to Cyrus the Great (ca. 500 or 576 B.C. to 530 B.C.), an Achaemenid ruler and the founder of the Great Persian Empire.

Leading figures of orientation knowledge are as useful as they are to a ruler mainly because they determine the moral norms, the criteria of truth, and the ideology and worldview favored by the ruler (see also Speth, 1997, p. 274). These shapers of the dominating ideology define which arguments, terms, and definitions are politically incorrect and which historical events may or may not be compared. They endorse patriotism, emotional identification with an institution or ideology, and the social cohesion of social systems. Religions, ideologies, and collective memories are an essential part of individual and collective identity. Orientation knowledge that meets with general social acceptance because it is traditionally considered true or correct in certain segments of the population or in particular territories can become enormously effective and dynamic. It can move people to go to great lengths, make immense sacrifice, persevere in difficult situations, or even die as martyrs for a "good cause." Because orientation knowledge is not subject to any scientific proof but instead is revealed by an authority and cannot be proven false, experts on orientation knowledge have far greater potential to influence, mobilize, and deceive people than natural scientists or engineers do. Elements of orientation knowledge are easier to convey to masses of people than scientific knowledge is.

Because orientation knowledge influences affectivity, it can also function as a filter for taking up information from the opposing side. External knowledge that shakes one's identity (self-image), exposes one's religious and political convictions to ridicule, or presents a history completely different from what one has experienced or has been taught by parents or other trustful persons is emotionally repudiated or repressed.

The Role of Orientation Knowledge in the Construction of the Self and the Other

Many conflicts and the tenacity with which they are dealt arise largely when two mutually incompatible bodies of orientation knowledge collide. In conflict regions such as the Balkans, Northern Ireland, or the Near East, each of the adversarial groups has created its own narratives, collective memory, and interpretation scheme. Each of them cultivates its own truth, basic consensus, and cultural memory through narratives, symbols, and figurative or metaphorical representations (e.g., monuments) and tries to keep its own region of influence free of the opposing party's interpretations.

The production of "geographical imaginations" (Gregory, 1994) or "imaginative geographies" (Gregory, 1995) is always articulated within a system of difference (Gregory, 1998, p. 20). The construction of collective identity (ethnicity, national consciousness, patriotism, basic consent) is all but inconceivable without demarcation between "Us" and the "Others." Distinguishing between the "sacred" and the "profane" or between the "self" and the "other" is inevitably tied to moral judgments, prejudices, and stereotypes.

Prejudices are presumably as easy to stoke as they are—and as difficult to eradicate—because associations between people and traits or between events and feelings form in an area of the human brain that lies beyond one's control. In the course of early human evolution, stereotypes and prejudices had a major part in survival. Habitualized schemata help the brain to accelerate its processing of information, to retrieve experiences and evaluations instantly with little cognitive effort, and to make decisions immediately (Brown, 1995; Leyens, 2001).

Dichotomies such as "we" and "the others," which are probably a foundation of nearly all ideological, religious, or ethnic conflicts, are imparted in early childhood within the family and other primary groups, that is, by the persons in whom one first establishes trust and to whom a close emotional relation exists. One learns in early childhood that strangers can pose a threat. Later, it is learned from the communicators of orientation knowledge that the "others" (the barbarians, nonbelievers, heathen, savages, or terrorists) have repeatedly inflicted injustice and violence on one's own group, that the others were the perpetrators and the members of one's own group were the innocent victims. Such attitudes and values are acquired in the primary community, not through conditioning in the sense discussed in modern learning theory but rather through emotional exchange with persons in whom one trusts and on whom one depends (Meier-Seethaler, 1999, p. 151). The fact that an individual recognizes mostly the good in his or her "own group" and tends to think the other side to be capable of inhumanity and atrocities partly springs from people's greater receptiveness to information that corresponds with their prejudices. Prejudices spare strenuous thinking. They enhance the cohesion of one's group and shield one's self-esteem (Leyens, 2001, p. 11987).

Orientation Knowledge and Moral Exclusion

An especially important task of people versed in orientation knowledge is to define and categorize situations, historical events, persons, and organizations and to construe their side's moral superiority. The same person in a given conflict can be defined as a terrorist or a freedom fighter, as a hero or a war criminal, as a patriot or a traitor. A messianic political agenda, a religious conviction of personal mission, and the firm conviction to be God's chosen people are particularly effective not only at creating identity but also at promoting moral ostracism and the double standards that stem from it.

As shown by the history of religion as a concept, the mechanism of moral ostracism has been practiced since ancient times. The Romans used the term *religio* only for their own cult. They did not grant the other cults in the empire the status of a religion but rather marginalized them as *superstitio* (Feil, 1986; Kerber, 1993). Later, Christianity appropriated the term *religio*, along with many other insignia of ancient Rome, exclusively for itself, with *superstitio* being used as a battle cry in the struggle against all other cults. The word *religion* was not used for non-Christian religions until after the Enlightenment.

Morally ostracizing the opponent depersonalizes, dehumanizes, and demonizes that person. The attendant propaganda aims to persuade people that their own group has a divine mission; embodies the only true religion; stands on the side of morality, historical necessity, or incontrovertible truth; or represents God's chosen land in the fight for freedom, world peace, human rights, and democracy. These teachings make the opponents out to be barbarians, subhumans, infidels, heretics, terrorists, criminals, or class enemies. Leaders of orientation knowledge ensure that their own side uses moral standards or legal codes different from those of the others. The laws of war, the Geneva Convention, and the International Court of Justice in The Hague apply to one side, and the others can ignore them with impunity. If opponents are depersonalized, demonized, and confined in a no-man's land devoid of rights and protection (e.g., Guantánamo Bay, the U.S. military prison camp at the southeastern end of Cuba) as terrorists, or unlawful combatants, they can be treated differently than prisoners of war, who are protected under the Geneva Convention. The dehumanization of the opponent and the dogma that one's own side embodies absolute good, and the other side absolute evil, go to rationalize and justify the use of force. They are also necessary for torturers to lose their inhibitions and sense of injustice.

> The enemy is demonic and the saints are perfectly pure, no matter what they may do in battle. These images have been presented in so many movies, stories, comic books, and newspapers that they have etched themselves firmly in the national consciousness. (Jewett & Lawrence, 2003, p. 222)

> This division of the world into good and bad people is a crucial component of the Captain America complex, visible in World War II, the Cold War of 1945–90, the Vietnam War, the Gulf War, and the current "war on terrorism." None of these struggles would have occurred as they did without such stereotypes. (Jewett & Lawrence, 2003, p. 215)

The dichotomy of good and evil always figures in conflicts. It is part of the oldest myths and is highly effective. Each of the parties to a conflict tries to present itself as a moral authority, the representative of a superior civilization, or an instrument of God ("God's own country," "the chosen people") that is fighting against the darkness, unbelievers, pagans, barbarians, or terrorists and seeking to spread the blessings of civilization (see Jewett & Lawrence, 2003; Chap. 7 by Jewett in this volume; Gregory, 2004). Abrams (1969), Jewett and Lawrence, Weinberg (1935), and many others describe how the concept of Manifest Destiny (one nation under God) determined expansionist American policy from the outset. Derived by the Puritans from the Old Testament and popularized around 1840, it rested on the premise that the United States was the "holy nation" referred to in Exodus 19:6 and that it had a divine mission. John Winthrop the Younger (1606–1676), a prominent Puritan leader, convinced his followers that their nation would become "a guiding light," an "example to the whole world," and a "bulwark against the kingdom of Anti-Christ," meaning the Jesuits (Winthrop, 1869, vol. I, pp. 309–311). Whoever had such a divine mission could, according to Jewett and Lawrence, take ruthless action against enemies.

> The unquestioned premise was that a victorious crusade would truly make the world safe[,] ... that the destruction of the demonic Beast would automatically bring the world under the control of the saints (p. 74).

> The biblical tradition of redemptive violence was popularized in Western culture by the Crusades, and it was then taken up by the Reformation in England. . . . Puritanism developed the crusading impulse of the Old Testament to the logical extreme (p. 250).

The massacre of the Native Americans was justified in the United States with similar concepts and arguments, as were all subsequent wars. "The bloodthirsty savages had to be radically decontaminated for inclusion in the kingdom of the saints; and if they refused, annihilation was the logical solution" (pp. 253–254). The following passages present only a few examples illustrating the rhetoric of Manifest Destiny passed along by newspapers, films, school texts, novels, and comic books (e.g., *Captain America*), beginning with a statement John Adams wrote to Thomas Jefferson on November 13, 1813: "Many hundred years must roll away before we shall be corrupted. Our pure, virtuous, public spirited, federative republic will last forever, govern the globe and introduce the perfection of man" (quoted in Jewett & Lawrence, 2003, p. 221). The United States was repeatedly portrayed as a chosen land in American literature. As Herman Melville wrote in his novel *White Jacket* (1850/1970), for instance,

> we Americans are the peculiar, chosen people—the Israel of our time; we bear the ark of the liberties of the world. . . . Long enough have we been sceptics with regard to ourselves, and doubted whether, indeed, the political Messiah had come. But he has come in us, if we would but give utterance to his promptings. (p. 151)

In January 1900 historian and Senator Albert J. Beveridge supported the Spanish-American War in the Senate with the words, "Almighty God . . . has marked the American people as the chosen nation to finally lead in the regeneration of the world. This is the divine mission of America . . . We are the trustees of the world's progress, guardians of the righteous peace" (Congressional record, 56th Cong., 1st Session, vol. 33, p. 711, as quoted in Jewett & Lawrence, 2003, p. 3). Addressing the Harvard Club in 1917, Theodore Roosevelt stated upon the U.S. declaration of war on the Axis Powers that year: "If ever there was a holy war, it is this war" (Jewett & Lawrence, p. 73). The entry of the United States into World War I was favored by preacher Randolph H. McKim, too. In Washington, D.C., he declared,

> It is God who has summoned us to this war. It is his war we are fighting . . . This conflict is indeed a crusade. The greatest in history—the holiest. It is in the profoundest and truest sense a Holy War. . . . Yes, it is Christ, the King of Righteousness, who calls us to grapple in deadly strife with this unholy and blasphemous power (Abrams, 1969, p. 55).

Speaking at the annual convention of the National Association of Evangelicals on August 3, 1983, Ronald Reagan called the Soviet Union an "evil empire." And on

January 28, 1991, George H. W. Bush characterized the first Gulf War in terms such as "good versus evil, right versus wrong, human dignity and freedom versus tyranny and oppression" (Jewett & Lawrence, 2003, pp. 2, 328). Likewise, the second Gulf War against Iraq was described by George W. Bush as a "monumental struggle of good versus evil" (Sandalow, 2001, p. A7). Similarly impressive moral self-aggrandizement and demonization of the opponent exist in numerous other countries as well.

It is the job of the cultural-awareness industry to mount targeted campaigns for a government, political party, business organization, or some other entity to spread rumors, withhold news, defame individuals, and supply public media with sanitized data and manipulated images in order to focus on particular topics and capture people's hearts and minds (see Maresch, 2002, p. 250). The result is that each party to a conflict can assume that it is serving a worthwhile aim, that it is acting in the name of God, morality, justice, or world peace. Along the way, religious convictions are used for political ends and political ideologies are supercharged into political religions (Tanner, 2005, p. 7).

> [P]owerful institutions and dominant social groups in modernity have been able to establish a hegemonic position whereby conceptions of what is good, true, real and universal have taken on the appearance of natural laws which bind us to a specific and seemingly inevitable social order. (Wilkin, 1997, p. 11)

Even in conflicts of a purely economic or imperialistic nature, the exponents of orientation knowledge have the task of morally justifying what their own side does. They are expected to use arguments suggesting, for example, that imperialism and colonialism are a dissemination of civilization or that a dispute over oil reserves is a struggle for democracy, freedom, and human rights. Propaganda, persuasion, psychological warfare, disinformation, camouflage, and manipulation are central features of politics, hegemonic practices, and warfare.[27] Truth is always the first victim of war. Propaganda "serves the purpose of disseminating a range of values, beliefs, and codes of behavior with which to develop and maintain popular support for the existing social order" (Wilkin, 1997, p. 122; see also Herman & Chomsky, 1988). At the heart of propaganda lies a core ideology uniting elite groups of a political system. Propaganda is about persuading and influencing the human being's inner self, heart, or conscience. "Whoever has power over people's hearts finds a following" (Tanner, 2005, p. 7).

As early as the beginning of the twentieth century, Mauthner (1910, p. 235) regarded language as the most important means of orientation for human beings. Because it is possible to affect the thought structures, discourses, and emotions of people and the legitimacy of actions through selected language regulations or prescribed terminologies, most conflicts involve clashes over the substance and "correct" use of terms. The use and misuse of language, or the relation between language, ideology, and power, are topics in which many authors are engaged (Arendt, 1951; Felder, 2013; Girnth, 2002; Herman & Chomsky, 1988; Imhof, 1996; Jäger, 2013; Marxhausen, 2010; Radeiskis, 2013; Schelsky, 1975; Wilkin, 1997). The party

---

[27] The famous Chinese military general and philosopher Sun Tzu (544–496 B.C.) stated, "All warfare is based on deception." Retrieved November 24, 2013, from http://www.military-quotes.com/Sun-Tzu.htm

whose vocabulary manages to win out in the public discussion has already won half the battle. Those embroiled in a conflict therefore do all they can within their sphere of influence to monopolize the use of words, the interpretation of texts and images, and the "power to define reality" (Imhof, 1996, p. 217), that is, to claim supreme authority over the discourse. It is about determining what is talked about and where; what kinds of labels are used; what gets pushed aside; what forms of power are intrinsically linked with the forms of speech; and in which institutions, mechanisms, and structures of global power the speech praxis is structurally embedded (Detel, 1998, p. 33).

Even tiny differences in vocabulary can have serious legal and political consequences, so the use made of words may be of the highest political significance. A historical event will be classified as a war crime, a crime against humanity, a violation of international law, or simply a migration, depending on whether it is seen as expulsion, ethnic cleansing, deportation, relocation, resettlement, or transfer. The term *wall* triggers political associations other than what the words *fence* or *peace border* suggest. In the German Democratic Republic the designation *new citizen* was an attempt to repress the historical experience of expulsion after World War II. The public impact of the word *torture* differs from that of *enhanced interrogation techniques*. Rather than say *war of aggression*, people find it more palatable today to refer to a *preemptive strike*. Whereas the U.N. High Commissioner for Human Rights, Louise Arbour, described waterboarding as torture, the Bush administration reclassified that technique and others as "alternative interrogation procedures" (Head, n.d., par. 2 & 5; "U.N. says," 2008, par. 1).

As far as the vocabulary of ideology is concerned, it is nomination theory, or the theory of language, that distinguishes between symbol words, whose function is to describe complex reality in condensed form, and words of demarcation, which ballyhoo a stance taken by a political party (see Marxhausen, 2010, pp. 222–223). Positively loaded symbol words or Miranda words ("that which must be admired," from *mirari*—to admire) include *peace*, *freedom*, and *justice*. Examples of negatively loaded symbol words (anti-Miranda) are *dictatorship*, *racism*, *torture*, and *terror*. Words of demarcation differentiate between positively loaded flag words, which raise the status of one's own group (*freedom fighters*), and negatively loaded stigma words (e.g., *terrorist*), which are intended to defame the opponent (see Girnth, 2002, pp. 53–54; Marxhausen, p. 223).

One of the most effective forms of language regulation is the coinage of *new* words that stress the unique nature of an event and thereby preclude comparisons with similar events. Another method of manipulation is to individualize and generalize events. In these cases torture and war crimes committed by one's own side are usually attributed to culpable individuals, and comparable crimes of the opponent are blamed on their system (e.g., nation, ethnic group, or political party). The torture and sexual abuse of Iraqis in the Abu Ghraib prison was attributed to individual soldiers and branded as un-American, for such practices were said to be inconsistent with American national character. But use of the generalizing term *national character*[28] suggests that such crimes can indeed be linked to the national character of other countries.

---

[28] Application of the term is predicated on a spatial dissemination of particular character traits, so the territory's residents—individually screened or unscreened—can be labeled with certain traits.

Yet another method of manipulation is the personalization and anonymization of victims. A personalization of one's own sacrifices is calculated to engender sympathy and to provoke outrage at the opponent. Making the opponent anonymous or invisible is supposed to desensitize or disengage one's conscience. Under totalitarian regimes executed opponents and the people killed through ethnic cleansing have usually been buried in anonymous mass graves. The dead have been deliberately stripped of their names and, if possible, their grave sites kept secret to prevent memories from surviving at those places. Whereas one's own war dead are treated as individuals, each with a name and a biography, the victims on the opposing side are called collateral damage allegedly unavoidable in strikes against military targets even in a clean war.

The point of following such language rules is to bring about an inequality in perception, evaluation, tolerance, grievance, and outrage. It is the foundation of every double standard and is one of the most important instruments of power in the information society. Semantics and the use of symbol words allow "inferences about the thinking and action of a speech community" (Girnth, 2002, p. 52), a topic that ought to be examined more vigorously in human geography than it has been.

Moral ostracism is nearly always linked with spatial exclusion. The "good" and "bad" are each localized and become the stuff of imaginative geographies (Said, 1978). For without precise localization, the bad cannot be attacked. "Imaginary geographies . . . are constructions that fold distance into difference through a series of spatializations . . . by multiplying partitions and enclosures that serve to demarcate 'the same' from 'the other'" (Gregory, 2004, p. 17). "Geography is inextricably linked to the architecture of enmity" (Shapiro, 1997, p. xi).

With Which Methods Do Power Centers Influence the Creation and Spread of Knowledge?

Manipulation of Epistemic Perspectives

From the very beginning of human history, the powers that be have always tried to legitimate and retain power and engender loyalties. They have done so by controlling the spread of information and generation of knowledge and by ensuring that the values, cultural norms, and interpretations of historical events conducive to their power prevail in public opinion. These components constitute what Gerhardt (1992) calls the "epistemic perspective": "Just as each act of seeing quite naturally entails an optical perspective, each act of cognition is linked to an accompanying epistemic perspective" (p. xii; see also Fellmann, 1992; Kaulbach, 1990). Tanner (1999) explains further that "all action predicates interpretive understanding and familiarity with systems of symbols. It is with such systems that our image of the world as a whole and our place in it arise" (p. 237).

Orientation knowledge, collective memories, traditions, and symbols that fashion a cultural, ethnic, or national identity must constantly be reorganized, practiced, and imparted. "Culture does not exist of things, people, behavior, or emotions, but in the forms or organization of the things in the minds of the people" (Goodenough, 1957, pp. 167–168). If power is defined in part as the ability to change the epistemic perspective and motivational structure of other people selectively (Detel, 1998, p. 21), then those in power either have to work closely with media experts or control them as much as possible. State institutions therefore try to "manufacture consent" (Herman & Chomsky, 1988; Lippmann, 1955; Rehberg, 2005; Wilkin, 1997) and to dominate or monopolize media coverage and interpretations of events. To do so, they resort to methods ranging from shutting down radio and TV stations, imposing censorship, embedding journalists, intimidating opponents, enforcing secrecy, and creating forgeries to exerting subtle influence through language and artwork.

> For this purpose they have a range of mechanisms from the use of experts and state officials to legitimize the state's response to an event, the planting of stories in the media, the bribery of journalists, the setting up of newspapers, magazines, radio stations and such like, through the more straightforward forms of propaganda such as lying, deception and misinformation. (Wilkin, 1997, p. 126)

One of the most radical historical examples of the manner in which political power could determine orientation knowledge within the territory it controls was the principle known as *cuius regio, eius religio* (Who rules, his religion). Anchored in the Peace of Augsburg (1555), this principle permitted the sovereign princes of the Holy Roman Empire to stipulate the confession (Lutheran, Calvinist, or Roman Catholic) to which the population of their respective domains were to belong.[29] In later centuries the communist systems, National Socialism, and fundamentalist theocracies also attempted to force such homogeneity upon orientation knowledge and to make their ideologies dominant.

Control of Access to Information, Censorship of Information, Bibliocide, and Memorycide[30,31]

Censorship, forgery, deception, disinformation, and memorycide have been among the instruments of rulers for more than two thousand years (see Post, 1998). Collective memories or shared knowledge call for specification of what is considered worth knowing. "The dispute over the admissibility of bodies of knowledge ran through the entire Middle Ages" (Kintzinger, 2003, p. 89). In the medieval monasteries the abbot determined what kind of knowledge was regarded as wholesome or "good"

---

[29] Persons not wishing to accept the sovereign's decision were granted a grace period during which they could resettle in a region allowing them to practice the religion of their own choice. The Electors residing in Heidelberg switched their religious affiliation seven times between 1556 and 1716 (Baar-Cantoni & Wolgast, 2012, p. 67). On each of these occasions, the professors at Heidelberg University had to choose between changing their confession or leaving the university.

[30] Some of the thoughts and arguments in this section have been published in other works of mine as well (Meusburger, 2005, 2007a, 2007b, 2011).

[31] The culture of memory is also manipulated through graphic representations and placement of monuments. Manipulation through pictures is even more effective than that through words (for a thorough discussion see vol. 4 of this series, *Cultural Memories*).

for the monks (whatever promoted the salvation of the soul), what kind was "detrimental" or wrong (divergent doctrine, heresy, whatever detracted from the salvation of the soul; pp. 61–62).

> The abbot decided for the monks of his monastery . . . which texts and books . . . were permitted to be borrowed from other cloisters for copying and which of its own were allowed to be passed on to those places. What was read by or to the monks as a community and what the monks read individually was subject to strict control. (p. 61)

The distinction between good and disapproved knowledge and between wrong and right sources of knowledge is also fundamental to many political ideologies. Censorship and secrecy are practiced especially if awareness of certain information would threaten one's self-image or the moral exaltation of one's social system. The most extreme manipulation of information arguably occurred under twentieth-century totalitarian systems and hegemonic democracies, which mastered techniques of faking photographs and documents. King (1997) thoroughly documented the manipulation of photography and art in the Soviet Union, showing that political purges not only liquidated Stalin's opponents but had to erase all memory of them in publications as well. As soon as members of the ruling apparatus fell out of favor, their images were deleted from encyclopedias, history books, and school texts.

It is also possible to manipulate information by eradicating published books (bibliocide) and preventing manuscripts from being printed. From the time the Bolsheviks seized power in Russia in 1917 to the collapse of the Soviet Union in 1991, approximately 100,000 book titles were added to the index of banned publications, and more than one billion printed books were destroyed there (Ingold, 2005). In Czechoslovakia, where the *Guide for Monitoring Book Inventories in Libraries of All Types* was published as late as 1953 (Míšková, 2005, p. 237), around 27.5 million books were destroyed by the end of the 1950s (Pešek, 2005, p. 247). A central administration for publications was set up in 1966 to "protect the interests of socialist society." This institution was responsible for ensuring "that no information contradicting other interests of society is published in the mass media" (Malý, 2005, p. 230). In Germany the National Socialist propaganda machine managed to control media reporting almost completely. "Wherever totalitarianism possesses absolute control, it replaces propaganda with indoctrination and uses violence not so much to frighten people (this is done only in the initial stages when opposition still exists) as to realize constantly its ideological doctrines and its practical lies" (Arendt, 1951, p. 333).

Democracies, too, control and manipulate sensitive information through an array of measures. In some of these countries, for instance, certain documents are locked away in archives longer than is legally required, and the history books used in schools are particularly graphic examples of manipulation. In the United Kingdom a system introduced in 1912 as the Defence Notice (D-Notice) is still in force in 2014 (renamed in 1993 as the Defence Advisory Notice, or DA-Notice). Under its provisions the British government may request on grounds of national security that news editors not broadcast specified information or disseminate it in any other way.[32] In June 2013 a DA-Notice was issued asking the media to refrain from running further stories related to the US PRISM spy program and British involvement therein (DA-Notice, n.d., final par. under "United Kingdom"). Most British media (except for *The Guardian*) reported in a very different way about the scandal involving the U.S. National Security Agency than did prominent newspapers of other European countries.

As far back as the Middle Ages, books and manuscripts have been burned to eliminate memory. The aim is to cause the loss of cultural memory and of the potential to remember, that is, to eschew any future memory and to foster collective amnesia. It was a proven method of silencing heretics (Werner, 1995, pp. 149–150). A milder approach to memorycide was to sequester undesirable books completely or to compile registers naming them as "errant"—"the most important instrument of censorship in the late Middle Ages" (p. 171). The Papal palace in the Vatican had its own library where the blacklisted books were deposited (p. 170). Error and its legally binding condemnation were intended to become an element of cultural memory if possible.

Manipulation of Public Attention

To function as experts of factual knowledge or conveyors of salvation and meaning, the relevant persons must have a platform granting them the spotlight and guaranteeing their presence in the media. This attention, a general term for selectivity in perception, determines what is learned and remembered and what is excluded. Because attention is an increasingly scarce resource amid the information overloads of our times (Franck, 1998), various sophisticated techniques have been developed to attract, manipulate, or divert it and to control access to the platforms that afford it. Directing the public's attention to certain objects, persons, or concerns has become a major business and a powerful device for shrewdly steering learning processes, value systems, identities, collective memories, and, yes, consumer behavior.

Most methods of manipulation help turn the public's attention and interest to particular aspects and away from others. As in a theater, the audience is to watch only those parts of the scene illuminated by the spotlights, with the other actors and events remaining in the dark. The careful choice of information and topics (agenda-setting) can manipulate the media's public without resort to distortions or lies. A conflict may end not because the problem is solved but because the public debate may have shifted to another problem.

Considering today's flood of information, the content or societal usefulness of a message is often less important for its broad diffusion than the platform on which it is presented. The locality in which new knowledge is declared mainly determines

---

[32] "The objective of the DA-Notice System is to prevent inadvertent public disclosure of information that would compromise UK military and intelligence operations and methods, or put at risk the safety of those involved in such operations, or lead to attacks that would damage the critical national infrastructure and/or endanger lives" (DA-Notice System, n.d., par. 1).

the relevance, visibility, and credibility of the associated knowledge claims. This ability to shift public attention to selected subjects, persons, objects, or places and to draw it away from unlit areas is one of the most effective instruments of power in the twenty-first century. However, most persons and institutions exerting this kind of power remain anonymous.[33]

Subjectivity and Credibility of Experts

Crisis of Expertise?

Beck (1986, 1992a, 1992b, 2007), Pfister and Stehr (2013), Schimank (1992), Stehr (1992), Stehr and Ericson (1992), van den Daele (1992), and many others have pointed out that knowledge societies are simultaneously also risk societies for the very reason that acceleration of social and technological change and escalation of complexity and interdependencies have made them ever more fragile and vulnerable.

> The increasing spread of knowledge in society and the attendant growth in alternatives for action produce societal uncertainty. Science cannot deliver any truths (in the sense of conclusive causal chains or universal laws) but only more or less well-founded hypotheses and probabilities. Instead of being a source of bedrock knowledge and certainty, science is thus essentially a *source of uncertainty* and sociopolitical problems. (Pfister & Stehr, 2013, p. 17)

Pfister and Stehr's (2013) arguments focus on the decay of the authority of experts, the loss of respect for the know-how embedded in public administration, and the undermining of the epistemological monopoly held by gatekeepers of the scientific disciplines (p. 16). The fragility of the knowledge society consists above all in the decision-makers' escalating dependence on outside specialists (Bauman, 1992), the subjectivity of people in the latter group, the resulting multiplicity of their differing opinions,[34] and ever-increasing disciplinary specialization, which ultimately obscures the inadvertent consequences of decisions. The current crisis of technical competence thus lies less in a scarcity of experts than in the continuing fragmentation and narrowness of what they know, overspecialization, the politicization of particular scientific domains, the conscious acquiescence of experts to the interests

---

[33] In most cases it is not the host of the talk-show who selects his or her guests, but anonymous members of the editorial staff of the TV channel.

[34] Baghel and Nüsser (2010) offer a striking illustration of the subjectivity of expertise: "The guidelines proposed in the World Commission on Dams (WCD) final report were vehemently rejected by several Asian governments, and dam building has continued apace in most Asian countries. This reaction is in line with the simplistic dam debate, where dam critics offer laundry lists of socioeconomic and environmental costs, and dam proponents highlight the benefits while underestimating associated costs. Whereas the WCD sought to evaluate dams in terms of 'costs and benefits', this approach is self-defeating due to the very subjectivity of such measurements" (p. 231).

of their clients, the partiality and subjectivity of some experts, the coalitions between bad science and media power, and technocratic arrogance toward critics. The following episode is a prime example of the situation.

> An erroneous prediction of the disappearance of Himalayan glaciers by 2035, took on a life of its own, being repeated periodically with greater credence, until it entered the report of the IPCC.[35] This one paragraph in a 980 page report was then used to garner public attention and support for action on climate change. As it became clear that this prediction was erroneous, and the original date had been 2350, it became sufficient to discredit the entire report of the premier body of experts on climate change. . . . The chairperson of the IPCC at first dismissed questioning of the 2035 date as "voodoo science", however as the error became clear, an apology eventually became necessary. (Baghel, 2012, p. 1)

There are various reasons why the credibility of experts today is much more contested than it was in earlier periods. Studies by some authorities have quickly proven to be wrong or useless. Others have bogusly claimed to have forecasting ability. The half-life period of technological innovations and, therefore, of specialized knowledge is steadily decreasing, leaving experts to contradict each other on many questions. They "frequently emphasize some aspects of a problem but overlook others, and . . . even if we could find the right experts, they may not have the answers" (Evans & Collins, 2008, p. 609). Some of them fail to see or acknowledge the significance of the spatial context; still others regard best practice in one place or socioeconomic setting as the single best solution, neglecting the fact that the optimal response to a problem also depends on the geographic and social framework involved. Solutions shown to be reliable or cost-effective in cities can lead to unintended, highly inimical consequences in rural areas on the periphery.

An increasingly serious problem is the "media power of bad science" (Grossarth, 2014, p. 17). In media-saturated democracy, lobby groups and politicians seek scientific knowledge that confirms their own preconceived opinions or political goals. This appetite of policy- and decision-makers is often fed by third-class scientists hoping that a big bang in the media will garner them the attention withheld from them in the scientific community. By the same token, journalists do not want to ruin a "good story" by including too much complexity, such as an analysis of the data's reliability, an exhaustive report on the statistical and methodological approach, or a detailed description of the laboratory experiments underlying a study (p. 17). In most media a good headline is more important than the methodological quality of a scientific investigation. Because of that attitude, results of a study that professional researchers have criticized or ignored as unscientific can reach millions of readers and listeners through talk shows and other forms of media. This power that bad science has in the media naturally erodes the credibility of experts in the eyes of the public.

The growing doubt about the competence of experts strengthens the position of nonexperts. This outcome may be one of several reasons for the ever greater demands of the affected population to have a say in vital decisions, especially on environmental issues and infrastructural planning. "The link between expertise and participation remains the Achilles heel in the relationship between [science and technology studies] and wider decision-making" (Evans & Collins, 2008, p. 612).

---

[35] Intergovernmental Panel on Climate Change.

The more expertise and scientific results are contested, the more uncertainty and risks increase (for details see Evans & Collins, 2008, p. 612). Uncertainty reinforces the need for new knowledge (Beck, 1992a, 1992b; Böhme, 1992), making this upward spiral, or the "will to knowledge" (Foucault, 1990), virtually inexorable. People never know enough. However, it is necessary to distinguish social choice (political bargaining, societal negotiations) from scientific or technical analysis. When solving a technical or scientific problem, empirical research or professional analysis cannot be replaced by political bargaining or mass participation. No social system can afford to bypass expertise or the explanatory power of science altogether, but it is possible to mobilize counterexpertise based on higher competence or new scientific results (Schimank, 1992, pp. 218–219; van den Daele, 1992, p. 336).

> Although the role of experts in risk regulation has been challenged over the last decades, the experts have survived. Issues of social choice implied in the regulation of technologies can be shifted to political processes of conflict and consensus formation. Issues of technical analysis, of prognosis and explanation cannot be shifted. They remain the domain of professional judgement. Experts may disagree, but their controversies belong, so to speak, to the profession. Outsiders can suspect that experts are biased, partial or even corrupt. But they cannot declare matters of fact to be matters of social choice. Nor can they as a rule substitute professional expertise with commonsense judgement (van den Daele, 1992, p. 337).

There are a number of crucial questions: How can we distinguish the capable (competent) expert from the nonexpert? What role does political and economic power play in defining who is a "real" expert? Which decision-making processes are a matter of social choice and political bargaining and which need scientific analysis first of all? In which domains and issues of decision-making are mass participation and citizen science[36] helpful and efficient and where are they inappropriate?[37] How and to whom is expert status attributed or denied in various societies (Evans & Collins, 2008, p. 609)? To what extent are solutions or best practices universally applicable and in which cases does site-specific or local expertise yield superior results?

Scholars as Instruments of Politics[38]

Academics, scientists, and other experts of factual knowledge serve power not only by describing situations as realistically or truthfully as possible, scientifically

---

[36] "Citizen science . . . is scientific research conducted, in whole or in part, by amateur or nonprofessional scientists" ("Citizen science," 2014). It has been defined as "the systematic collection and analysis of data; development of technology; testing of natural phenomena; and the dissemination of these activities by researchers on a primarily avocational basis" ("Finalizing," 2011). Citizen scientists often partner with professional scientists to achieve common goals. Large volunteer networks often allow scientists to carry out tasks that would be too expensive or time-consuming by other means.

[37] It is unlikely that laypersons are permitted to perform surgery.

[38] "In Jonathan Swift's *Gulliver's Travels*, when Lemuel Gulliver arrived in May 1707 in the land of Laputa on the continent of Balnibarbi, the king of Laputa received his guest at the royal palace. Concerned about how his power would be perceived by a foreigner, the king proudly showed Gulliver his loyal scientists, astronomers, and musicians, all of whom were dedicated to enlightened governance. Gulliver, however, was a traveler with a keen eye. He observed how

analyzing contexts, explaining interactions and interdependencies, and finding solutions to technical problems. Some of them support the power structure by pretending to provide objective or scientific arguments that justify political and military actions, by manipulating facts, or by serving their nation in military intelligence (see Heffernan, 1996, 2002), as a diplomatic weapon (Doel & Harper, 2006), or as an instrument of colonialism and imperialism. When engaging in ideological disputes, politicians and other figures in power like to turn to the academic community when seeking to create the appearance that their arguments and actions are rational, their actions legitimate, their analyses scientific, or their evaluations objective (see Gregory, 1978). Gyuris (2014) shows how Marxist thinkers frequently emphasized that their findings were "scientific," "objective," and factual (p. 115) and that they branded the ideas of their political rivals as politically biased and unscientific.

> [Lenin] contrasted "official science" (Lenin, 1964, p. 200) with the Marxist approach. In his interpretation, "official science" produced findings that contradicted empirical evidence, but still aggressively tried to destroy all competing concepts. The Marxist approach was, however, verified by "facts" in his eyes. Thus, Lenin interpreted the scientific field as being configured by the dichotomy between those having political power, and those who are right. He consequently referred to the former as "bourgeois science" or "science" in quotation marks, while about the latter he simply wrote as science, without quotation marks. For him, "bourgeois science . . . strives to obscure the essence of the matter, to hide the forest behind the trees" (p. 216). He considered this by no means as accidental since, as he put it, "bourgeois scholars . . . are all apologists of imperialism and finance capital" (p. 226). That is why "official science tried, by a conspiracy of silence, to kill the works of Marx" (p. 200). However, what Marx had written about capitalism and its tendency to end up in monopolies, "has become a fact", and "facts are stubborn things", so "they have to be reckoned with, whether we like it or not" (ibid.). It was in the light of these tendencies that Lenin judged the findings of Marx a "precise, scientific analysis" (p. 304). . . . [A]ll kinds of knowledge not based on Marxist-Leninist, or rather Stalinist, grounds were automatically exiled from the domain of science. . . . Stalin not only upgraded the already existing legitimate authority of science, but also positioned himself as the leading expert and representative of this "mode of knowledge". . . . Stalin again and again stressed the objective, scientific and unambiguous nature of his statements. Here, he claimed social laws to have the same explanatory power as natural laws. (Gyuris, 2014, pp. 116–117)

In democracies and totalitarian states alike, some scientists, experts, and legal advisors have been ready to sign a "devil's pact" (Cornwell, 2003) with those in power in order to acquire the resources and opportunities they need to accomplish their research or burnish their reputation (see Chap. 9 by Barnes in this volume; Szöllösi-Janze, 2004). In the widely known "torture memos" of 2005 and 2006, legal advisors of the Bush administration (John Yoo, Jay Bybee, and Stephen

---

the netherland population of Lindalino, part of the kingdom of Laputa, dwelled below a floating island which the king could order his scientists to manipulate. The scientists could use magnetic levitation to move the floating island, thus preventing sunlight and rain from getting to the population of Lindalino. If insurrection broke out, the sovereign could demand that Lindalino be bombarded with rocks, or destroy Lindalino by lowering the floating island on top of the population" (Seegel, 2012, p. 1).

Bradbury) redefined torture and argued that torturers acting under presidential orders could not be prosecuted (Head, n.d., par. 2 & 5; "U.N. says," 2008, par. 1).

Another variety of courtship between the intelligentsia and the power structure was apparent in the way scholars celebrated communist leaders as scientific geniuses.

[Stalin's] scientific "genius" was frequently praised by leading members of the scientific hierarchy. For instance, the then President of the Academy, Sergey Vavilov, otherwise an internationally acknowledged physicist, often referred to Stalin as "the genius of science" (Vavilov, 1950[1949], p. 11) or "the coryphaeus of science" (Pollock, 2006, p. 1). (Gyuris, 2014, p. 117)

There are also numerous historical instances of scientists' involvement in forgeries in order to please someone in power. They have produced data, images, and arguments utterly divorced from reality solely to realize political goals or pursue the interests of their government leaders. It is well known that maps, for example, have figured as important tools of politics, nationalism, colonialism, and imperialism (Black, 1997; Cosgrove & della Dora, 2005; Heffernan, 1996, 2002; Seegel, 2012; Stone, 1988; Wilkinson, 1951). "[C]artography was a means of control used by governments to conquer and then engineer territorial space. Cartography was a representational language of power and protest . . . [and] a tool of imperial governance" (Seegel, 2012, p. 2). Maps have been used "as a geopolitical artefact; as an image of political space, both actual and potential, and as a military and strategic device that both reflected and challenged the objectives of the major nation-states" (Heffernan, 2002, p. 207; for further details see Seegel, 2012; Wilkinson, 1951).

After World War I, for instance, geographers in several countries turned to the problem of "just" or "natural" borders and provided "scientific" arguments for corrections in territorial boundaries. For the peace negotiations culminating in the Treaty of Trianon, the French geographer Emmanuel de Martonne (1973–1955)[39] submitted a largely faked ethnographic map entitled "Distribution of nationalities in regions dominated by Romanians"–published in 1919 by the Service Géographique de l'Armée–that blithely ignored official data from several national censuses.[40] De Martonne's cartographic manipulations and distortions of information were documented in detail by Boulineau (2001), Bowd (2011), Palsky (2002), and others. Presented as objective and authentic facts, of course, these manipulations were used to justify the decisions of the Allies to attach large parts of Hungary to Romania at the peace negotiations of Trianon. Such subterfuge and methodological manipulations of ethnic maps had started in Romania already in the 1890s. Twenty-nine of

[39] During the Paris Peace Conference after World War I, de Martonne was an adviser to French Minister of Foreign Affairs André Tardieu and French Prime Minister Georges Clemenceau.

[40] He declared national census data on the mother tongue of the Romanian population as unreliable and "corrected" them by data about the religious denomination of the population. In addition he applied a number of cartographic techniques and tricks that masked the real distribution of minorities. The census data on the mother tongue of the population in Transylvania are still retrievable at http://www.kia.hu/konyvtar/erdely/erd2002.htm. Kocsis (1994, 2007), Kocsis and Kocsisné (1998), and others describe the distribution of minorities that is based on mother tongue and the background of the ethnic conflicts in the Carpathian Basin. Jordan (2010) gives a general overview about methods used to manipulate maps showing the distribution of ethnic awareness.

the most impertinent manipulations of maps and atlases published between 1894 and 1941 in Romania have been documented in detail by Staatswissenschaftliches Institut (1942) in Budapest.

For the peace negotiations in St. Germain, the Italian geographer and philologist Ettore Tolomei (1865–1952)–later a politician and member of the Fascist party– produced a map of South Tyrol in which all German names of places, rivers, mountains, and landscapes had been replaced with Italian names, although more than 95 % of the population there was German-speaking at that time. Tolomei made a few embarrassing mistakes in his map because he did not know what some of the German terms meant, or he translated them incorrectly. This map was meant to justify the fact that South Tyrol had been annexed by Italy after World War I. The American geographer Isaiah Bowman, who was also involved in the peace negotiations, described the situation as follows:

> Each one of the Central European nationalities had its own bagful of statistical and cartographical tricks. When statistics failed, use was made of maps in color. It would take a huge monograph to contain an analysis of all the types of map forgeries that the war and the peace conference called forth. A new instrument was discovered—the map language. A map was as good as a brilliant poster, and just being a map made it respectable, authentic. A perverted map was a life-belt to many a foundering argument. It was in the Balkans that the use of this process reached its most brilliant climax. (Bowman, 1921, p. 142)

It turned out that assessment criteria or scientific rationales for corrections of borders were randomly interchanged at the peace negotiations after World War I. In South Tyrol the watershed was regarded by the victorious powers as a legitimate or "natural" border, but they rejected the watershed concept in Istria and the Carpathian Basin. The victorious powers considered rivers to be ideal, natural boundaries when it suited them; in other cases, rivers were called a connective element that united rather than separated regions.

Academics (e.g., historians, geographers, literary scholars, archeologists, and anthropologists) interacting with museums and producing school textbooks have also advanced the nation-building process and the objectives of their respective national policies by espousing the interpretations and views of history taken by their governments or by rewriting the history of a region's settlement. To claim the right to rule or to justify wars, "experts" have even been commissioned to forge documents[41] or corroborating evidence (as with the alleged weapons of mass destruction in Iraq).

---

[41] To aggrandize the legitimacy and influence of the House of Habsburg and its Austrian lands, Rudolph IV (1339–1365) ordered the creation of a forged document called the Privilegium Majus ("the greater privilege") in the winter of 1358–1359. It consisted of five faked deeds, some of which had supposedly been issued by Julius Caesar and Nero to the historic Roman province called *regnum Noricum*, whose borders ran a course similar to those of modern Austria. The Privilegium Majus was modeled on the Privilegium Minus (a grant of special privileges and a reduction of obligations toward the empire, issued by Holy Roman Emperor Frederick I Barbarossa when Austria was raised to a duchy). The original of the latter document, however, "got lost" at the same time, and the Privilegium Majus was identified as a fake even by contemporaries, such as the Italian scholar Francesco Petrarch.

Conclusion

This chapter explains why human action is a blend of factual and orientation knowledge and why both categories of knowledge are needed for the acquisition and retention of power. The art of exercising power appears to lie in finding the right balance between the two epistemological categories for each task, situation, and set of competitive conditions.

The distinction between factual knowledge and orientation knowledge, however, is not just academic; it has great existential significance for goal-oriented, social systems. For if orientation knowledge is so dominant that it impairs the decision-maker's perception and faculty of judgment, or if orientation knowledge is applied where factual knowledge is primarily needed, the eventual result is faulty analyses, wrong objectives, and decisions that impair the performance and viability of the social system involved. Orientation knowledge can instill tremendous motivation and strengthen a social system's cohesion, but in its exaggerated form as religious fundamentalism and political fanaticism it restricts the ability to judge a situation impartially and realistically, to foresee unintended long-term consequences of actions, to distinguish between representation and reality, and to make the right decisions for accomplishing objectives. Many policy-makers and business leaders have failed to reach their goals because they believed in their own propaganda, which was originally intended only to keep their system intact and had no claim to truth.

Self-reflection, necessary corrections, processes of learning and adaptation, and continued dynamic development of organizations can take place only if uncomfortable information is not repressed and if public discourse avoids preference falsification.[42] The people holding political power or controlling the media must desist from trying to thwart public expression of views that do not conform to political correctness or the opinions of the "intellectual theocratic caste" (Schelsky, 1975). Self-censorship and preference falsification have a number of adverse impacts on the social system or society in question. According to Kuran (1995):

> [P]reference falsification generates inefficiencies, breeds ignorance and confusion, and conceals social possibilities. (p. 6)

> [In communist systems] individuals routinely applauded speakers they disliked, joined organizations whose mission they opposed, ostracized dissidents they admired, and followed orders they considered nonsensical, unjust, or inhuman. (p. 119)

> [T]he distortion of public discourse paralyzed the critical faculties of individual citizens, making them accept lies as unquestionable truths and hollow slogans as profound wisdom. (p. 206)

The relations between knowledge, power, and space are pivotal in myriad issues and theoretical approaches. However, as meritorious as it may be to interpret Arendt, Chomsky, Gramsci, and Foucault repeatedly from new angles and to discuss minute details of hegemonic practices, surveillance, and governmentality, these discourses cover only a fraction of the total complex known as the relations between power,

---

[42] Kuran (1995) takes preference falsification to mean "the act of misrepresenting one's genuine wants under perceived social pressure" (p. 3).

knowledge, and space. Further research should direct attention to at least four issues. First, it is necessary to increase the integration of the findings of communication and organization theory, management studies, social psychology, network studies, political geography, and other fields focusing on the role of power in the task of organizing social systems in space. Organization studies in the tradition of Mintzberg (1979) have much to offer when the spatial distribution of power and knowledge has to be explained, for they focus on the relations between the stability (or instability) of an organization's tasks, the organization's environmental uncertainties, its autonomy, and its internal structure (architecture).[43] If organizations are dealing with simple tasks and a stable environment (low degree of uncertainty), then decision-making, problem-solving, research, development, and planning will shift to the upper levels (the center) of the system's hierarchy. Consequently, the lower levels will predominantly keep routine activities and workplaces for the low-skilled person. This type of organization can be called *bureaucratic*. If organizations are dealing with complex tasks and a dynamic environment (high degree of uncertainty) and are confronted by constantly changing, unpredictable, one-time transactions, then decentralization of competence and authority within the system is more effective. This type of organization is called *organic* (for details see Mintzberg, 1979, pp. 86–87; 188–202; 271–273; Meusburger, 1998, pp. 131–152).

The second issue that has been widely disregarded in studies about the interplay between power and knowledge is the role of secrecy, disinformation, leaks, camouflage, and deception. Having exclusive knowledge, disclosing secrets of adversaries in due time, keeping essential information secret as long as necessary, manipulating information, and dominating media are among the most important instruments of power. Some authors have observed that the state exercises power, among other things, by conducting censuses or introducing land registers and personal registration (Hannah, 1997, 2000). Conversely, the public can be duped and manipulated if a government does not gather or share particular data that would damage its image or if it switches to publishing that information only in an aggregate form that henceforth obscures social or regional disparities.[44]

A third topic worthy of more research attention than it has hitherto received is the sociospatial implications of surveillance. Boyne (2000), Deleuze (1992), Klauser (2009, 2013, 2014), Lyon (2001, 2003), Murakami Wood (2007), and others argue that the information society is also a control or surveillance society. Examples of such work are the studies by Klauser, who analyzes the importance of space as the locus, object, and tool of surveillance; the relationship between space and surveillance in different institutional contexts (cities, airports, major sports events); and the

[43] The architecture of an organization is defined as an ordered arrangement of different functions (workplaces with different tasks). It can be described by the hierarchical arrangement of units fulfilling line and staff functions; by the distribution of expertise, responsibilities, and control functions; by the centralization or decentralization of decision-making; by the channels of formal communication; and many other attributes (Meusburger, 2007b, p. 119).

[44] Poverty, crime, income inequalities, and gender inequalities purportedly did not exist in communist countries. The relevant data were not collected or not published (Meusburger, 1997).

logics, functioning, and effects of control and regulation in particular geographical locales. What effects do the Internet and the new social media have on processes of wielding power, on grass-root participation in policy decisions, and on resistance to political propaganda? To what extent are the new technological possibilities for the surveillance of digital communication by secret services changing the national and global asymmetries of power and the definition of privacy?

A fourth engaging field of research relates to the question of the spatial scale—the global, national, regional, or local—at which solutions to salient problems tend to emerge and where expertise and power will have to be wielded. How much power has the nation-state surrendered to regions, corporations, and the financial sector? How much will it have to surrender in the future?

Acknowledgements I would like to thank Gabriela Christmann (The Leibniz Institute for Regional Development and Structural Planning (IRS) in Erkner) for the inspiring talks and exchange of ideas that we had while writing this chapter.

References

Abel, G. (2008). Forms of knowledge: Problems, projects, perspectives. In P. Meusburger, M. Welker, & E. Wunder (Eds.), *Clashes of knowledge: Orthodoxies and heterodoxies in science and religion* (Knowledge and space, Vol. 1, pp. 11–33). Dordrecht, The Netherlands: Springer. doi:10.1007/978-1-4020-5555-3_1.

Abrams, R. H. (1969). *Preachers present arms: The role of the American churches and clergy in World Wars I and II, with some observations on the war in Vietnam* (Rev. ed.). Scottdale, PA: Herald Press.

Ahmed, M. D. (1988). Traditionelle Formen der Erziehung in der Islamischen Welt [Traditional forms of education in the Islamic world]. *Zeitschrift für Kulturaustausch, 38*, 332–337.

Allen, A. (1999). *The power of feminist theory: Domination, resistance, solidarity.* Boulder, CO: Westview Press.

Anacker, M. (2004). Wissen VI. 19. und 20. Jahrhundert [Knowledge, subsection VI: Nineteenth and twentieth centuries]. In J. Ritter, K. Gründer, & G. Gabriel (Eds.), *Historisches Wörterbuch der Philosophie: Vol. 12. W–Z* [CD-Rom] (pp. 891–900). Basel, Switzerland: Schwabe.

Arendt, H. (1951). *The origins of totalitarianism.* New York: Harcourt Brace and Company.

Arendt, H. (1970). *On violence.* New York: Harcourt Brace and Company.

Arndt, A. (2004). Wissen V. Von Kant bis zum Nachidealismus [Knowledge V: From Kant to postidealism]. In J. Ritter, K. Gründer, & G. Gabriel (Eds.), *Historisches Wörterbuch der Philosophie: Vol. 12. W–Z* [CD-Rom] (pp. 884–891). Basel, Switzerland: Schwabe.

Avelino, F., & Rotmans, J. (2009). Power in transition: An interdisciplinary framework to study power in relation to structural change. *European Journal of Social Theory, 12*, 543–569.

Baar-Cantoni, R., & Wolgast, E. (2012). Migration of professors between 1550 and 1700. In P. Meusburger & T. Schuch (Eds.), *Wissenschaftsatlas of Heidelberg University. Spatio-temporal relations of academic knowledge production* (pp. 66–69). Knittlingen, Germany: Bibliotheca Palatina.

Bacon, F. (1863). *The new organon or: True directions concerning the interpretation of nature* [Text based on the standard translation of *Novum Organum Scientiarum* by J. Spedding, R. L. Ellis, & D. D. Heath in *The Works: Vol. VIII.* Boston: Taggard and Thompson.] (Original work published 1620). Retrieved April 15, 2014, from http://ebooks.adelaide.edu.au/b/bacon/francis/organon/ and from http://www.constitution.org/bacon/nov_org.htm

Baghel, R. (2012). Knowledge, power and the environment: Epistemologies of the Anthropocene. *Transcience, 3*, 1–6.

Baghel, R., & Nüsser, M. (2010). Discussing large dams in Asia after the World Commission on Dams: Is a political ecology approach the way forward? *Water Alternatives, 3*, 231–248.

Barton, T. S. (1994). *Power and knowledge: Astrology, physiognomics, and medicine under the Roman Empire*. Ann Arbor, MI: University of Michigan Press.

Baum, M., & Kron, T. (2012). Von Gärtnern und Jägern—Macht und Herrschaft im Denken Zygmunt Baumans [Of gardeners and hunters—Power and politics in the thinking of Zygmunt Bauman]. In P. Imbusch (Ed.), *Macht und Herrschaft. Sozialwissenschaftliche Theorien und Konzeptionen* (pp. 335–356). Wiesbaden, Germany: Springer Fachmedien. doi:10.1007/978-3-531-93469-3_16.

Bauman, Z. (1992). Life-world and expertise: Social production of dependency. In N. Stehr & V. R. Ericson (Eds.), *The culture and power of knowledge: Inquiries into contemporary societies* (pp. 81–106). Berlin, Germany: Walter de Gruyter.

Bauman, Z. (2000). *Liquid modernity*. Cambridge, UK: Polity Press.

Bauman, Z., & Haugaard, M. (2008). Liquid modernity and power: A dialogue with Zygmunt Bauman. *Journal of Power, 1*, 111–130.

Beck, U. (1986). *Risikogesellschaft. Auf dem Weg in eine andere Moderne* [Risk society: On the way to a different modernity]. Frankfurt am Main, Germany: Suhrkamp.

Beck, U. (2007). *Weltrisikogesellschaft. Auf der Suche nach der verlorenen Sicherheit* [World risk society: Seeking lost security]. Frankfurt am Main, Germany: Suhrkamp.

Beck, U. (1992a). Modern society as a risk society. In N. Stehr & V. R. Ericson (Eds.), *The culture and power of knowledge: Inquiries into contemporary societies* (pp. 199–214). Berlin, Germany: Walter de Gruyter.

Beck, U. (1992b). *Risk society*. London: Sage.

Bell, D. (1973). *The coming of the post-industrial society: A venture in social forecasting*. New York: Basic Books.

Belyaev, D. (2008). *Geographie der alternativen Religiosität in Russland. Zur Rolle des heterodoxen Wissens nach dem Zusammenbruch des kommunistischen Systems* [Geography of alternative religiousness in Russia. About the role of heterodox knowledge after the collapse of the communist system] (Heidelberger Geographische Arbeiten: Vol. 127). Heidelberg, Germany: Selbstverlag des Geographischen Instituts der Universität Heidelberg.

Berger, P. L., & Luckmann, T. (1966). *The social construction of reality: A treatise in the sociology of knowledge*. Garden City, NY: Doubleday.

Berthold, L. (1997). Transitive Macht, intransitive Macht und ihre Verbindung: Hermann Hellers Begriff der Organisation [Transitive power, intransitive power, and their connection: Hermann Heller's concept of organization]. In G. Göhler (Ed.), *Institution—Macht—Repräsentation. Wofür politische Institutionen stehen und wie sie wirken* (pp. 349–359). Baden-Baden, Germany: Nomos.

Black, J. (1997). *Maps and politics*. London: Reaktion Books.

Böhme, G. (1992). The techno-structures of society. In N. Stehr & V. R. Ericson (Eds.), *The culture and power of knowledge: Inquiries into contemporary societies* (pp. 39–50). Berlin, Germany: Walter de Gruyter.

Borch, C. (2005). Systemic power: Luhmann, Foucault and analytics of power. *Acta Sociologica, 48*, 155–167.

Boulineau, E. (2001). Un géographe traceur de frontières: Emmanuel de Martonne et la Roumanie. *L'Espace géographique, 30*, 358–369.

Bowd, G. (2011). Emmanuel de Martonne et la naissance de la Grande Roumaine. *Revue Roumaine de Géographique/Romanian Journal of Geography, 55*, 103–120.

Bowman, I. (1921). Constantinople and the Balkans. In E. M. House & C. Seymour (Eds.), *What really happened at Paris: The story of the Peace Conference, 1918–1919* (pp. 140–175). New York: Charles Scribener's Sons.

Boyer, D. (2003). Censorship as a vocation: The institutions, practices, and cultural logic of media control in the German Democratic Republic. *Comparative Studies in Society and History, 45*, 511–545.

Boyne, R. (2000). Post-Panopticism. *Economy and Society, 29*, 285–307. doi:10.1080/03085140360505.

Brown, R. (1995). *Prejudice: Its social psychology*. Oxford, UK: Blackwell.

Brunés, T. (1967). *The secrets of ancient geometry—and its use* (Vol. 1). Copenhagen, Denmark: Rhodos.

Buber, M. (1965). *Daniel: Dialogues on realization* (M. S. Friedman, Trans.). New York: McGraw-Hill. (Original work published 1913)

Bublitz, H. (2008). Macht [Power]. In C. Kammler, R. Parr, & U. J. Schneider (Eds.), *Foucault-Handbuch. Leben—Werk—Wirkung* (pp. 273–277). Stuttgart, Germany: Metzler.

Burt, R. (1998). (Un)censoring in detail: The fetish of censorship in the early modern past and the postmodern present. In R. Post & Getty Research Institute for the History of Art and the Humanities (Eds.), *Censorship and silencing: Practices of cultural regulation* (pp. 17–41). Los Angeles: Getty Research Institute for the History of Art and the Humanities.

Cassirer, E. (1922). *Das Erkenntnisproblem in der Philosophie und Wissenschaft der neueren Zeit* [The cognition problem in philosophy and science in recent times]. Vol. 2 (1994 reprint of 3rd ed.). Darmstadt, Germany: Wissenschaftliche Buchgesellschaft. (Original work published 1907)

Chomsky, N. (1987). *Turning the tide*. Montreal, Canada: Black Rose Books.

Christmann, G. B. (Ed.). (2013). *Zur kommunikativen Konstruktion von Räumen. Theoretische Konzepte und empirische Analysen* [About the communicative construction of spaces. Theoretical concepts and empirical analyses]. Wiesbaden, Germany: Springer VS.

Cialdini, R. B. (2008). Turning persuasion from an art into a science. In P. Meusburger, M. Welker, & E. Wunder (Eds.), *Clashes of knowledge: Orthodoxies and heterodoxies in science and religion* (Knowledge and space, Vol. 1, pp. 199–209). Dordrecht, The Netherlands: Springer. doi:10.1007/978-1-4020-5555-3_12.

Cicourel, A. V. (1974). *Methode und Messung in der Soziologie* [Method and measurement in sociology]. Frankfurt am Main, Germany: Suhrkamp.

Citizen science. (2014, April 17). Retrieved May 2, 2014, from http://en.wikipedia.org/wiki/Citizen_science

Cornwell, J. (2003). *Hitler's scientists: Science, war and the devil's pact*. London: Viking.

Cosgrove, D. E., & Dora, V. (2005). Mapping global war: Los Angeles, the Pacific, and Charles Owens's pictorial cartography. *Annals of the Association of American Geographers, 95*, 373–390. doi:10.1111/j.1467-8306.2005.00465.x.

Dan, Y. (2007). *Die Kabbala. Eine kleine Einführung* [The kabbala: A small introduction]. Stuttgart, Germany: Reclam.

DA-Notice System. (n.d.). Retrieved May 1, 2014, from http://www.dnotice.org.uk/

DA-Notice. (n.d.). Retrieved May 1, 2014, from http://en.wikipedia.org/wiki/DA-Notice

Deleuze, G. (1992). Postscript on the societies of control. *October, 59*, 3–7.

Descartes, R. (2001). *Discours de la méthode pour bien conduire sa raison et chercher la vérité dans les sciences. Bericht über die Methode, die Vernunft richtig zu führen und die Wahrheit in den Wissenschaften zu erforschen* [Discourse on the method of rightly conducting one's reason, and of seeking truth in the sciences]. French/German. Translated and edited by H. Ostwald. Stuttgart, Germany: Reclam. (Original work published 1637)

Detel, W. (1998). *Macht, Moral, Wissen. Foucault und die klassische Antike* [Power, morality, knowledge: Foucault and classical antiquity]. Frankfurt am Main, Germany: Suhrkamp.

Doel, R. E., & Harper, K. C. (2006). Prometheus unleashed: Science as a diplomatic weapon in the Lyndon B. Johnson administration. *Osiris, A Research Journal Devoted to the History of Science and Its Cultural Influences, 2nd Series, 21*, 66–85.

Drucker, P. F. (1969). *The age of discontinuity: Guidelines to our changing society*. New York: Harper & Row.

Elias, N., & Scotson, J. L. (1994). *The established and the outsiders: A sociological enquiry into community problems* (2nd ed.). London: Sage.

Evans, R., & Collins, H. (2008). Expertise: From attribute to attribution and back again? In E. J. Hackett, O. Amsterdamska, M. Lynch, & J. Wajcman (Eds.), *The handbook of science and technology studies* (3rd ed., pp. 609–630). London: MIT Press.

Feil, E. (1986). *Religio. Die Geschichte eines neuzeitlichen Grundbegriffs vom Frühchristentum bis zur Reformation* [Religio: The history of an early modern concept from early Christianity to the Reformation]. (Forschungen zur Kirchen- und Dogmengeschichte, Vol. 36). Göttingen, Germany: Vandenhoeck & Ruprecht.

Felder, E. (2013). Faktizitätsherstellung mittels handlungsleitender Konzepte und agonaler Zentren. Der diskursive Wettkampf um Geltungsansprüche. [Creating facticity in discourses through action-guiding concepts and agonal centers: The discursive competition over claims of validity]. In E. Felder (Ed.), *Faktizitätsherstellung in Diskursen: Die Macht des Deklarativen* (pp. 13–28). Berlin, Germany: De Gruyter.

Fellmann, F. (1992). Perspektivismus und symbolischer Pragmatismus [Perspectivism and symbolic pragmatism]. In V. Gerhardt & N. Herold (Eds.), *Perspektiven des Perspektivismus. Gedenkschrift zum Tode Friedrich Kaulbachs* (pp. 235–249). Würzburg, Germany: Königshausen und Neumann.

Fichte, J. G. (1845–1846). *Sämmtliche Werke* [Complete works] (Vol. 1, (I. H. Fichte, Ed.)). Berlin, Germany: Veit.

Finalizing a definition of "citizen science" and "citizen scientists." (2011, September 3). *OpenScientist.* Retrieved May 2, 2014, from http://www.openscientist.org/2011/09/finalizing-definition-of-citizen.html

Foucault, M. (1980). *Power/knowledge: Selected interviews & other writings, 1972–1977* (C. Gordon, Ed.; C. Gordon, L. Marschall, J. Mepham, & K. Soper, Trans.). New York: Pantheon Books.

Foucault, M. (2007). *Security, territory, population: Lectures at the Collège de France, 1977–1978* (G. Burchell, Trans.). London: Palgrave Macmillan.

Foucault, M. (1979). *Discipline and punish: The birth of the prison.* New York: Vintage Books.

Foucault, M. (1990). *The will to knowledge: Vol. 1. The history of sexuality.* London: Penguin Books.

Franck, G. (1998). *Ökonomie der Aufmerksamkeit. Ein Entwurf* [The economics of attention: An outline]. Munich, Germany: Hanser.

Gadenne, V. (1999). Haben wir Erkenntnis von einer unabhängigen Welt? [Do we have knowledge of an independent world?]. In J. Mittelstraß (Ed.), *Die Zukunft des Wissens* (pp. 89–95). Konstanz, Germany: Universitätsverlag Konstanz.

Gehring, P. (2004). Wissen VII [Knowledge, subsection VII]. In J. Ritter, K. Gründer, & G. Gabriel (Eds.), *Historisches Wörterbuch der Philosophie: Vol. 12. W–Z* [CD-Rom] (pp. 900–902). Basel, Switzerland: Schwabe.

Gerhardt, V. (1992). Die Perspektive des Menschen [The perspective of the human being]. In V. Gerhardt & N. Herold (Eds.), *Perspektiven des Perspektivismus. Gedenkschrift zum Tode Friedrich Kaulbachs* (pp. v–xv). Würzburg, Germany: Königshausen und Neumann.

Gimbel, J. (1990). *Science, technology, and reparations: Exploitation and plunder in postwar Germany.* Stanford, CA: Stanford University Press.

Girnth, H. (2002). *Sprache und Sprachverwendung in der Politik: Eine Einführung in die linguistische Analyse öffentlich-politischer Kommunikation* [Language and language use in politics: An introduction to the linguistic analysis of public political communication]. Tübingen, Germany: De Gruyter.

Göhler, G. (2011). Macht [Power]. In G. Göhler, M. Iser, & I. Kerner (Eds.), *Politische Theorie. 25 umkämpfte Begriffe* (2nd enlarged ed., pp. 224–240). Wiesbaden, Germany: VS Verlag.

Göhler, G. (1997). Der Zusammenhang von Institution, Macht und Repräsentation [The relation between institution, power, and representation]. In G. Göhler (Ed.), *Institution—Macht—Repräsentation. Wofür politische Institutionen stehen und wie sie wirken* (pp. 11–62). Baden-Baden, Germany: Nomos Verlag.

Goodenough, W. (1957). Cultural anthropology and linguistics. In P. L. Garvin (Ed.), *Reports of the seventh annual round table meeting on linguistics and language study* (pp. 167–173). Washington, DC: Georgetown University Press.

Gordon, N. (2002). On visibility and power: An Arendtian corrective of Foucault. *Human Studies, 25,* 125–145.

Gramsci, A. (1971). *Selection from the prison notebooks.* London: Lawrence and Wishart.

Gregory, D. (1978). *Ideology, science and human geography.* New York: St. Martin's Press.

Gregory, D. (1994). *Geographical imaginations.* Oxford, UK: Basil Blackwell.

Gregory, D. (1995). Imaginative geographies. *Progress in Human Geography, 19,* 447–485.

Gregory, D. (1998). Power, knowledge and geography. In *Explorations in critical human geography* (Hettner-lecture, Vol. 1, pp. 9–40). Heidelberg, Germany: Department of Geography, Heidelberg University.

Gregory, D. (2004). *The colonial present: Afghanistan, Palestine, Iraq.* Malden, MA: Blackwell.

Grossarth, J. (2014, February 17). Die Medienmacht schlechter Wissenschaft [Media power of bad science]. *Frankfurter Allgemeine Zeitung,* p. 17.

Gyuris, F. (2014). *The political discourse of spatial disparities: Geographical inequalities between science and propaganda.* Heidelberg, Germany: Springer. doi:10.1007/978-3-319-01508-8.

Halbertal, M. (2007). *Concealment and revelation. Esotericism in Jewish thought and its philosophical implications.* Princeton, NJ: Princeton University Press.

Hanegraaff, W. J. (2008). Reason, faith, and gnosis: Potentials and problematics of a typological construct. In P. Meusburger, M. Welker, & E. Wunder (Eds.), *Clashes of knowledge: Orthodoxies and heterodoxies in science and religion* (Knowledge and space, Vol. 1, pp. 133–144). Dordrecht, The Netherlands: Springer. doi:10.1007/978-1-4020-5555-3_7.

Hannah, M. (1997). Space and the structuring of disciplinary power: An interpretive review. *Geografiska Annaler: Series B, Human Geography, 79,* 171–180.

Hannah, M. (2000). *Governmentality and the mastery of territory in nineteenth-century America.* Cambridge, UK: Cambridge University Press.

Hardy, J., & Meier-Oeser, S. (2004). Wissen I A. Terminologie [Knowledge, subsection 1A: Terminology]. In J. Ritter, K. Gründer, & G. Gabriel (Eds.), *Historisches Wörterbuch der Philosophie: Vol. 12. W–Z* [CD-Rom] (pp. 855–856). Basel, Switzerland: Schwabe.

Harmssen, G. W. (1951). *Am Abend der Demontage. Sechs Jahre Reparationspolitik (mit Dokumentenanhang)* [On the eve of the dismantling program: Six years of reparations policy]. Bremen, Germany: Friedrich Trujen.

Haugaard, M. (Ed.). (2002). *Power: A reader.* Manchester, UK: Manchester University Press.

Head, T. (n.d.). Is torture justified? *About.com News and Issues: Civil liberties.* Article retrieved February 26, 2014, from http://civilliberty.about.com/od/tortureandrendition/p/is_torture_just. htm

Heffernan, M. (1996). Geography, cartography and military intelligence: The Royal Geographical Society and the First World War. *Transactions of the Institute of British Geographers, New Series, 21,* 504–533.

Heffernan, M. (2002). The politics of the map in the early twentieth century. *Cartography and Geographic Information Science, 29,* 207–226. doi:10.1559/152304002782008512.

Herman, E. S., & Chomsky, N. (1988). *Manufacturing consent: The political economy of the mass media.* New York: Pantheon Books.

Imbusch, P. (2012a). Macht und Herrschaft in der wissenschaftlichen Kontroverse [Power and domination in scientific controversy]. In P. Imbusch (Ed.), *Macht und Herrschaft. Sozialwissenschaftliche Theorien und Konzeptionen* (pp. 9–35). Wiesbaden, Germany: Springer Fachmedien. doi:10.1007/978-3-531-93469-3_1.

Imbusch, P. (2012b). Machtfigurationen und Herrschaftsprozesse bei Norbert Elias [Power configurations and processes of domination in the work of Norbert Elias]. In P. Imbusch (Ed.), *Macht und Herrschaft. Sozialwissenschaftliche Theorien und Konzeptionen* (pp. 169–193). Wiesbaden, Germany: Springer Fachmedien. doi:10.1007/978-3-531-93469-3_9.

Imhof, K. (1996). Intersubjektivität und Moderne [Intersubjectivity and modernity]. In K. Imhof & G. Romano (Eds.), *Die Diskontinuität der Moderne. Zur Theorie des sozialen Wandels* (pp. 200–292). Frankfurt am Main, Germany: Campus.

Ingold, F. P. (2005, November 2). *Zaristisch-bolschewistisch. In Russland wächst die Sehnsucht nach Zensur* [Czarist-Bolshevist: The desire for censorship is growing in Russia]. *Frankfurter Allgemeine Zeitung,* p. N3.

Jäger, L. (2013). Erinnern und Vergessen. Zwei transkriptive Verfahrensformen des kulturellen Gedächtnisses [Remembering and forgetting: Two transcriptive procedures of cultural memory]. In E. Felder (Ed.), *Faktizitätsherstellung in Diskursen: Die Macht des Deklarativen* (pp. 265–285). Berlin, Germany: De Gruyter.

Jewett, R., & Lawrence, J. S. (2003). *Captain America and the crusade against evil: The dilemma of zealous nationalism.* Grand Rapids, MI: W. B. Eerdmans.

Jordan, P. (2010). Methodik und Objektivität von Karten des nationalen/ethnischen Bewusstseins [Methodology and objectivity of maps about national/ethnic awareness]. In J. Happel & C. von Werdt (Eds.), *Osteuropa kartiert—Mapping Eastern Europe* (Osteuropa, Vol. 3, pp. 175–185). Vienna, Berlin: Lit Verlag.

Kammler, C. (2008). Wissen [Knowledge]. In C. Kammler, R. Parr, & U. J. Schneider (Eds.), *Foucault-Handbuch. Leben—Werk—Wirkung* (pp. 303–306). Stuttgart, Germany: Metzler.

Kant, I. (1996). What does it mean to orient oneself in thinking. In I. Kant, *Religion and rational theology* (A. W. Wood & G. Di Giovanni, Trans. & Eds., pp. 7–18). Cambridge, UK: Cambridge University Press. (Original work published 1786)

Kaulbach, F. (1990). *Philosophie des Perspektivismus: Vol. 1. Wahrheit und Perspektive bei Kant, Hegel und Nietzsche* [Philosophy of perspectivism: Vol. 1. Truth and perspective in Kant, Hegel, and Nietzsche]. Tübingen, Germany: Mohr.

Keller, R. (2013). Kommunikative Konstruktion und diskursive Konstruktion [Communicative constructivism and discursive construction]. In R. Keller, H. Knoblauch, & J. Reichertz (Eds.), *Kommunikativer Konstruktivismus. Theoretische und empirische Arbeiten zu einem neuen wissenssoziologischen Ansatz* (pp. 69–94). Wiesbaden, Germany: Springer VS.

Keller, R., Knoblauch, H., & Reichertz, J. (Eds.). (2013). *Kommunikativer Konstruktivismus. Theoretische und empirische Arbeiten zu einem neuen wissenssoziologischen Ansatz* [Communicative constructivism: Theoretical and empirical papers to a new approach of sociology of knowledge]. Wiesbaden, Germany: Springer VS.

Kerber, W. (Ed.). (1993). *Der Begriff der Religion* [The concept of religion]. Munich, Germany: Kindt.

King, D. (1997). *The commissar vanishes: The falsification of photographs and art in Stalin's Russia*. Edinburgh, UK: Canongate Books.

Kintzinger, M. (2003). *Wissen wird Macht. Bildung im Mittelalter* [Knowledge becomes power: Education in the Middle Ages]. Darmstadt, Germany: Jan Thorbecke.

Klauser, F. R. (2014). Introduction: Foundations for a political geography of surveillance. In F. R. Klauser, *Governing the everyday in the information age: Towards a political geography of surveillance* (pp. 2–46). Postdoctoral dissertation, University of Berne, Switzerland.

Klauser, F. R. (2009). Interacting forms of expertise in security governance: The example of CCTV surveillance at Geneva International Airport. *British Journal of Sociology, 60,* 279–297.

Klauser, F. R. (2013). Spatialities of security and surveillance: Managing spaces, separations and circulations at sport mega events. *Geoforum, 49,* 289–298.

Knebel, S. (2004). Wahrheit, objektive [Objective truth]. In J. Ritter, K. Gründer, & G. Gabriel (Eds.), *Historisches Wörterbuch der Philosophie: Vol. 12. W–Z* [CD-Rom] (pp. 154–160). Basel, Switzerland: Schwabe.

Kneer, G. (2012). Die Analytik der Macht bei Michel Foucault [Analytics of power: Michel Foucault]. In P. Imbusch (Ed.), *Macht und Herrschaft. Sozialwissenschaftliche Theorien und Konzeptionen* (pp. 265–283). Wiesbaden, Germany: Springer Fachmedien. doi:10.1007/978-3-531-93469-3_13.

Knoblauch, H. (2013a). Grundbegriffe und Aufgaben des kommunikativen Konstruktivismus [Basic terms and tasks of communicative constructivism]. In R. Keller, H. Knoblauch, & J. Reichertz (Eds.), *Kommunikativer Konstruktivismus. Theoretische und empirische Arbeiten zu einem neuen wissenssoziologischen Ansatz* (pp. 25–47). Wiesbaden, Germany: Springer VS.

Knoblauch, H. (2013b). Über die kommunikative Konstruktion der Wirklichkeit [About the communicative structure of reality]. In G. B. Christmann (Ed.), *Zur kommunikativen Konstruktion von Räumen. Theoretische Konzepte und empirische Analysen* (pp. 29–53). Wiesbaden, Germany: Springer VS.

Kobusch, T., & Oeing-Hanhoff, L. (1980). Macht [Power]. In J. Ritter, K. Gründer, & G. Gabriel (Eds.), *Historisches Wörterbuch der Philosophie: Vol. 5. L–Mn* [CD-Rom] (pp. 585–588). Basel, Switzerland: Schwabe.

Kocsis, K. (Ed.). (2007). *South Eastern Europe in maps* (2nd, rev. and expanded ed.). Budapest, Hungary: Geographical Research Institute, Hungarian Academy of Sciences.

Kocsis, K. (1994). Contribution to the background of the ethnic conflicts in the Carpathian Basin. *GeoJournal, 32,* 425–433.

Kocsis, K., & Kocsisné, E. (1998). *Ethnic geography of the Hungarian minorities in the Carpathian Basin.* Budapest, Hungary: Geographical Research Institute, Hungarian Academy of Sciences.

Kuran, T. (1995). *Private truths, public lies: The social consequences of preference falsification.* Cambridge, MA: Harvard University Press.

Lappo, G., & Poljan, P. (1997). Transformation der geschlossenen Städte Rußlands [Transformation of the closed cities of Russia]. In Bundesinstitut für ostwissenschaftliche und internationale

Studien (Ed.), *Bericht des Bundesinstituts für ostwissenschaftliche und internationale Studien* (Vol. 6, pp. 3–29). Cologne, Germany: Bundesinstitut für ostwissenschaftliche und internationale Studien.

Lappo, G., & Poljan, P. (2007). Naoukograds, les villes interdites [Soviet science towns: The forbidden cities]. In C. Jacob (Ed.), *Lieux de savoir. Espaces et communautés* (pp. 1226–1249). Paris: Editions Albin Michel.

Lasby, C. G. (1971). *Project paperclip: German scientists and the Cold War*. New York: Atheneum.

Latour, B. (1987). *Science in action: How to follow scientists and engineers through society*. Milton Keynes, UK: Open University Press.

Leed, E. (1981). *No Man's Land: Combat and identity in World War I*. Cambridge, UK: Cambridge University Press.

Lenin, V. I. (1964). Imperialism, the highest stage of capitalism. In G. Hanna (Ed.), *V. I. Lenin. Collected works: Vol. 22. December 1915–July 1916* (Y. Sdobnikov, Trans., pp. 185–304). Moscow, Russia: Progress Publishers.

Lewis, J. R., & Hammer, O. (2011). *Handbook of religion and the authority of science*. Leiden, The Netherlands/Boston: Brill.

Leyens, J. P. (2001). Prejudice in society. In N. J. Smelser & P. B. Baltes (Eds.), *International encyclopedia of the social & behavioral sciences* (Pes–Pre, Vol. 17, pp. 11986–11989). Amsterdam, The Netherlands: Elsevier.

Liebenberg, L. (1990). *The art of tracking: The origin of science*. Claremont, South Africa: David Philip.

Lippmann, W. (1955). *The public philosophy*. New York: Mentor Books.

Löbl, R. (1997). *Texnh-Techne: Untersuchungen zur Bedeutung dieses Worts in der Zeit von Homer bis Aristoteles: Bd. 1. Von Homer bis zu den Sophisten* [Texnh—Techne: Inquiries into the meaning of this word from Homer to Aristotle: Vol. 1. From Homer to the sophists]. Würzburg, Germany: Königshausen & Neumann.

Löbl, R. (2003). *Texnh-Techne: Untersuchungen zur Bedeutung dieses Worts in der Zeit von Homer bis Aristoteles. Bd. 2: Von den Sophisten bis Aristoteles* [Texnh—Techne: Inquiries into the meaning of this word in the period from Homer to Aristotle: Vol. 2. From the sophists to Aristotle]. Würzburg, Germany: Königshausen & Neumann.

Luckmann, T. (1967). *The invisible religion, The problem of religion in modern society*. New York: Macmillan.

Lyon, D. (2001). *Surveillance society: Monitoring everyday life*. Buckingham, UK: Open University Press.

Lyon, D. (Ed.). (2003). *Surveillance as social sorting*. London: Routledge.

Malý, K. (2005). Presserecht und Zensur in der Tschechoslowakei in den Jahren 1945–1990 [Press law and censorship in Czechoslovakia, 1945–1990]. In M. Anděl, D. Brandes, A. Labisch, J. Pešek, T. Ruzicka, & A. Kutsch (Eds.), *Propaganda, (Selbst-)Zensur, Sensation. Grenzen von Presse- und Wissenschaftsfreiheit in Deutschland und Tschechien seit 1871* (Veröffentlichungen zur Kultur und Geschichte im östlichen Europa, Vol. 27, pp. 223–233). Essen, Germany: Klartext.

Mann, M. (1986). *The sources of social power: Vol. I. A history of power from the beginning to A.D. 1760*. Cambridge, UK: Cambridge University Press.

Maresch, R. (2002). Hard power/Soft power. Amerikas Waffen globaler Raumnahme [Hard power/soft power: America's weapons for seizing global space]. In R. Maresch & N. Werber (Eds.), *Raum—Wissen—Macht* (2nd ed., pp. 237–262). Frankfurt am Main, Germany: Suhrkamp.

Marxhausen, C. (2010). *Identität, Repräsentation, Diskurs. Eine handlungsorientierte linguistische Diskursanalyse zur Erfassung raumbezogener Identitätsangebote* [Identity, representation, discourse: A linguistic discourse analysis based on action theory for capturing spatial identity] (Sozialgeographische Bibliothek, Vol. 14). Stuttgart, Germany: Steiner.

Maul, S. M. (1994). *Zukunftsbewältigung: Eine Untersuchung altorientalischen Denkens anhand der babylonisch-assyrischen Löserituale (Namburbi)* [Coping with the future: A study of ancient oriental thinking on the basis of Babylonian–Assyrian apotropaic rituals (Namburbi)] (Baghdader Forschungen, Vol. 18). Mainz, Germany: von Zabern.

Maul, S. M. (2003). Omina und Orakel. A: In Mesopotamien [Omina and oracle: A. In Mesopotamia]. In D. O. Edzard & M. P. Streck (Eds.), *Reallexikon der Assyriologie und Vorderasiatischen Archäologie: Vol. 10. Oannes–Priesterverkleidung* (pp. 45–88). 1./2. Lieferung. Berlin, Germany: De Gruyter.

Maul, S. M. (2013). *Die Wahrsagekunst im Alten Orient. Zeichen des Himmels und der Erde* [The art of scrying in the old orient: Signs of heaven and earth]. Munich, Germany: Beck.

Maurer, A. (2012). Herrschaftsordnungen. Die Idee der rationalen Selbstorganisation freier Akteure von Hobbes über Weber zu Coleman [Authority structures: The idea of rational self-organization of free agents from Hobbes to Weber and Coleman]. In P. Imbusch (Ed.), *Macht und Herrschaft. Sozialwissenschaftliche Theorien und Konzeptionen* (pp. 357–378). Wiesbaden, Germany: Springer Fachmedien. doi:10.1007/978-3-531-93469-3_17.

Mauthner, F. (1910). *Wörterbuch der Philosophie. Neue Beiträge zu einer Kritik der Sprache* [Dictionary of philosophy: New contributions to a critique of language]. Leipzig, Germany: Felix Meiner-Verlag.

Meier-Oeser, S. (2004). Wissenschaft [Science]. In J. Ritter, K. Gründer, & G. Gabriel (Eds.), *Historisches Wörterbuch der Philosophie: Vol. 12. W–Z* [CD-Rom] (pp. 902–915). Basel, Switzerland: Schwabe.

Meier-Seethaler, C. (1999). Gefühle als moralische und ästhetische Urteilskraft [Emotions as moral and aesthetic power of judgment]. In J. Mittelstraß (Ed.), *Die Zukunft des Wissens* (pp. 147–152). Konstanz, Germany: Universitätsverlag Konstanz.

Melville, H. (1970). *White jacket; or, The world in a man-of-war*. Evanston, IL: Northwestern University Press (Original work published 1850).

Meusburger, P. (1997). Spatial and social inequality in communist countries and in the first period of the transformation process to a market economy: The example of Hungary. *Geographical Review of Japan, 70 (Series B)*, 126–143.

Meusburger, P. (1998). *Bildungsgeographie. Wissen und Ausbildung in der räumlichen Dimension* [Geography of education: Knowledge and education in the spatial dimension]. Heidelberg, Germany: Spektrum Akademischer Verlag.

Meusburger, P. (2005). Sachwissen und symbolisches Wissen als Machtinstrument und Konfliktfeld. Zur Bedeutung von Worten, Bildern und Orten bei der Manipulation des Wissens [Factual knowledge and symbolic knowledge as an instrument of power and a field of conflict: The meaning of words, images, and places for the manipulation of knowledge]. *Geographische Zeitschrift, 93*, 148–164.

Meusburger, P. (2007a). Macht, Wissen und die Persistenz von räumlichen Disparitäten [Power, knowledge, and the persistence of spatial disparities]. In I. Kretschmer (Ed.), *Das Jubiläum der Österreichischen Geographischen Gesellschaft. 150 Jahre (1856–2006)* (pp. 99–124). Vienna, Austria: Österreichische Geographische Gesellschaft.

Meusburger, P. (2007b). Power, knowledge and the organization of space. In J. Wassmann & K. Stockhaus (Eds.), *Experiencing new worlds* (pp. 111–124). New York: Berghahn Books.

Meusburger, P. (2008). The nexus of knowledge and space. In P. Meusburger, M. Welker, & E. Wunder (Eds.), *Clashes of knowledge: Orthodoxies and heterodoxies in science and religion* (Knowledge and space, Vol. 1, pp. 35–90). Dordrecht, The Netherlands: Springer. doi:10.1007/978-1-4020-5555-3_2.

Meusburger, P. (2011). Knowledge, cultural memory, and politics. In P. Meusburger, M. Heffernan, & E. Wunder (Eds.), *Cultural memories: The geographical point of view* (Knowledge and space, Vol. 4, pp. 51–69). Dordrecht, The Netherlands: Springer. doi:10.1007/978-90-481-8945-8_4.

Meusburger, P. (2013). Relations between knowledge and economic development: Some methodological considerations. In P. Meusburger, J. Glückler, & M. El Meskioui (Eds.), *Knowledge and the economy* (Knowledge and space, Vol. 5, pp. 15–42). Dordrecht, The Netherlands: Springer. doi:10.1007/978-94-007-6131-5_2.

Meusburger, P., Heffernan, M., & Wunder, E. (2011). Cultural memories: An introduction. In P. Meusburger, M. Heffernan, & E. Wunder (Eds.), *Cultural memories: The geographical point of view* (Knowledge and space, Vol. 4, pp. 3–14). Dordrecht, The Netherlands: Springer. doi:10.1007/978-90-481-8945-8_1.

Mintzberg, H. (1979). *The structuring of organizations: A synthesis of the research.* Englewood Cliffs, NJ: Prentice Hall.

Míšková, A. (2005). "Politische Säuberungen" der Bestände der wissenschaftlichen Bibliotheken der tschechoslowakischen Akademie der Wissenschaften in den 1950er Jahren ["Political cleansing" of the research library holdings of the Czechoslovakian Academy of Sciences in the 1950s]. In M. Anděl, D. Brandes, A. Labisch, J. Pešek, T. Ruzicka, & A. Kutsch (Eds.), *Propaganda, (Selbst-)Zensur, Sensation. Grenzen von Presse- und Wissenschaftsfreiheit in Deutschland und Tschechien seit 1871* (Veröffentlichungen zur Kultur und Geschichte im östlichen Europa, Vol. 27, pp. 235–243). Essen, Germany: Klartext.

Mitchell, T. (1990). Everyday metaphors of power. *Theory and Society, 19*, 545–577.

Mittelstraß, J. (1982). *Wissenschaft als Lebensform. Reden über philosophische Orientierungen in Wissenschaft und Universität* [Knowledge as way of life: Addresses on philosophical orientations in science and the university]. Frankfurt am Main, Germany: Suhrkamp.

Mittelstraß, J. (2001). *Wissen und Grenzen. Philosophische Studien* [Knowledge and borders: Philosophical Studies]. Frankfurt am Main, Germany: Suhrkamp.

Mittelstraß, J. (2010). The loss of knowledge in the information age. In E. De Corte & J. E. Fenstad (Eds.), *From information to knowledge; from knowledge to wisdom* (Wenner–Gren International Series, Vol. 85, pp. 19–23). London: Portland Press.

Moldaschl, M., & Stehr, N. (2010). Eine kurze Geschichte der Wissensökonomie [A brief history of the economics of knowledge]. In M. Moldaschl & N. Stehr (Eds.), *Wissensökonomie und Innovation. Beiträge zur Ökonomie der Wissensgesellschaft* (pp. 9–74). Marburg, Germany: Metropolis-Verlag.

Münkler, H. (1995). Die Visibilität der Macht und die Strategien der Machtvisualisierung [Visibility of power and strategies of visualizing power]. In G. Göhler (Ed.), *Macht der Öffentlichkeit— Öffentlichkeit der Macht* (pp. 213–230). Baden-Baden, Germany: Nomos Verlag.

Murakami Wood, D. (2007). Beyond the panopticon? Foucault and surveillance studies. In J. W. Crampton & S. Elden (Eds.), *Space, knowledge and power: Foucault and geography* (pp. 245–263). Aldershot, UK: Ashgate Publishing Company.

Mutschler, F.-H. (2005). Potestatis nihilo amplius habui quam ceteri. Zum Problem der Invisibilisierung der Macht im frühen Prinzipat [On the problem of rendering power invisible in the early principate]. In G. Melville (Ed.), *Das Sichtbare und das Unsichtbare der Macht. Institutionelle Prozesse in Antike, Mittelalter und Neuzeit* (pp. 259–282). Cologne, Germany: Böhlau.

Nietzsche, F. (1968). *The will to power* (W. Kaufmann & R. J. Hollingdale, Trans.; W. Kaufmann, Ed.; with commentary by W. Kaufmann). New York: Vintage Books.

Onnasch, E.-O. (2004). Wahrheit, absolute [Absolute truth]. In J. Ritter, K. Gründer, & G. Gabriel (Eds.), *Historisches Wörterbuch der Philosophie: Vol. 12. W–Z* [CD-Rom] (pp. 135–137). Basel, Switzerland: Schwabe.

Palsky, G. (2002). Emmanuel de Martonne and the ethnographical cartography of Central Europe (1917–1920). *Imago Mundi, 54*, 111–119. doi:10.1080/03085690208592961.

Parsons, T. (1967). *On the concept of political power: Sociological theory and modern society.* London: Free Press.

Penzlin, H. (2002). *Die Welt als Täuschung* [The world as illusion]. *Gehirn & Geist, 3*, 68–73.

Pešek, J. (2005). "Litterae et libri prohibiti" in der kommunistisch beherrschten Tschechoslowakei [Forbidden texts and books in communist Czechoslovakia]. In M. Anděl, D. Brandes, A. Labisch, J. Pešek, & T. Ruzicka (Eds.), *Propaganda, (Selbst-)Zensur, Sensation. Grenzen von Presse- und Wissenschaftsfreiheit in Deutschland und Tschechien seit 1871* (Veröffentlichungen zur Kultur und Geschichte im östlichen Europa, Vol. 27, pp. 245–252). Essen, Germany: Klartext.

Pestre, D. (2003). Regimes of knowledge production in society: Towards a more political and social reading. *Minerva, 41*, 245–261.

Pfister, T., & Stehr, N. (2013). Einführung: Fragile Welten aus Wissen [Introduction: Fragile worlds of knowledge]. In S. A. Jansen, E. Schröter, & N. Stehr (Eds.), *Fragile Stabilität—stabile Fragilität* (pp. 9–18). Wiesbaden, Germany: Springer VS.

Pitkin, H. (1972). *Wittgenstein and justice*. Berkeley, CA: University of California Press.
Pollock, E. (2006). *Stalin and the soviet science wars*. Princeton, NJ: Princeton University Press.
Popitz, H. (1992). *Phänomene der Macht* [Phenomena of power] (2nd enlarged ed.). Tübingen, Germany: Mohr.
Post, R. C. (Ed.). (1998). *Censorship and silencing: Practices of cultural regulation*. Los Angeles: Getty Research Institute for the History of Art and the Humanities.
Pulte, H. (2004a). Wahrheitsähnlichkeit [Verisimilitude]. In J. Ritter, K. Gründer, & G. Gabriel (Eds.), *Historisches Wörterbuch der Philosophie: Vol. 12. W–Z* [CD-Rom] (pp. 170–177). Basel, Switzerland: Schwabe.
Pulte, H. (2004b). Wissenschaft III. Ausbildung moderner Wissenschafts-Begriffe im 19. und 20. Jh. [Science, subsection III: Emergence of modern scientific terms in the nineteenth and twentieth centuries]. In J. Ritter, K. Gründer, & G. Gabriel (Eds.), *Historisches Wörterbuch der Philosophie: Vol. 12. W–Z* [CD-Rom] (pp. 921–948). Basel, Switzerland: Schwabe.
Puster, R. (1999). Die Endlichkeit des Wissens: Epistemologie zwischen Genese und Geltung [The finiteness of knowledge: Epistemology between inception and acceptance]. In J. Mittelstraß (Ed.), *Die Zukunft des Wissens* (pp. 96–103). Konstanz, Germany: Universitätsverlag Konstanz.
Radbruch, G. (1993). *Rechtsphilosophie II* [Philosophy of law II]. Gesamtausgabe (A. Kaufmann, Ed.). Vol. 2. Heidelberg, Germany: Müller, Juristenverlag.
Radeiskis, B. (2013). Erinnerungen an die DDR oder Erinnerungen an DDR-Propaganda? Exemplarische Überlegungen zur strukturellen Ähnlichkeit von Erinnerungs- und Propagandadiskursen [Memories of the GDR or memories of GDR propaganda? Exemplary considerations about structural similarity between discourses of remembering and discourses of propaganda]. In E. Felder (Ed.), *Faktizitätsherstellung in Diskursen: Die Macht des Deklarativen* (pp. 359–376). Berlin, Germany: De Gruyter.
Rehberg, K.-S. (2005). Sichtbarkeit und Invisibilisierung der Macht durch die Künste. Die DDR-"Konsensdiktatur" als Exemplum [The visibility and invisibility of power as rendered by the arts: The GDR's "consensual dictatorship"]. In G. Melville (Ed.), *Das Sichtbare und das Unsichtbare der Macht. Institutionelle Prozesse in Antike, Mittelalter und Neuzeit* (pp. 355–382). Cologne, Germany: Böhlau.
Richta, R. (1977). The scientific and technological revolution and the prospects of social development. In R. Dahrendorf (Ed.), *Scientific-technological revolution: Social aspects* (pp. 25–72). London: Sage.
Richta, R. (with research team). (1969). *Civilization at the crossroads: Social and human implications of the scientific and technological revolution* (3rd enlarged ed.). White Plains, NY: International Arts and Sciences Press.
Ricœur, P. (2005). *The course of recognition* (D. Pellauer, Trans.). Cambridge, MA: Harvard University Press.
Röttgers, K. (1980). Macht [Power]. In J. Ritter, K. Gründer, & G. Gabriel (Eds.), *Historisches Wörterbuch der Philosophie: Vol. 5. L–Mn* [CD-Rom] (pp. 588–603). Basel, Switzerland: Schwabe.
Rueschemeyer, D. (1986). *Power and the division of labor*. Stanford, CA: Stanford University Press.
Russell, B. (1958). *In praise of idleness and other essays*. London: G. Allen & Unwin.
Said, E. W. (1978). *Orientalism*. New York: Vintage.
Sandalow, M. (2001, September 13). War Footing/Bush promises to conquer a new kind of enemy/ President sees battle between 'good' and 'evil'. San Francisco Chronicle, p. A7. Retrieved January 22, 2014, from http://www.sfgate.com/default/article/WAR-FOOTING-Bush-promises-to-conquer-a-new-kind-2879805.php
Scheler, M. (1926). *Die Wissensformen und die Gesellschaft. Probleme einer Soziologie des Wissens* [The forms of knowledge and society: Problems of a sociology of knowledge]. Leipzig, Germany: Der Neue-Geist Verlag.
Schelsky, H. (1975). *Die Arbeit tun die Anderen. Klassenkampf und Priesterherrschaft der Intellektuellen* [The others do the work: Class struggle and intellectual theocratic caste]. Opladen, Germany: Westdeutscher Verlag.

Schimank, U. (1992). Science as a societal risk-producer: A general model of intersystemic dynamics, and some specific institutional determinants of research behavior. In N. Stehr & R. V. Ericson (Eds.), *The culture and power of knowledge: Inquiries into contemporary societies* (pp. 215–233). Berlin, Germany: Walter de Gruyter.

Schleiermacher, F. D. E. (1988). *Dialektik* (1814–1815) [Dialectics]. *Einleitung zur Dialektik* (1833) [Introduction to dialectics] (Edited by A. Arndt). Hamburg, Germany: Felix Meiner.

Schönrich, G. (2005). Machtausübung und die Sicht der Akteure. Ein Beitrag zur Theorie der Macht [The exercise of power and the view of the actors: On the theory of power]. In G. Melville (Ed.), *Das Sichtbare und das Unsichtbare der Macht. Institutionelle Prozesse in Antike, Mittelalter und Neuzeit* (pp. 383–409). Cologne, Germany: Böhlau.

Schütz, A., & Luckmann, T. (1973). *The structures of the life-world.* Vol. 1 (R. M. Zaner & H. T. Engelhardt, Trans.). Evanston, IL: Northwestern University Press.

Seegel, S. (2012). *Mapping Europe's borderlands: Russian cartography in the age of the Empire.* Chicago: University of Chicago Press.

Shapiro, M. (1997). *Violent cartographies: Mapping cultures of war.* Minneapolis, MN: University of Minnesota Press.

Simonds, A. P. (1989). Ideological domination and the political information market. *Theory and Society, 18,* 181–211.

Simpson, G. G. (1963). Biology and the nature of science. *Science, New Series, 139*(3550), 81–88.

Sorokin, P. A. (1985). *Social and cultural dynamics: A study of change in major systems of art, truth, ethics, law, and social relationships.* New Brunswick, NJ: Transaction Books.

Speth, R. (1997). Foucaults Theorie der Disziplinarmacht mit Bezug auf Nietzsche [Foucault's theory of disciplinary power with regard to Nietzsche]. In G. Göhler (Ed.), *Institution—Macht—Repräsentation. Wofür politische Institutionen stehen und wie sie wirken* (pp. 262–320). Baden-Baden, Germany: Nomos.

Speth, R., & Buchstein, H. (1997). Hannah Arendts Theorie intransitiver Macht [Hannah Arendt's theory of intransitive power]. In G. Göhler (Ed.), *Institution—Macht—Repräsentation. Wofür politische Institutionen stehen und wie sie wirken* (pp. 224–261). Baden-Baden, Germany: Nomos.

Staatswissenschaftliches Institut (Ed.). (1942). *Rumänische ethnographische Landkarten und ihre Kritik* [Romanian ethnographical maps and critique thereof]. Budapest, Hungary: Staatswissenschaftliches Institut.

Stegmaier, W. (2008). *Philosophie der Orientierung* [Philosophy of orientation]. Berlin, Germany: Walter de Gruyter.

Stegmaier, W. (1992). Wahrheit und Orientierung. Zur Idee des Wissens [Truth and orientation: About the idea of knowledge]. In V. Gerhardt & N. Herold (Eds.), *Perspektiven des Perspektivismus. Gedenkschrift zum Tode Friedrich Kaulbachs* (pp. 287–307). Würzburg, Germany: Königshausen and Neumann.

Stehr, N. (1992). Experts, counselors and advisers. In N. Stehr & V. R. Ericson (Eds.), *The culture and power of knowledge: Inquiries into contemporary societies* (pp. 107–155). Berlin, Germany: Walter de Gruyter.

Stehr, N., & Ericson, V. R. (1992). The culture and power of knowledge in modern society. In N. Stehr & V. R. Ericson (Eds.), *The culture and power of knowledge: Inquiries into contemporary societies* (pp. 3–19). Berlin, Germany: Walter de Gruyter.

Stenmark, M. (2008). Science and the limits of knowledge. In P. Meusburger, M. Welker, & E. Wunder (Eds.), *Clashes of knowledge: Orthodoxies and heterodoxies in science and religion* (Knowledge and space, Vol. 1, pp. 111–120). Dordrecht, The Netherlands: Springer. doi:10.1007/978-1-4020-5555-3_5.

Stone, J. (1988). Imperialism, colonialism and cartography. *Transactions of the Institute of British Geographers, New Series, 13,* 57–64.

Szöllösi-Janze, M. (2004). Politisierung der Wissenschaften—Verwissenschaftlichung der Politik. Wissenschaftliche Politikberatung zwischen Kaiserreich und Nationalsozialismus [Politicization of the sciences—Scientification of politics: Scienific policy consulting between the empire and National Socialism]. In S. Fisch & W. Rudloff (Eds.), *Experten und Politik. Wissenschaftliche Politikberatung in geschichtlicher Perspektive* (pp. 79–100). Berlin, Germany: Duncker & Humblot.

Tanner, K. (1999). Ethik und Religion [Ethics and religion]. In R. Anselm, S. Schleissing, & K. Tanner (Eds.), *Die Kunst des Auslegens. Zur Hermeneutik des Christentums in der Kultur der Gegenwart* (pp. 225–241). Frankfurt am Main, Germany: Peter Lang.

Tanner, K. (2005). Die unsichtbare Dimension der Macht. Ekklesiologie als Exemplum der Analyse des Institutionellen [The invisible dimension of power: Ecclesiology as exemplum of the analysis of the institutional]. In G. Melville (Ed.), *Das Sichtbare und das Unsichtbare der Macht. Institutionelle Prozesse in Antike, Mittelalter und Neuzeit* (pp. 3–17). Cologne, Germany: Böhlau.

Taylor, P., Hoyler, M., & Evans, D. M. (2010). A geohistorical study of "the rise of modern science": Mapping scientific practice through urban networks, 1500–1900. In P. Meusburger, D. N. Livingstone, & H. Jöns (Eds.), *Geographies of science* (Knowledge and space, Vol. 3, pp. 37–56). Dordrecht, The Netherlands: Springer. doi:10.1007/978-90-481-8611-2_3.

U.N. says waterboarding should be prosecuted as torture. (2008, February 8). Reuters Edition UK. Retrieved February 26, 2014, from http://uk.reuters.com/article/2008/02/08/uk-usa-torture-un-idUKN0852061620080208

van den Daele, W. (1992). Scientific evidence and the regulation of technical risks: Twenty years of demythologizing the experts. In N. Stehr & V. R. Ericson (Eds.), *The culture and power of knowledge: Inquiries into contemporary societies* (pp. 323–340). Berlin, Germany: Walter de Gruyter.

Vavilov, S. I. (1950). Sztálin a tudomány géniusza [Stalin, the genius of science]. In A. Szovjetunió Tudományos Akadémiája (Ed.), *Sztálin és a szovjet tudomány* (pp. 11–25). Budapest, Hungary: Szikra.

Weber, M. (1964). *The theory of social and economic organization* (A. M. Henderson & T. Parsons, Trans.). Glencoe, Ill: The Free Press. (Original work published 1922)

Weber, M. (1978). *Economy and society: An outline of interpretive sociology.* 2 Vols. (G. Roth & C. Wittich, Eds.; E. Fischoff, H. Gerth, A. M. Henderson, F. Kolegar, C. Wright Mills, T. Parsons, M. Rheinstein, G. Roth, E. Shils, & C. Wittich, Trans.). Berkeley, CA: University of California Press. (Original work published 1922)

Weinberg, A. K. (1935). *Manifest destiny: A study of nationalist expansionism in American history.* Baltimore: Johns Hopkins Press.

Welker, M. (2008). The demarcation problem of knowledge and faith: Questions and answers from theology. In P. Meusburger, M. Welker, & E. Wunder (Eds.), *Clashes of knowledge: Orthodoxies and heterodoxies in science and religion* (Knowledge and space, Vol. 1, pp. 145–153). Dordrecht, The Netherlands: Springer. doi:10.1007/978-1-4020-5555-3_8.

Werner, T. (1995). Bücherverbrennungen im Mittelalter [Book burnings in the Middle Ages]. In O. G. Oexle (Ed.), *Memoria als Kultur* (Veröffentlichungen des Max-Planck-Instituts für Geschichte, Vol. 121, pp. 149–184). Göttingen, Germany: Vandenhoeck & Ruprecht.

Westwick, P. J. (2000). Secret science: A classified community in the national laboratories. *Minerva, 38,* 363–391.

Wieland, W. (1982). *Platon und die Formen des Wissens* [Plato and the categories of knowledge]. Göttingen, Germany: Vandenhoeck & Ruprecht.

Wilkin, P. (1997). *Noam Chomsky: On power, knowledge and human nature.* New York: St. Martin's Press.

Wilkinson, H. R. (1951). *Maps and politics: A review of the ethnographic cartography of Macedonia.* Liverpool, UK: University Press.

Winthrop, R. C. (Ed.). (1869). *Life and letters of John Winthrop: Governor of the Massachusetts-Bay Company at their emigration to New England, 1630.* 2 vols. Boston, MA: Little, Brown.

Zachhuber, J. (2004). Wahrheit, praktische bzw. moralische [Truth, practical and moral]. In J. Ritter, K. Gründer, & G. Gabriel (Eds.), *Historisches Wörterbuch der Philosophie: Vol. 12. W–Z* [CD-Rom] (pp. 164–167). Basel, Switzerland: Schwabe.

Zeller, B. (2011). New religious movements and science. *Nova Religio: The Journal of Alternative and Emergent Religions, 14*(4), 4–10. doi:10.1525/nr.2011.14.4.4.

Enabling Knowledge

3

Nico Stehr

> *A democratic system in which knowledge is made the focus of continuing public concern is the only basis, under modern conditions, for government which is both effective and responsible.*
>
> Sanford A. Lakoff, *Knowledge, Power, and Democratic Theory* (1971, p. 12)

In an essay in the *New York Review of Books*, the molecular biologist Richard Lewontin (2004, p. 38) maintains that "the knowledge required for political rationality, once available to the masses, is now in the possession of a specially educated elite, a situation that creates a series of tensions and contradictions in the operation of representative democracy." Has, therefore, as Jones (2004, pp. 16–63) has suggested, the optimism of the philosophers of the French Enlightenment, particularly the Marquis de Condorcet's view of the role of knowledge in overcoming poverty, violence, and ignorance, as well as in building a sustainable democratic society, been destroyed?

By the same token, English chemistry Nobel laureate Harry Kroto, in an opinion piece in the *Guardian* (Kroto, 2007, p. 1), denounces the UK government for wrecking British science and science education, despite the fact that the "need for a general population with a satisfactory understanding of science and technology has never been greater." Kroto, who has left England and is now researching and teaching in the United States, adds that "we live in a world economically, socially, and culturally dependent on science not only functioning well, but being wisely applied."

Moreover, in light of the growing specialization of the production of scientific knowledge, a social scientist observes that all but a few individuals are deprived of the "capacity for individual rational judgment either about the quality of the evidence

N. Stehr (✉)
Cultural Studies, Institute for Political and Social Sciences, Zeppelin University,
Am Seemoser Horn 20, 88045, Friedrichshafen, Germany
e-mail: nico.stehr@zu.de

© Springer Netherlands 2015
P. Meusburger et al. (eds.), *Geographies of Knowledge and Power*,
Knowledge and Space 7, DOI 10.1007/978-94-017-9960-7_3

proffered or about the tightness of the theoretical reasoning applied to the analysis of the data. The 'harder' the science, the truer this is" (Wallerstein, 2004, p. 8).[1]

But it is not only the deficient capacity to engage in "rational" discourse—for example with the carriers of expert knowledge, on account of a lack of relevant skills in comprehending and contesting specialized knowledge claims—that is at issue in the claims about the restrictions on participation of many citizens in an emerging expert society. The lack of cognitive skills is also seen to have an impact on traditional material, economic opportunities and hence the true nature of the inequality regime in modern society (see Stehr, 1999). As Mancur Olson (1982) has observed, "Individuals in a few special vocations can receive considerable rewards in private goods if they acquire exceptional knowledge of public goods … Withal, the typical citizen will find that his or her income and life chances will not be improved by zealous study of public affairs, or even of any single collective good" (p. 26).[2]

Richard Lewontin, Harry Kroto, Immanuel Wallerstein, and Mancur Olson are representative of a far greater number of individuals in the science community who skeptically view the increasing use of contemporary, and in particular natural scientific, knowledge, not only by governments but also as a tool in politics (cf. Pielke, 2007). According to these scholars, this tendency has led to the massive increase, paralleling the widening gap between rich and poor in many developed societies, in the inability of large segments of the population to take part in democratic decision-making. Given these circumstances, "ordinary" citizens apparently are also robbed of the ability to rationally enter into discourse about modern science and technology and its social consequences.[3]

A less strident and less frequent perspective in the debate on the relations between scientific knowledge, governance, civil society, and participation of the public in policy matters would question the very premise of the observers I just cited. Yaron Ezrahi (2004, p. 273), for example, notes that science has declined in significance in contemporary politics; scientists are much less in demand by politicians; science is no longer as important a component of modern state authority as in the past; and scientific

---

[1] The historian James Harvey Robinson (1923, p. 76) stresses, in a treatise entitled *The Humanizing of Knowledge*—that is, of ensuring that the "scientific frame of mind" and specialized scientific knowledge are not an esoteric enterprise confined to a small number of members of the scientific community—that the "divisions of knowledge … form one of the most effective barriers to the cultivation of a really scientific frame of mind in thee young and thee public at large." The solution therefore lies in "re-synthesizing" and "re-humanizing" knowledge.

[2] Applying an economic logic to the question of the lack of knowledge among ordinary citizens with respect to policy issues or, for that matter, any issue in everyday life, Olson's observations raise the question of the incentive of ordinary citizens to learn enough about issues that require a decision from them. If the acquisition of relevant knowledge entails costs and the consequences of their decisions or informed participation are obscure, citizens presumably have little incentive to acquire relevant knowledge—at least according to the "economic theory of democracy" proposed by Anthony Downs (1957). However, it would be shortsighted to limit the possible incentives to acquire knowledge to mere economic considerations.

[3] Aside from the *ability* to enter a field of discourse, there is also the question of the *desire* to enter a field of discourse in an active manner. Ability and desire likely interact on a psychological level, and desire and ability do vary from person to person as well as from issue to issue (cf. Mulder, 1971).

knowledge actually is no longer the "resource it once was, with which policies and public choices could be legitimated as impersonal, objective and technical."

The frequent lament about the extent to which specialized knowledge disenfranchises the majority of citizens in modern societies conveniently sums up the questions about the multiple linkages between knowledge and democracy that I wish to explore in this chapter.[4] Is it indeed the case that we cannot escape the dilemma of deferring our judgments to self-selected communities of experts? And can it be, for example, that most members of modern society do not know enough to participate intelligently in policy discourse?[5]

Knowledgeability and Democracy

On the surface, questions of the relations between the knowledgeability of broad segments of the population and democratic government are not a widely or explicitly discussed set of issues in contemporary social science. Much higher on the agenda of the social sciences are lively discussions of the notion that "democratic theory cannot be articulated in satisfactory terms today without looking in detail at the politics of science and technology" (Jasanoff, 2005, p. 6). Such a conclusion is of course a measure of the practical political significance of scientific and technical developments in modern societies. However, if one extends one's perspective to *mediated* relations between knowledge, the economy, civil society, and democratic regimes, one constantly encounters the relations of knowledgeability and democracy. To name but a few of the issues on the agenda of social science and politics today, we encounter cultural capital and political franchise, access to educational institutions and the social distribution of knowledge, accountability and citizen participation, the competitiveness of nations, and social identities and political inclusiveness.

I shall begin with a rather broad set of questions and claims. As Max Horkheimer emphasized—in contrast to Karl Marx—justice or equity and freedom do not mutually support each other. Does Horkheimer's assertion also apply to democracy and knowledge? Or is knowledge a democratizer? Is the progress of knowledge, especially rapid advances in knowledge, a burden on democracy, civil society, and the capacity of the individual to assert his or her will? If there is a contradiction between knowledge and democratic processes, is it a *new* development, or is the advance of liberal democracies codetermined by the joint force of knowledge and democratic political conduct that enable one to claim that civil society, if not democracy, is the daughter of knowledge?

---

[4] My observations in this chapter draw on a study in which I explore the origins, formations, and sustainability of the linkages between knowledge, knowledgeability, and democracy (see Stehr, 2013). I have also relied on a paper jointly written with Jason Mast (Stehr & Mast, 2011).

[5] Russell Hardin (2002, p. 214) argues that the answer to this question requires, first of all, what he calls a "street-level epistemology". Unlike standard philosophical epistemology, street-level epistemology attends to what counts as knowledge among ordinary citizens, not to what justifies truth claims. For Hardin, a street-level epistemology is essentially an economic theory of knowledge since it is not about justification but about usefulness; relevant consequences that are part of such an economic theory of knowledge include all of the costs and benefits of acquiring knowledge and deploying it.

Overview

I shall advance my exploration of the multiple linkages between civil society, governance, and democracy in a number of steps, and ask whether these linkages are codetermined by a growing knowledgeability of modern actors. This approach focuses on the growing opportunities for reflexive cooperation in civil society organizations, for social movements, and perhaps for growing influence from greater segments of society on democratic regimes through the actors' improved knowledgeability. Access to and the command of knowledge are of course stratified. However, in addition to the often underestimated knowledgeability of many citizens in public affairs, there is, of course, the still growing role of scientific knowledge in the capacity for action in politics.

Initially, I explore a few barriers to access to knowledge and ask the following questions: Is it possible to reconcile expertise and civil society? Is it conceivable to reconcile civil society and knowledge as a private good? And to what extent does discussion about the role of expertise and knowledge as a private good apply to the social sciences and the humanities? Are the social sciences and the humanities sources of enabling knowledge in contemporary society? Are they creating new capacities for action by generating novel as well as practical policy advice?

Each of the central terms I introduced in my brief overview is an essentially contested concept, the meanings of which give rise to unending debates (cf. Gallie, 1955–1956). I will therefore attempt to clarify how I plan to use these concepts, especially the notion of knowledge in general and the enabling of knowledge in particular.

The Terms

Knowledge may be defined as a *capacity for action*.[6] Use of the term "knowledge" as a capacity for action is derived from Francis Bacon's famous observation that knowledge is power (*scientia est potentia*). Francis Bacon suggested that knowledge derives its utility from the capacity to set something in motion, for example, new communication devices, new forms of power, new regulatory regimes, new chemical substances, new political organizations, new financial instruments, and new illnesses. In my view, science not only strives for comprehending or understanding

---

[6]To begin with, I use the term "knowledge" as a generic term that encompasses all forms of knowledge and not only knowledge produced in the scientific community. In the context of the discussion of "enabling knowledge," knowledge generated in science is in the forefront of the discussion, last but not least because the scientific community in highly differentiated societies is the main institution in charge of generating "additional knowledge" (including negating or destroying knowledge claims) rather than merely reproducing existing and often widely shared forms of knowledge (i.e., common knowledge). In other words, aside from other differences in the forms of knowledge in different social institutions (e.g., with respect to dissemination, accessibility, formalization), what counts in most modern social institutions, such as those of education, religion, and the economy, are the processes designed to transmit common knowledge.

in the sense of developing *models of reality* but also, in a practical sense, is interested in how to accomplish things and therefore becomes a *model for reality*.[7]

I refer to *civil society* not in the traditional sense, as a political society or a state, but as the public arena of active citizens interposed between the state and intimate forms of life. Civil society comprises a wide variety of social groups, associations, and movements that are not in the business of production, in government, or part of the family. Possession of knowledge enhances *agency*, which is at the heart of civil society. Agency is the ability of citizens to set goals, develop commitments, pursue values, and succeed in realizing them. Valuing agency is at the heart of subsidiary government and self-government. By asking about the varying command of knowledge by actors in modern societies, I am applying the issue of differential access to knowledge to the question of mastering one's own life through the help of knowledge as a resource.

Theories of Democracy and Civil Society

There is, of course, a large number of more or less rival hypotheses giving reasons for the emergence and persistence of democratic regimes and the strength of civil societies within such social systems. For example, in presenting his thesis about the end of competing ideologies in the last century, Francis Fukuyama (1992) stresses that "there are fundamental economic and political imperatives pushing history in one direction, towards greater democracy" (p. 72). Other scholars argue that democracies can take hold in countries that are poor and that democracy therefore does not follow economic development. But as justifications for the war in Iraq have shown, democracy is also expected to follow from the barrel of a gun.

In contrast to these relatively recent claims, John Stuart Mill in "The Spirit of the Age" (1831/1986), published after his return to England from France, affirms his conviction that the *intellectual accomplishments* of his own age make social progress inevitable. However, progress in the improvement of social conditions is not, Mill argues, the outcome of an "increase in wisdom" or of the collective accomplishments of science. It is linked, rather, to a *general diffusion of knowledge*. Mill, observing the moral and political transitions of the mid-nineteenth century, predicts that increased individual choice and emancipation from "custom" will follow broad diffusion of knowledge and education. This theme strongly resonates in the social structure that is emerging today as the industrial society gives way to a *knowledge society*.

John Stuart Mill was a great admirer of the classic study of American society by Alexis de Tocqueville. As a matter of fact, Mill (1832/1977) wrote a review of *Democracy in America* which was published almost at the same time as his "Spirit

---

[7] Today, we are constantly engaged—whether deliberately or in response to the unintended consequences of deliberate conduct—in remaking not only our social but most significantly our natural environment. It follows that the boundaries between social and natural constructs are constantly and deliberately shifting in favor of social constructs.

of the Age." However, there are decisive differences between Mill and De Tocqueville in their assessment of democracy, especially regarding the role of citizens' knowledge for and in democratic regimes. De Tocqueville closes his study of American society with the observation that the educational attainment of its citizens is an influential factor in maintaining democracy in America. Whereas Mill has considerable confidence in the independent capacity of enlightenment, seeing education, knowledge, and intellectual skills as *necessary* conditions for the strength of democratic regimes, De Tocqueville views knowledge as a *sufficient* condition for democracy.

From Mill's estimation, it follows that intellectuals and scientists will play a significant political role in democracies. In the case of De Tocqueville, it is the ordinary citizen, the enlightened public, and his or her immediate political practice that strengthen democratic political systems and check political power. Without taking sides in the specifics of the dispute between De Tocqueville and Mill, I generally concur with their emphasis on the social role that distribution of knowledge plays in civil society and democracy.[8]

I therefore reject the microphysics of power theory, as elaborated by Michel Foucault. In Foucault's genealogical work, he describes the one-sided shaping of the individual by scientific disciplines such as penology and psychoanalysis, and the enormous micromanaged power of regimentation and measurement of major social institutions. Foucault's (1972) observations on "the undoing of the subject" are based on a view that assigns too much power to the agencies that deploy knowledge. Knowledge, as described in *The Archaeology of Knowledge*, is an anonymous discourse that exercises control over a powerless individual.[9] Foucault thereby underestimates the malleability of knowledge, the extent to which knowledge is contested, and the capacity of individuals as well as civil society organizations to deploy knowledge in order to *resist*, *oppose*, and *restrain* the oppression that may be exercised by major social institutions in modern society.

Various societal restraints curb the broad dissemination of knowledge in society and thus hinder the efficacy of knowledge in a democracy. I shall refer to some of these barriers when posing the following questions: Is it possible to reconcile democracy and expertise? And is it possible to reconcile democracy and knowledge as property?

Reconciling Democracy and Expertise

As we have seen, many observers are convinced that the gap between the expert skills of powerful agents and the knowledge of laypersons has dramatically and irreversibly widened recently. However, it is evident that the social deference and

---

[8] For the record, Jean-Jacques Rousseau, Thomas Hobbes, and Karl Marx do not share the positive assessment of the role of (scientific) knowledge in rationalizing political action and enhancing democracy, let alone with regard to happiness or controlling human passions.

[9] Foucault's assertion about the affinity between the powerful and knowledge brings to mind the thesis that an increase in collective human capital, though it "raises the people's ability to resist oppression," also "raises the ruler's benefits from subjugating them" (Barro, 1999, p. 159).

unquestioned respect paid to the knowledge of professions such as teachers, doctors, and lawyers has declined since the 1960s at the latest, at least in modern Western society. Nonetheless, there is still widespread support for the "scientistic" (instrumental) perspective on the societal impact of knowledge claims, or the enlightenment model—namely, that knowledge is universal and universally useful and that there is a one-way flow of knowledge from the experts to the lay public.

With the rapidly growing volumes and speed of new information, a growing cleavage between those who directly participate in the process of knowledge production and the lay public has become apparent. As the large majority of the public is excluded, the asymmetry between expert knowledge and the knowledge of the general public is seen to have serious consequences for the nature of civil society.[10] In this section I describe the "enlightenment" or "deficit" model in somewhat greater detail. The ease with which one delegates to what the economists call "principles" (of course aside from one's own specialty)—that is, to the judgment of experts—has become widespread in all social institutions in modern society, not only in science. At the same time, it is widely assumed in the field of the "public understanding of science" that scientific illiteracy decreases citizens' capacity for democratic participation, including the democratic governance of science.

Large segments of the public have been disadvantaged in and disenfranchised from effective involvement in democratic processes. Exercise of citizenship today requires increasing scientific literacy. This loss of contact and of epistemic deference is not only the result of the growing cognitive distance between science and everyday knowledge; it is also affected by the rapid expansion of knowledge through the growing division of labor in science (e.g., on account of scarce cognitive resources even in science) and by the deployment of knowledge in a productive capacity.

Decreasing cognitive proximity increases the political distance from science, for example, by restricting public reflection on both the anticipated and the unanticipated transformations of social and cultural realities that might result from the application of new knowledge. The scientific community shares responsibility for this diminishing intellectual proximity, as the preferred self-image of science as a consensual, albeit a monolithic and monologic, enterprise, conflicts with both its public role and its own internal struggles regarding research priorities, including the generation of data and their interpretation.

However, on both political and moral grounds, many groups, constituencies, and institutions must be consulted before decisions are made about issues that affect the regulation of new knowledge and therefore indirectly affect the development of

---

[10]The justification for concern about a gap between expertise and democratic governance in the political sphere is of course based on the premise that the right to democratic governance should not be restricted and that the expert should be no more influential than the layperson. In the case of organizational governance (e.g., in industrial or governmental organizations), other norms such as employee satisfaction or productivity gains may be employed to legitimate and assess broad participation in organizational decision-making processes. In other words, is it feasible to equalize power in organizations? After reviewing a range of empirical studies of organizations, Mohr (1994, p. 55) concluded that such a goal is a utopian ambition on account of the attendant costs of power equalization.

science and technology. It would be misleading to think that the loss of contact and the considerable scientific illiteracy found in modern societies is somehow a "potentially fatal flaw in the self-conception of the people today," as Gerald Holton (1992, p. 105) suggests, or that it signals the possibility of a dramatic decline in public support for science. It is more accurate to speak of a precarious balance that affects the autonomy and dependence of science in modern society. Loss of close intellectual contact between science and the public is perfectly compatible with both broad support for science and an openness to legal and political efforts to control the impact of science and technology.

In another sense, however, the loss of cognitive contact is almost irrelevant when "contact" is meant to refer to *close cognitive proximity* as a prerequisite for public participation in decisions affecting scientific and technological knowledge. Such a claim is practically meaningless because it virtually requires public engagement in science-in-progress.

To make a judgment about expertise and civil society, one needs to take specific contexts into account. As a matter of fact, the "solution" to the social role of knowledge and democracy in modern society does not require a general answer but one that can only be solved on a case-by-case basis (cf. Bohman, 1999, p. 190).

The conditions under which different groups make sense of specialized knowledge vary considerably. For example, we live in an age in which science "no longer enjoys the uncontested esteem it had for two centuries as the most certain form of truth—for many the only certain form of truth" (Wallerstein, 2004, p. 7). Thus, rather than treating the relationship between expertise and the public as a series of fixed events involving individual, isolated actors, we need to think of that interaction as being mediated by cultural identities and changing conceptions of the social benefits of science and technology. The resourcefulness with which civil society organizations reconstruct science and technology so distinctly is affected by both political and economic circumstances.

In an age of *knowledge politics*, with efforts to regulate and police new knowledge and technical artifacts, it no longer makes sense to view the public as naively resistant to new capacities to act; it is more useful and accurate to view them as cautious, uncertain, and curious about the possible consequences of new information (cf. Stehr, 2005). Scientific and technology-based innovations are judged by members of civil society against the background of their worldviews, value preferences, and beliefs. Take stem cell research, medical genetics, and genetically modified foods as cases in point. In short, within the context of knowledge, politics, and public discourse about authorizing innovative capacities to act, the balance of power between science and civil society is now shifting toward civil society.

Nonetheless, without some element of impersonal trust (see Shapiro, 1987) toward experts, expertise would vanish. Today's experts are frequently involved in a remarkable number of controversies. The growing policy field of setting limits to the presence of certain ingredients in foodstuffs, safety regulations, risk management, and the surveillance and control of hazards have had the side effect of ruining the reputation of experts. As long as an issue remains a contested matter, particularly when it is a publicly contentious matter, the power and influence of experts and counter-experts are limited; but once a decision has been made and closure achieved, the authority of experts becomes virtually uncontested.

For the scientific community, the lack of cognitive proximity to the general public has both advantages and disadvantages. The separation between science and the public can perhaps explain, at least in part, why the scientific community, in view of its usefulness for corporations, the military, and the state, has been able to preserve a considerable degree of intellectual autonomy. Such autonomy, nonetheless, is contingent on a host of factors within and without the scientific community. It signals symbolic detachment and independence that can be translated into an asset vis-à-vis the state and other social institutions. Science can become an authoritative voice in policy matters and represent the openness of society in ideological and material struggles with other political systems. However, cognitive distance limits the immediate effectiveness of the "voice of science" in both civil society organizations and policy matters.[11] Too much independence of science may also result in excessive celebration of "normal" scientific activity and lead to a lack of innovativeness.

Reconciling Democracy and Knowledge as Property

In testimony before the U.S. Congress more than a century ago, John Powell, a pioneer in the field of the earth sciences, put his finger on one of the most intriguing features of knowledge, namely, the possession of property is exclusive; possession of knowledge is not exclusive. Contrary to Powell's thesis, some forms of knowledge are exclusive and become private goods by means of legal restraints, such as patents or copyright restrictions attached to knowledge.

Whether knowledge is treated as a public or a private good has many noteworthy consequences; for example, it is most likely incremental or new knowledge that is protected. In the context of economic systems and also in science, this situation raises a serious dilemma. The basis of the growth of knowledge is knowledge. If knowledge is protected, the growth of knowledge is hampered. However, if knowledge is not protected, economists argue, the incentive to invest in new knowledge disappears; monopoly rights are essential for the growth of knowledge and inventions. In contrast to incremental knowledge, the general mundane and routinized stock of knowledge consists mostly of knowledge that is not characterized by rivalry or excludability; that is, this type of knowledge may very well constitute public goods.

Scientific knowledge constitutes one of the most important conditions for the possibility of modernization in the sense of a persistent extension and enlargement of social and economic action that is generated by science, and not by any social system in modern society. I do not wish to discuss the contentious issue of trade-offs that may exist between assigning proprietary rights to knowledge and realizing

---

[11] Despite the lack of detailed scientific and technical knowledge held by the public, Collins and Evans (2007, p. 138) stress that the public's interest and involvement in any regulation of genetically modified technologies, for example, remain unaffected; what "we should be celebrating is this political right in a democratic society, not the spurious technical abilities of the public."

gains in the overall welfare of society, or the trade-off between treating knowledge as a public good and sanctioning loss of revenue for those who cannot reap the benefits from their inventions and discoveries.

Economists, legal scholars, and major international organizations such as the World Bank make the case that knowledge must be a global public asset. From an economic viewpoint, this approach would mean that knowledge should lack the characteristics otherwise typical of economic assets, namely, rivalry and excludability. The fact that some forms of knowledge are public goods is not likely to advance the case for additional knowledge, and it is this new knowledge that turns a profit. Thus, the age-old dilemma of whether property generates power and thereby fashions human relations or whether it is the other way around continues to be played out even in knowledge societies.

Discussion about the relationship between scientific knowledge and democracy has been science-centered, be it talk about the role of experts or the contested idea that knowledge is property. The discussion has concentrated exclusively on the social role of natural scientific or technical knowledge. In my concluding remarks, I shall focus instead on knowledge claims of the social sciences and their impact on modern society.

Enabling Knowledge?

The social sciences and the humanities have generated two models for dealing with scientific knowledge claims. The *model of instrumentality* resonates with much of the previous discussion and asserts a wide knowledge gap between science and society. Science speaks to society and does so not only with considerable authority but also with significant success, whereas society has little if any opportunity to talk back.[12] In short, using the instrumental model as a standard, social science knowledge itself is the author of its success (or failure) in society. More specifically, the instrumentality model stipulates that the practical usefulness associated with social science is linked solely to the solid "scientificity" of such knowledge.

The alternative approach to the social pathways of social science knowledge (but not only social science knowledge) is the *capacity model*. Because the capacity model views the practical influence of science as a process driven by the impact of *ideas* on society and its actors, it stresses the *conditions* for the social sciences and the humanities to have considerable influence on society. In this sense, the social sciences and the humanities operate as meaning producers. The social sciences and the humanities do not primarily offer the instrumental knowledge described by the model of instrumentality, a form of enabling knowledge

---

[12] The alleged dominance of scientific knowledge in society and the respect granted to scientific knowledge to the exclusion of other forms of knowledge provoked Paul Feyerabend (2006) to ask how society can be defended against science. His answer is with the help of an education system that is more inclusive in its intellectual pursuits.

that originates mainly from the natural sciences and technology. The social sciences are—borrowing a term from the historian James Harvey Robinson (1923, p. 16)—"mind-makers."[13]

The social sciences, even if considered a major, if not growing, reservoir of meaning that disseminates through various social "pipelines" (such as the media, teachers, priests, and writers) into society, do not have a monopoly on meaning production. But in contrast to the model of instrumentality, the capacity model stresses that the agents who "employ" social science knowledge are *active* agents who transform, re-issue, and otherwise redesign social science knowledge. This active attribute of the "mind-seekers" speaks against a straightforward "social scientification" of mundane worldviews by social science discourse.

The capacity model stipulates that social science knowledge is an intellectual resource that is open and complex and thus can be molded in the course of "travel" from the social science community to society. This model further assumes that neither the production nor the application of this knowledge involves identical reproduction. The capacity model therefore accepts that people may critically engage social science knowledge using local knowledge resources and thus make social science accountable to the public.

Concluding Remarks

The Marquis de Condorcet, a philosopher of the French Enlightenment, was convinced that "the argument that the citizen could not take part in the whole discussion and that each individual's argument could not be heard by everyone can have no force" (as cited in Urbinati, 2006, p. 202).

For Condorcet, the issue was not one of competency with respect to the issue at hand but of good rules and settings within which individuals would be able to deliberate jointly. Aside from the normative or even constitutional empowerment of ordinary citizens to be heard on policy matters, even when they involve highly specialized knowledge claims, Condorcet reminds us that collective deliberation and involvement benefit from rules, settings, and opportunities conducive to such reflection. This idea is one side of the issue of the relation between democracy and specialized knowledge. The other side argues that the public consideration of

---

[13] Robinson (1923, pp. 16–17) refers to a list of occupations and professions serving as mind-makers in modern society: "mind-seekers" are the questioners (of the taken-for-granted or the commonplace) and the seers. We classify them roughly as poets, religious leaders, moralists, storytellers, philosophers, theologians, artists, scientists, and inventors. But Robinson (p. 17) also raises a significant follow-up question: "What determines the *success* of a new idea; what establishes its currency and gives it social significance by securing its victory over ignorance and indifference or older rival and conflicting beliefs?" In this context, he stresses that the "*truth* of a new idea proposed for acceptance plays an altogether secondary role" (p. 20). Robinson's question about the conditions for the success of a new idea must of course be extended to the question of why new ideas are incapable of displacing the commonplace and the taken-for-granted, or what ideas established by "social labor" exactly accomplish, and under what circumstances.

specialized knowledge claims is a futile enterprise from the beginning because of the inability of ordinary citizens to engage in public deliberation of such forms of knowledge.

Drawing on my observations of the current situation, I suggest that the evolvement of modern societies as knowledge societies increasingly extends to the *democratization and negotiation of knowledge claims*. We are slowly moving away from what has been the case of expert rule to a much broader, shared form of knowledge claims governance (cf. Leighninger, 2006; Stehr, 2005). And indeed, one of the virtues of liberal democracies is citizen involvement in political decisions. Such participation, whatever formal basis it may assume, does not hinge, as a prerequisite, on the degree of technical or intellectual competence that citizens may or may not command.

In addition, I maintain that scientific and technical knowledge is more malleable and accessible in practice than has been suggested in the "enlightenment model" and other classical approaches to the relations between science and society (cf. Irwin, 1999). Moreover, the new sociology of scientific knowledge has contributed to an understanding that the production of scientific knowledge is in many ways very similar to other social practices, and that the barriers between science and society are lower than frequently assumed, albeit by no means eradicated. In short, the boundaries between expertise and everyday knowledge are much less fixed than is often surmised, particularly with respect to the alleged growing distance between expert knowledge and public knowledge.

In addition, what is increasingly problematic in modern society is not that we may now know enough but that we may know too much. The social negotiation of novel capacities for knowledge (generated in science and in technology) is not as dependent on specialized natural scientific and technical knowledge as on the enabling knowledge generated by the social sciences and the humanities.

The general access of civil society to enabling knowledge produced in the social sciences faces fewer hurdles than access to knowledge in the natural sciences. Knowledgeability has gained in social reach and accessibility through the development of a more participatory democracy and citizenship, which above all benefit civil society organizations. Altogether this situation produces particular challenges not only in terms of access to social science knowledge but also in the form of new modes of participation. It is in this respect that civil society organizations will be challenged.

Social realms for communication between science/social science and the public already exist. The possibility for democratic negotiation and scientific practice must be seen as part of a larger social enterprise and a larger social context in which both professional scientists as experts and the lay public engage in discussion. Science is an effective social force because it, in turn, can engage and rely on civil society organizations and institutions. The cases for climate change and AIDS activism are rich examples of social processes in which the boundaries of expert and lay public are quite malleable (cf. Bohman, 1999). Finally, one should not be too harsh about the lack of *scientific* foundation in much of what we as members of society treat as knowledge in ordinary life, because we tend to get on quite well with such knowledge, at least most of the time (cf. Hardin, 2003, p. 5; Schutz, 1946). As Wittgenstein (1969, p. 344) observed, "My life consists in my being content to accept many things."

References

Barro, R. (1999). The determinants of democracy. *Journal of Political Economy, 107*, 158–183. doi:10.1086/250107.

Bohman, J. (1999). Citizenship and norms of publicity: Wide public reason in cosmopolitan societies. *Political Theory, 27*, 176–202. doi:10.1177/0090591799027002002.

Collins, H., & Evans, R. (2007). *Rethinking expertise*. Chicago: University of Chicago Press. doi:10.7208/chicago/9780226113623.001.0001.

Downs, A. (1957). *An economic theory of democracy*. New York: Harper.

Ezrahi, Y. (2004). Science and the political imagination in contemporary democracies. In S. Jasanoff (Ed.), *States of knowledge: The co-production of science and the social order* (pp. 254–273). London: Routledge.

Feyerabend, P. (2006). How to defend society against science. In E. Selinger & R. P. Crease (Eds.), *The philosophy of expertise* (pp. 358–369). New York: Columbia University Press.

Foucault, M. (1972). *The archaeology of knowledge*. New York: Pantheon Books.

Fukuyama, F. (1992). *The end of history and the last man*. New York: Free Press.

Gallie, W. B. (1955–1956). Essentially contested concepts. *Proceedings of the Aristotelian Society New Series, 56*, 167–198.

Hardin, R. J. (2002). Street-level epistemology and democratic participation. *The Journal of Political Philosophy, 10*, 212–229. doi:10.1111/1467-9760.00150.

Hardin, R. J. (2003). If it rained knowledge. *Philosophy of the Social Sciences, 33*, 3–24. doi:10.1177/0048393102250280.

Holton, G. (1992). How to think about the "anti-science" phenomenon. *Public Understanding of Science, 1*, 103–128. doi:10.1088/0963-6625/1/1/012.

Irwin, A. (1999). Science and citizenship. In E. Scanlon, E. Whitelegg, & S. Yates (Eds.), *Communicating science: Contexts and channels* (Reader 2, pp. 14–36). London: Routledge.

Jasanoff, S. (2005). *Designs on nature: Science and democracy in Europe and the United States*. Princeton, NJ: Princeton University Press.

Jones, G. S. (2004). *An end to poverty? A historical debate*. New York: Columbia University Press.

Kroto, H. (2007, May 22). The wrecking of British science. *The Guardian*, Education section, pp. 1–2. Retrieved from http://www.guardian.co.uk/science/2007/may/22/highereducation.education

Lakoff, S. A. (1971). Knowledge, power, and democratic theory. *Annals of the American Academy of Political and Social Science, 394*, 4–12. doi:10.1177/000271627139400102.

Leighninger, M. (2006). *The next form of democracy: How expert rule is giving way to shared governance and why politics will never be the same*. Nashville, TN: Vanderbilt University Press.

Lewontin, R. C. (2004, November 18). Dishonesty in science. *New York Review of Books*. Retrieved from http://www.nybooks.com/articles/archives/2004/nov/18/dishonesty-in-science/

Mill, J. S. (1977). Democracy in America. In J. M. Robson (Ed.), *The collected works of John Stuart Mill, volume XVIII – Essays on politics and society, part I*. Toronto: University of Toronto Press. (Original work published 1832)

Mill, J. S. (1986). The spirit of the age. Parts I–VII. In A. P. Robson & J. M. Robson (Eds.), *The collected works of John Stuart Mill, volume XXII – Newspaper writings, December 1822–July 1831, part I*. Toronto: University of Toronto Press. (Original work published 1831)

Mohr, L. B. (1994). Authority in organizations: On the reconciliation of democracy and expertise. *Journal of Public Administration Research and Theory, 4*, 49–65.

Mulder, M. (1971). Power equalization through participation. *Administrative Science Quarterly,* *16*, 31–39. doi:10.2307/2391284.

Olson, M. (1982). *The rise and decline of nations: Economic growth, stagflation, and social rigidities*. New Haven, CT: Yale University Press.

Pielke, R. A., Jr. (2007). *The honest broker: Making sense of science in policy and politics*. Cambridge, UK: Cambridge University Press. doi:10.1017/CBO9780511818110.

Robinson, J. H. (1923). *The humanizing of knowledge*. New York: George H. Doran.

Schutz, A. (1946). The well-informed citizen. *Social Research, 13*, 463–478.

Shapiro, S. (1987). The social control of personal trust. *American Journal of Sociology, 93*, 623–658. doi:10.1086/228791.

Stehr, N. (1999). The future of social inequality. *Society, 36*, 54–59.

Stehr, N. (2005). *Knowledge politics: Governing the consequences of science and technology*. Boulder, CO: Paradigm Publishers.

Stehr, N. (2013). *Die Freiheit ist eine Tochter des Wissens* [Liberty is a daughter of knowledge]. Weilerswist, Germany: Velbrück Wissenschaft.

Stehr, N., & Mast, J. (2011). The modern slaves: Specialized knowledge and democratic governance. *Society, 48*, 36–40. doi:10.1007/s12115-010-9391-6.

Urbinati, N. (2006). *Representative democracy: Principles & genealogy*. Chicago: University of Chicago Press. doi:10.7208/chicago/9780226842806.001.0001.

Wallerstein, I. (2004). *The uncertainties of knowledge*. Philadelphia: Temple University Press.

Wittgenstein, L. (1969). *On certainty*. Oxford, UK: Basil Blackwell.

Gabriel's Map: Cartography and Corpography in Modern War

4

Derek Gregory

> *As the balloon calmed, the major looked down once more at the Belgian soil they had recently vacated…*
> *Foot by foot, yard by yard, the war was heaving into view…*
> *"Believe you me, Major, this is the only way you can make sense of what's down there. Once you are in the trenches, you keep your head down and the world shrinks…"*
>
> Robert Ryan, *Dead Man's Land*

I Would Rather Be in France …

In the winter of 1980–1981 William Boyd was researching his second novel in the Bodleian Library's collections at Rhodes House. *An Ice-Cream War* opens in June 1914, but Boyd was at Rhodes House because his eyes were fixed not on the killing fields of the Western Front—these would appear in later novels—but on a little known colonial conflict in British East Africa and German East Africa. This was one of the most remote theaters of World War I and, as Boyd said himself, in many ways the very opposite of the war in Europe: a war of movement, of skirmish and pursuit through desperately difficult bush country "on a scale unimaginable to soldiers on

I am extremely grateful to audiences at the University of Kentucky (Committee on Social Theory), King's College London, and the Peter Wall Institute for Advanced Studies in Vancouver for their comments on earlier presentations of these arguments. I also owe Trevor Barnes a great debt, not least for introducing me to Tom McCarthy's novel *C*.

D. Gregory (✉)
Department of Geography, The University of British Columbia,
1984 West Mall, Vancouver, BC V6T 1Z2, Canada

Peter Wall Institute for Advanced Studies University Centre, University of British Columbia,
6331 Crescent Road, Vancouver, BC V6T 1Z2, Canada
e-mail: derek.gregory@geog.ubc.ca

© Springer Netherlands 2015
P. Meusburger et al. (eds.), *Geographies of Knowledge and Power*,
Knowledge and Space 7, DOI 10.1007/978-94-017-9960-7_4

the Western Front—two armies pursuing each other for 4 years over territory five times the size of Germany." Boyd found it difficult to get its measure too—which is why 25 years later he welcomed a new history (Boyd, 2007)—and made the most of one of its better documented and most dramatic episodes: the Battle of Tanga.

The battle took place in the first few days of November 1914 in and around the small German-held port of Tanga. It was an epic disaster for a British expeditionary force that had sailed from India with orders "to bring the whole of German East Africa under British authority" (for detailed accounts, see Anderson, 2001; Godefroy, 2000). The officers in command made a series of miscalculations that allowed the much smaller opposing force to seize the advantage and ultimately to draw out the campaign for another 4 years. Many of them revolved around inadequate planning and incomplete intelligence, and in *An Ice-cream War* Boyd renders this fatal combination in vividly cartographic terms.[1] A young subaltern, Gabriel, joins a group of officers on board a lighter from the troop ship, "all peering at copies of the map by the light of torches" (Boyd, 1982, p. 144).

> "What's this mark?" someone asked. "It's a railway cutting," Major Santoras replied. "Between the landing beaches and the town." He went on less confidently: "There'll be bridges over it, I think . . . Should be, anyway." (p. 144)

The morning of the next day the men plunge ashore and climb the low, scrub-covered cliffs before advancing on the town:

> Gabriel tried to visualise the advance as if from a bird's-eye view—3,000 men moving on Tanga—but found it impossible. (p. 158)

> He wondered if they'd wandered off course in the coconut plantation. But what lay beyond the maize field? Gabriel waved his men down into a crouch and got out his map. It made no sense at all. (p. 160)

As the fighting continues, and the British are forced into an ignominious retreat, Gabriel has the epiphany that provides me with my title:

> "It's all gone wrong," Bilderbeck said . . . He took out his map from his pocket and smoothed it on the ground. *Gabriel thought maps should be banned. They gave the world an order and reasonableness it didn't possess.* (p. 169; my emphasis)

The contrast between what Clausewitz called more generally "paper war" and "real war" bedevils all conflicts, but the lack of what today would be called geospatial intelligence proved to be catastrophic for the East Africa campaign (Lohman, 2012, p. 21). It was a peculiarly brutal affair, described by one official historian as "a war of attrition and extermination which [was] without parallel in modern times"

---

[1] What I have described as "cartographic anxiety" (Gregory, 1994) is advertised in Boyd's epigraph to *An Ice-Cream War*, which comes from Rudyard Kipling's (1910) *The Brushwood Boy*:

"He hurried desperately, and islands slipped and slid under his feet, the straits yawned and widened, till he found himself utterly lost in the world's fourth dimension with no hope of return. Yet only a little distance away he could see the old world with the rivers and mountain chains marked according to the Sandhurst rules of map-making."

Those "Sandhurst rules of map-making" were unbuttoned in the war in East Africa and, as I will show, were simultaneously enforced and confounded in the war in Europe.

(Sandes, 1933, p. 498, cited in Paice, 2007, p. 3). In his more recent history Paice (2007) sharpens the point with an extraordinary vignette:

> In 1914 Lieutenant Lewis had witnessed the slaughter of every single man in his half-battalion on the Western Front and had experienced all the horrors of trench warfare. Yet sixteen months later, in a letter sent to his mother from the East African "front," Lewis wrote "I would rather be in France than here." (p. 7)

In this essay I turn back to British experience on the Western Front in 1914–1918 and, for all the distance and difference from East Africa, re-locate Gabriel's despair at the ordered geometry of the map in the no less surreal and slippery landscapes of war-torn Belgium and France. In contrast to the East African campaign, war here was fought with increasingly sophisticated, highly detailed geospatial intelligence. In the next sections I describe a combination of mapping and sketching, aerial reconnaissance and sound ranging that transformed the battlefield into a highly regulated, quasi-mathematical space: the abstract space of a military Reason whose material instruments were aircraft, artillery, and machine-guns. In the sections that follow I counterpose this cartography and its intrinsically optical-visual logic to the muddy, mutilated and shell-torn slimescapes in which the infantry were immersed month after month. I call the radically different knowledges that the war-weary soldiers improvised as a matter of sheer survival a corpography: a way of apprehending the battle space through the body as an acutely physical field in which the senses of sound, smell and touch were increasingly privileged in the construction of a profoundly haptic or somatic geography.[2] I conclude with some reflections on the shadows cast by this analysis over war in our own troubled present.

The Optical War and Cartographic Vision

It has become commonplace to identify World War I with a crisis of perception that was, through its intimate connections with modernist experimentation, also a crisis of representation (Kern, 1983; see also Eksteins, 1989).[3] And yet—or rather, in consequence—it was also what Saint-Amour (2003, p. 354) calls "the most optical war yet" that depended on a rapidly improvised and then swiftly professionalized techno-military assemblage whose political technology of vision not only brought the war into view but also ordered its conduct through a distinctive scopic regime whose parameters I must now sketch out.[4]

---

[2] I thought I had made the word up—I discuss its filiations below—but I have since discovered that Pugliese (2013) uses "geocorpographies" to designate "the violent enmeshment of the flesh and blood of the body within the geopolitics of war and empire"(p. 86). My intention is to use the term more directly to confront the optical privileges of cartography through an appeal to the corporeal (and to the corpses of those who were killed in the names of war and empire).

[3] More specifically, Jay (1994) describes this as a crisis of *ocularcentrism* (pp. 192–217).

[4] Saint-Amour (2003, p. 354) describes this as a "technological matrix" but I use "assemblage" to emphasize both its heterogeneity and its materiality.

When the British Expeditionary Force set sail in August 1914 it was assumed that tried and tested methods of geospatial intelligence would suffice.[5] In the War Office's collective view, existing maps of the combat zone would be perfectly adequate, and the Ordnance Survey was instructed to provide General Headquarters (GHQ) with copies of two medium-scale topographic maps of Belgium and northeastern France (1:100,000) and of France (1:80,000). Any updates would be made by traditional means, and in July General Douglas Haig made it clear that the only useful reconnaissance would be conducted by the cavalry: "I hope none of you gentlemen is so foolish as to think that aeroplanes will be usefully employed for reconnaissance purposes in war" (Sykes, 1942, p. 105).[6] Neither assumption survived the first encounters with enemy forces. During the chaotic retreat from Mons in the last week of August one subaltern recalled that "maps were non-existent. We had been issued with maps for an advance, and we soon walked off those!" (Lt. B. K. Young, in Barton, Doyle, & Vandewalle, 2010, p. 19). For many days he and his fellows relied on a road map confiscated from a fleeing motorist.

Confronted with a cascade of unforeseen events, GHQ demanded regular updates for its (as it turned out, wholly inadequate) maps, and turned to the fledgling Royal Flying Corps for reconnaissance. The first results were not encouraging; the pilots flew without observers and the official historian admits that "the machines lost their way and lost each other" (Raleigh, 1922, p. 300). The officers had no training for these missions, and during the first Battle of Ypres in October observers from No. 6 Squadron "mistook long patches of tar on macadamized roads for troops on the move, and the shadows cast by gravestones in a churchyard for a military bivouac" (Raleigh, p. 304). But a system was already beginning to emerge. Reports were made in narrative-tabular form, under three standard headings—Time, Place, Observation—and as soon as the aircraft landed the pilot and observer would report to GHQ where they were debriefed and the base maps updated and annotated (Sykes, 1922).

After the battles in Flanders in October the Western Front stabilized and the conflict turned into a war of attrition with the armies, in William Brodrick's (2008) splendidly evocative phrase, "scratching behind the skirting boards of France and Belgium" (p. 27). The first (1:50,000) British trench maps showing the position of the German lines had been produced in great haste for the first battle of the Aisne in September, but it was now clear that many more and still larger scale maps would be required—the sooner the better—and that they would need to be regularly updated and overprinted with the latest, fine-grained tactical intelligence.[7] A small

---

[5] For details of the various offensives, see Hart (2013). My own account is largely confined to the British experience, but Hart restores the French to the prominent place from which they have been evicted in too many English-language accounts of the war.

[6] Sykes served as Chief of Staff for the Royal Flying Corps in 1914–1915.

[7] My discussion of military cartography and its ancillary practices has two principal limitations. First, it is confined largely to the practice of the British Army, though this may not be as restrictive as it appears. Chasseaud (2002) shows that, for all the differences between them, "in almost every aspect of war survey and mapping" the British, French, and German armies "developed remarkably similar organisations and methods, suggesting that problems were clear and solutions obvious"

Ranging and Survey section of military surveyors arrived in France in November 1914, and by March 1915 their field sheets had been submitted to the Ordnance Survey for the production of a new series of 1:20,000 maps. These improved the accuracy of artillery fire, but their very success generated demands for even more detailed maps.[8] By the end of November a new series of 1:10,000 trench maps had been distributed by the Ordnance Survey from its printing house in Southampton.

The presses rolled through the night, sometimes printing as many as 20,000 sheets in a day, and by the end of the war 34 million sheets had been supplied to Britain's armed forces. The sheets were shipped to Le Havre and then taken by train to forward distribution points from which Ordnance Survey "map cars" would eventually make daily runs to General, Corps, and Divisional Headquarters.

This was a formidable feat of production and distribution that required the military to overcome two major challenges. The first was to map occupied territory that lay beyond the scope of field survey, while the second was to update the database in line with a fluid battle space. In fact, for all the apparent authority of the printed map, it was always provisional; it always belonged to a past that was rapidly receding. In 1915 it took 2 weeks to produce a finished map at Southampton, and by 1916 this had doubled, so that by the time the map arrived at the Front it was already out of date (Chasseaud, 1999, p. 87).[9] In these difficult circumstances four techniques were used to consolidate and refine cartographic vision. Aerial photography and field sketching apprehended the battlefield as a space of objects, locating trenches and troop dispositions, while aerial observation and sound ranging animated the battlefield as a space of events, tracking troops advancing and guns firing (Table 4.1).

First (and foremost), aerial photography proved to be indispensable for what was, by the standards of day, near real-time mapping.[10] In the view of one observer, the camera was "a means for recording, with relentless precision, the multitudinous

---

(p. 201). Second, it is primarily concerned with the production of topographic maps and their trench overlays. As the conflict developed other geo-technical maps were required, based on the topographic series. Supplying water for troops, horses, and mules was a major problem—some estimates put the daily requirement at 45 liters per man or animal—and from 1915 water supply maps at various scales were used to identify likely sources and plan new boreholes. The development of tunneling and mining relied on geological maps and the production of meticulous mine plans (see Barton, Doyle, & Vandewalle, 2010; Doyle & Bennett, 1997; Rose & Rosenbaum, 1993). Towards the end of the war enterprising intelligence officers prepared terrain maps indicating the suitability (or otherwise) of the ground for tanks, but these "goings" maps were not always appreciated by staff officers. Haig's Chief of Intelligence intercepted one of them, which showed how limited the safe ("white") areas were, and returned it to its author with the curt instruction: "Pray do not send me any more of these ridiculous maps" (Macdonald, 1993, p. 116).

[8] The canonical account of British military cartography is Chasseaud (1999, 2013); see also Murray (1988) and Forty (2013).

[9] This increased the urgency for printing in theater, and by 1917 every Field Service Company was provided with powered printing presses for limited distribution, time-critical ("hasty") runs. By then, fears of attacks on Channel shipping had also prompted the Ordnance Survey to open an Overseas Branch in a disused factory in northern France.

[10] The definitive account is Finnegan (2011), but see also Slater's (n.d.) highly informative series on "British Aaerial photography and photographic interpretation on the Western Front" at http://tim-slater.blogspot.ca

Table 4.1 Cartographic vision and the battlefield

	Space of objects	Space of events
Air	Aerial photography	Aerial observation
Ground	Field sketching	Sound ranging

changes that take place within the restless area of an army at war" (H. A. Jones, 1928, p. 87). It was that capacity to track changes—to set the printed map in motion—that gave aerial photography its power. "Every day there are hundreds of photographs to be taken," one pilot explained, "so that the British map-makers can trace each detail of the German trench positions and can check up on any changes in the enemy zone" (Bishop, 1918, p. 22.). Still, it had a slow start; the Royal Flying Corps took only one official camera to France in 1914, and the first plates were unimpressive. But by January 1915 one enterprising observer had on his own initiative assembled a photomosaic that sufficiently impressed the General Staff for it to establish an experimental photographic section whose first sorties took place in early March.[11] The reconnaissance flights photographed the German lines to a depth of 700–1,500 yards, and the plates were used to overprint the existing 1:50,000 maps with an outline of the enemy trench system. This was the first trench map to be augmented by aerial photographs, and Haig used it to plan the first large-scale offensive by the British Army, the Battle of Neuve Chapelle, which took place a week or so later between 10 and 13 March 1915 (Fig. 4.1).[12]

Reconnaissance flights soon became so routine that one pilot compared them with "going to the office daily, the aeroplane being substituted for the suburban train" (H. A. Jones, 1928, p. 82), although once the power of aerial reconnaissance was recognized the commute became much more dangerous and often deadly. The reference to the relentless rhythm of the workaday world became ever more appropriate as the interval between reconnaissance and reproduction decreased. In the summer of 1916, in preparation for the Battle of the Somme, the Royal Flying Corps (RFC) conducted a series of "speed tests" in which less than an hour—and sometimes as little as 30 minutes—elapsed between taking a photograph and delivering the print to Corps HQ.[13] The tempo of reconnaissance increased too,

---

[11] The first "A" camera was handheld and required the observer to perform 11 separate operations "in thick gloves or with numbed fingers" to expose the first plate; its limitations were obvious, and by the summer a semi-automated "C" camera was fixed to the aircraft (Slater, n.d., Part 8; H. A. Jones, 1928, pp. 89–90).

[12] Finnegan (2011, p. 55) calls this "the first imagery-planned battle" but the newly detailed map was not sufficient to turn aerial photography from a novelty into a necessity. Slater (n.d., part 10) argues that it was the critical shortage of ammunition for the artillery—which Sir John French also blamed for the military failure at Neuve Chapelle—that drove the search for more accurate and efficient methods of targeting that aerial photography promised to provide.

[13] Slater (n.d.) claims that it was the Battle of the Somme that marked aerial photography's admission to the very center of operational planning; for a vivid account of the RFC's wider role

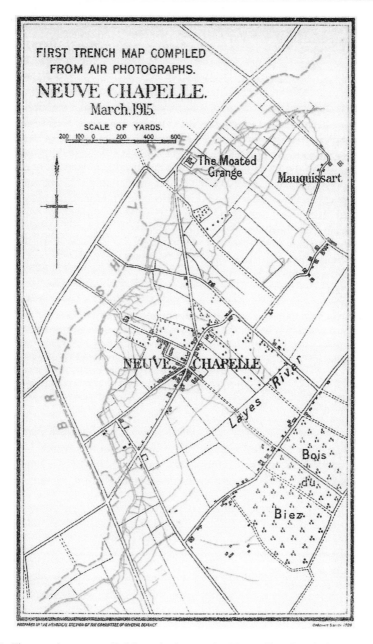

Fig. 4.1 First trench map compiled from air photography. Neuve Chapelle, 1915 (From *The War in the Air*, (Vol. 2, p. 91), by H. A. Jones, 1928, Oxford, UK: Clarendon Press. Reprinted with permission)

particularly during a major offensive. In 1916 the RFC planned to photograph the German lines to a depth of 3,000 yards every five days and the counter-battery area to the rear every ten days, but during the preparatory barrage for the Battle of Messines in July 1917 the German lines were being photographed every day. The production of photographic prints, like so much else on the Front, was becoming thoroughly industrialized. In 1915 the standard British production cycle had called for photographic plates to arrive at 2130 in the evening and for 100 copies to be ready for distribution to headquarters down the line by 0600 the following morning (Finnegan, 2011, p. 56). But this system was rapidly overtaken by events; photographic sections were decentralized to meet the growing demand for near real-time prints so that imagery also flowed up the command hierarchy. The stream of images rapidly accelerated, and was given a further boost once the United States entered the war in April 1917. What Sekula (1975, p. 27) called the "instrumental collage" of aircraft, camera, and artillery was central to modern, industrialized war, and the middle term was crucial. By November *Scientific American* could describe the camera as "a deadly instrument" that was "many times deadlier than its equivalent weight of high explosive" ("The Camera at the Front," 1917, p. 389). Its payload was most effectively delivered through assembly-line production. Sekula (1975) again:

> The establishment of this method of production grew out of demands for resolution, volume, and immediacy. No method of reproduction but direct printing from the original negative would hold the detail necessary for reconnaissance purposes. Large numbers of prints from a single negative had to be made for distribution throughout the hierarchy of command. In addition, the information in prints dated very rapidly. Under these circumstances, efficiency depended on a thorough-going division of labor and a virtually continuous speedup of the work process. Printers worked in unventilated, makeshift darkrooms; 20 workers might produce as many as 1,500 prints in an hour, working 16-hour shifts.[14] (p. 28)

Like the economic model from which it derived, the system was the product of a synergy between industrial innovation and scientific advance. As soon as semi-automation made it possible for pilots to produce series of overlapping photographs the analysis of stereoscopic pairs made the art of photo-interpretation equally scientific. Some staff officers no doubt still believed that the raw photograph spoke for itself, but careful interpretation was essential to *make* the image speak. "Reconnaissance images are highly encoded," Amad (2012) insists, "non-literal, non-transparent and opaque documents" (p. 83; see also Saint-Amour, 2011).[15] Reading them was an exacting business, and their capacity to disclose the battlefield was complicated as militaries not only integrated aerial reconnaissance into their

---

in that offensive, see Hart (2012).

[14] One of the principal managers of these production methods was Edward Steichen, who commanded the photographic division of the American Expeditionary Forces. He organized the 55 officers and 1,111 men under his command into what Virilio (1989) described as "a factory-style output of war information" that "fitted perfectly with the statistical tendencies of this first great military-industrial conflict" (p. 201).

[15] Hüppauf (1993) emphasizes how the photograph worked to project order onto a disordered landscape "by reducing the abundance of detail to restricted patterns of surface texture." In his view, "the morphology of the landscape of destruction, photographed from a plane, is the visual order of an abstract pattern" (p. 57).

operations but also sought to confound its use by their adversaries. There were two developments of particular significance. Most—or perhaps least—obviously, the refinement of aerial photography stimulated the development of the counter-science of camouflage. By 1916 the British had developed a sort of "net work warfare," using specially designed scrimmed netting throughout their section to subvert the enemy's photographic gaze. This required the *camoufleurs* to "see like a camera," and Shell (2012) suggests that the netting effectively "seeded itself into the emulsive space both within and between the photographic frames," so that the viewed became "active agents, operating to conceal themselves within regulated and serially photographed time" (p. 77; see also Forsyth, 2013).[16] This in turn required the photo-interpreters to see like a sort of reverse camera, and to peel away the deceptive layers of the frames. The problems were not only on the surface, however, because once the trenches had more or less stabilized part of the battlefield they disappeared deep underground. Both sides dug tunnels beneath the opposing lines to detonate enormous explosions (mines), and these too relied on detailed mapping for their success. Although the excavations were not visible from the air—the point was to take the enemy by surprise—the spoil was highly vulnerable to aerial reconnaissance. In consequence, large numbers of troops were employed at night to remove and distribute the spoil far from the mine head, and it was no simple task to detect traces of these operations on the photographic plates in time for countermeasures to be taken (Barton et al., 2010, p. 94; S. Jones, 2010).

All the way down the distribution chain aerial photographs were scrutinized, annotated, and used to construct makeshift maps modified from the printed sheets. But on the front lines direct observation from the ground was also indispensable and here a second, heterogeneous set of techniques came into its own: sketching of both maps and terrain. Thus Edmund Blunden was ordered "to produce an enlargement of the trench map showing our front line and the German front line at a chosen point" (in preparation for a raid), and later crawled along a disused sap towards a suspected German observation post, all the while "pretty certain that German topographers were crawling from their end in like fashion" (Blunden, 1928/2000, p. 39). The knowledge obtained from these sorts of expedition was typically recorded on annotated sketch maps (Fig. 4.2).

These were supplemented by formal field sketching carried out by military draftsmen. This was artisan rather than factory production, art more than science, but it maintained significant connections with both cartography and photography: some draftsmen revised their initial drawings with the aid of aerial photographs, and while the perspective of the field sketch was horizontal—unlike the vertical frame

---

[16] Even this could be undone by the violence of war. One artillery officer at Ypres in 1917 worried that "the ground was so devastated and wrecked that the usual camouflage netting might give you away. So we would make the [battery] position look as untidy as the surroundings. . . . We were told to do this by the RFC pilots. They said, 'For God's sake don't have any kind of order'" (Arthur, 2002, p. 214).

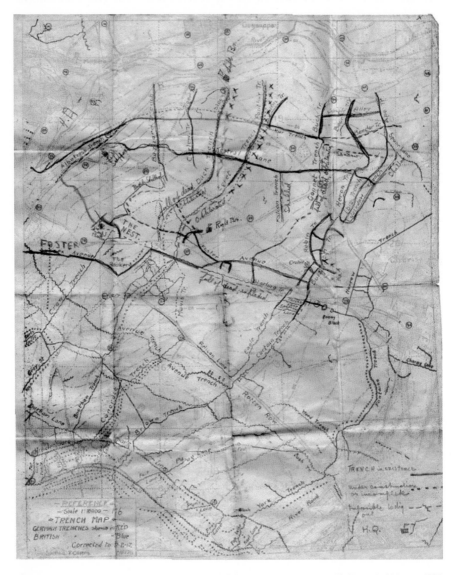

Fig. 4.2 Annotated trench map (Source: http://britishtrenchmaps.co.uk/pdfs/Trench%20maps%20 A4%20leaflet.pdf. Copyright unknown)

of the map or the photograph—Gough (1998, 2009, p. 244) has demonstrated that field sketches were almost always "heavily dressed in the idiom of map-making."[17]

The demand for field sketches increased throughout the war, and Gough (2009) claims that the panorama became "a surrogate view for the distant artillery" (p. 238). That was probably more important in the first phase of the war, when gun batteries

---

[17] See also Mattison's discussion of the work of British-Canadian military topographer Walter Draycot(t) in "Representations of war as autobiographical media" at http://www.walterdraycot.com

relied on direct fire; since this made them highly vulnerable to counterattack—because the line of sight could readily be reversed—indirect fire from concealed positions against unseen targets soon became the norm, and then other means had to be used to register the locations of enemy batteries. A third set of techniques thus relied on direct observation from the air, and in his history of the air war Raleigh (1922) insisted that

> Reconnaissance, or observation, can never be superseded; knowledge comes before power; and the air is first of all a place to see from. It is also a place to strike from, but, speaking historically, offensive action in the air, on any large scale, began, as had been anticipated, in the effort of the conflicting forces to deprive each other of the opportunity and means of vision. (p. 446)

Bombing played its role in the war, both on and off the battlefield, but much of the time the most vital vector of military violence was the artillery, and although its shells flew high into the air the ground—its ground—remained "the place to strike from".[18] Raleigh's sharp point was that effective artillery fire depended on air-to-ground coordination. The use of balloons and aircraft for direct observation and ranging allowed near real-time communication with gun batteries, and overrode the delay between reconnaissance, reproduction, and dissemination that remained no matter how fast a Steichen could spin the aerial photography cycle.

Unlike other belligerents, the British Expeditionary Force arrived without a single observation balloon, and the first British Kite Balloon Section was not deployed until May 1915. Its balloons were set 12–15 miles apart, usually 3 miles behind the front line trenches so that they were beyond the range of small arms and artillery fire, and tethered to a truck-mounted winch. They could rise to a height of 3,000–4,000 ft, which provided a field of view (see Fig. 4.3) that could extend 15 miles or so beyond the enemy's front line, and although they were static they had significant advantages over reconnaissance aircraft.

The balloons provided more or less persistent presence since, apart from changing observers, they could remain aloft all day and all night so long as they were not attacked by enemy aircraft; in ideal conditions the motion of the basket suspended beneath the balloon was so slight that observers could use high-magnification field glasses to conduct detailed surveillance; and the telephone line incorporated into the cable gave them two-way voice communication with the ground (H. A. Jones, 1928, p. 115; Kennett, 1991, p. 25).

Still, most historians seem to agree that balloons were much better for providing general situational awareness, and while at first the artillery was reluctant to have its guns "run by the Flying Corps" (Hart, 2012, p. 36) aircraft soon became the preferred platform for ranging guns on to specified targets.[19] An aircraft was assigned to work with a particular battery, but the first communications were hit-or-miss

---

[18] On some estimates artillery fire accounted for 58 % of all combat deaths during the war. On the role of artillery see Marble (2008) and Strong and Marble (2011).

[19] Batteries were not wholly reliant on aircraft, but also used forward observers, flash spotting, and sound ranging.

Fig. 4.3 Balloon view of bombardment, Roclincourt, 23 September 1915. Imperial War Museum, photograph Q42236 (Retrieved from http://www.gutenberg-e.org/mas01/images/mas03a.html. Also listed at http://www.iwm.org.uk/collections/item/object/205276700 as ©IWM)

affairs, involving written messages dropped to the ground ("You hit them—We must go home—No petrol"), very lights, and signaling lamps. Wireless communication was soon introduced, and in short order "a wireless aeroplane was as popular as an opera-singer" (Raleigh, 1922, p. 340) . But it had its own shortcomings: the equipment was so heavy that aircraft could only carry a transmitter and not a receiver, reception was often scrambled and disrupted by weather conditions, and there were teething problems when different aircraft used the same wavelength (Raleigh, 1922, p. 343). The first transmissions were verbal, as in this wireless communication on 24 September 1914:

4.02 p.m.   A very little short. Fire. Fire.
4.04 p.m.   Fire again. Fire again.
4.12 p.m.   A little short; line O.K.
4.15 p.m.   Short. Over, over and a little left.
4.20 p.m.   You were just between two batteries. Search 200 yards each side of your last shot. Range O.K.
4.22 p.m.   You have them.
4.26 p.m.   Hit. Hit. Hit.

In early 1915 this impressionistic system was replaced by a standard clock-code: a clock face was superimposed over a target identified on an aerial photograph, with 12 indicating north; radial distances were lettered from Y and Z (10 and 25 yards) out to E and F (400 and 500 yards), and the aircraft transmitted the location of each salvo (Y4, say) to the battery in Morse code. "With a good battery," one pilot reckoned, "you should get them right on target at about the third salvo" (Hart, 2012, p. 35; see also pp. 107–108).[20] The next year zone calls were introduced, in which guns were ranged on to quartered grid squares ("zones") on a 1:40,000 map (the "Artillery Board"). These refinements were moments in the abstraction of the battle space. As one reporter noted,

> The affair is not like shooting at anything. A polished missile is shoved into the gun. A horrid bang—the missile has disappeared, has simply gone. Where it has gone, what it has done, nobody in the hut seems to care. There is a telephone close by, but only numbers and formulae—and perhaps an occasional rebuke—come out of the telephone, in response to which the perspiring men make minute adjustments in the gun or in the next missile.
>
> Of the target I am absolutely ignorant, and so are the perspiring men. (Bennett, 1915, p. 97)

The same can be said of a fourth set of techniques known as sound ranging, which Liddle (1998) hailed as "the 'Manhattan Project' of the 1914–18 war" (p. 120). This involved locating an enemy battery by calculating its distance and direction from the sound-wave generated by its shell. The usual configuration had six low-frequency microphones stationed at carefully surveyed intervals along an arc 4,000 yards behind the front line with two observation posts in front of them, all linked to a recording station in the rear. When the forward observers saw a gun flash or heard its boom they sent a signal that activated an oscillograph and film recorder in the recording station (MacLeod, 2000).[21] The British established their first sound-ranging section in October 1915, following the French example, and by the end of the year they could locate an enemy gun within 500 yards. In the course of 1916 another seven sections were established, each plotting battery positions on printed Ordnance Survey sheets. In ideal conditions (which were rare) the operation could be completed for a single battery within 3 minutes and with an accuracy of 25–100 yards. Tom McCarthy's (2010) novel *C* provides a vivid reconstruction of the process:

> This [hut]'s wall has a large-scale map taped to it; stuck in the map in a neat semi-circle are six pins. Two men are going through a pile of torn-off, line-streaked film-strips, measuring the gaps between the kicks with lengths of string; then, moving the string over to the map slowly, careful to preserve the intervals, they transfer the latter onto its surface by fixing one end of the string to the pin and holding a pencil to the other, swinging it from side to side to mark a broad arc on the map. "Each pin's a microphone," the slender-fingered man explains. "Where the arcs intersect, the gun site must be." "So the strings are time, or space?" Serge asks. "You could say either," the man answers with a smile. "The film-strip knows no

---

[20] There is an imaginative description of artillery ranging from the pilot's point of view in McCarthy (2010, pp. 177–178).

[21] For a more informal account that describes the everyday routine of the sound rangers, see Innes (1935).

difference. The mathematical answer to your question, though, is that the strings represent the asymptote of the hyperbola on which the gun lies." (p. 195)

We might draw three conclusions from all this. First, the war was—in this register and in this place at least—a profoundly visual-optical affair. Even sound ranging relied on "filming sound," as McCarthy's protagonist realizes, and he inhabits a world of arcs and parabolas and gridded space. "I don't think of it as mathematics," he says at one point, "I just see space: surfaces and lines" (McCarthy, 2010, p. 152). Second, that space was in constant motion. In fact, the seeming stasis of trench warfare was Janus-faced, produced by myriad movements—advances and retreats, raids and repulses—whose effectiveness depended not on the fixity of the map or the photograph but on their more or less constant updating. This capacity is not the unique achievement of twenty-first century digital navigation, and Chasseaud (2002) is not exaggerating when he describes the results of the entanglements between aerial reconnaissance, photography, and cartography on the Western Front as the production of "a new geographical information system" that provided "a sophisticated three-dimensional fire-control database or matrix of the battlefield." "In effect," he claims, "the battlefield had been digitized" (p. 172). Third, this matrix was performative, producing the quick-fire succession of events that it represented. McCarthy captures this to great effect in his account of a pilot working with a gun battery:

> Serge feels an almost sacred tingling, as though he himself had become godlike, elevated by machinery and signal code to a higher post within the overall structure of things, a vantage point from which the vectors and control lines linking earth and heaven . . . have become visible, tangible even, all concentrated at a spot just underneath the index finger of his right hand which is tapping out, right now, the sequence C3E MX12 G . . .
>
> Almost immediately, a white rip appears amidst the wood's green cover on the English side. A small jet of smoke spills up into the air from this like cushion stuffing; out of it, a shell rises. It arcs above the trench-meshes and track-marked open ground, then dips and falls into the copse beneath Serge, blossoming there in vibrant red and yellow flame. A second follows it, then a third. The same is happening in the two-mile strip between Battery I and its target, and Battery M and its one, right on down the line: whole swathes of space becoming animated by the plumed trajectories of plans and orders metamorphosed into steel and cordite, speed and noise. Everything seems connected: disparate locations twitch and burst into activity like limbs reacting to impulses sent from elsewhere in the body, booms and jibs obeying levers at the far end of a complex set of ropes and cogs and relays. (p. 177)

Serge is using the clock code to range the guns on to their target, but the passage is remarkable for McCarthy's imagery of "machinery and signal code" and "ropes and cogs and relays." Some of those who survived the war used the same mechanical imagery, perhaps nobody more effectively than Ernst Jünger (1920/2003):

> The modern battlefield is like a huge, sleeping machine with innumerable eyes and ears and arms, lying hidden and inactive, ambushed for the one moment on which all depends. Then from some hole in the ground a single red light ascends in fiery prelude. A thousand guns

roar out on the instant, and at a touch, driven by innumerable levers, the work of annihilation goes pounding on its way.[22] (p. 107)

What distinguishes McCarthy (2010), I think, is his realization that more is happening than lights setting levers in motion. Later he has Serge recognize that he is the messenger of death but insists that

> He doesn't think doesn't think of what he's doing as a deadening. Quite the opposite: it's a quickening, a bringing to life. He feels this viscerally, not just intellectually, every time his tapping finger draws shells up into their arcs, or sends instructions buzzing through the woods to kick-start piano wires for whirring cameras, or causes the ground's scars and wrinkles to shift and contort from one photo to another: it's an awakening, a setting into motion. (p. 200)

When "the ground's scars and wrinkles . . . shift and contort from one photo to another" the cycle is complete: the image becomes the ground which becomes the image. A clockwork war is set in motion through the changing contours of the map.

"Clockwork War" and the Mathematics of the Battlefield

The new face of industrialized warfare with its intricate co-ordination of military forces along the Front required time and space to be choreographed with unprecedented precision. Various methods were used to synchronize time, but the wristwatch (or trench watch) was the indispensable mechanism, as the *Stars and Stripes* made plain in 1918 in an essay entitled "The Wrist Watch Speaks." The wristwatch was "at the heart of every move in this man's war."

> On the wrist of every line officer in the front line trenches, I point to the hour, minute and second at which the waiting men spring from the trenches to the attack. I . . . am the final arbiter as to when the barrage shall be laid down, when it shall be advanced, when it shall case, when it shall resume. I need but point with my tiny hands and the signal is given that means life or death to thousands upon thousands.

Synchronizing watches was a two-step process. Time-signals were transmitted from the Observatoire de Paris to the French military's radio-telegraphic station at the Eiffel Tower and broadcast twice a day in three bursts in the morning and again in the evening. Signals Officers or orderlies would be summoned to Brigade Headquarters to receive the official time and set their rated watch to Eiffel Tower Time, and they would then redistribute the synchronized watches to the officers.[23]

---

[22] The book was first published in German in 1920 but this passage was omitted by Jünger in subsequent revisions (which continued until 1961), and so does not appear in the (superior) English translation of the 1961 edition by Michael Hofmann (London, UK: Penguin, 2003). Unless otherwise noted, all subsequent references are to the Hofmann translation.

[23] Hence, for example, this "synchronisation instruction" contained in Operation Order (no 233) from the 112th Infantry Brigade on 10 October 1918: "O.C. No.2 Section, 41st Divisional Signal Company, will arrange for EIFFEL TOWER Time to be taken at 11.49 on 'J' minus one day ['J' was

By these means, as Stephen Kern (1983) has it, "the war imposed homogeneous time" (p. 288)—or at any rate attempted to do so.[24]

The relentless timetabling of the war was partly a product of the scale of the conflict, the sheer numbers of men and machines that had to be maneuvered across the battlefield, but it was also necessitated by the difficulty of real-time communication between infantry, artillery and aircraft. It also required a no less rigid mathematization of the battlespace. "We are to go over from tapes laid by the Engineers," wrote A. M. Burrage (1930). "The whole thing must be done *with mathematical precision*, for we are to follow a creeping barrage which is to play for 4 min only a hundred yards in front of the first 'ripple' of our first 'wave'." (p. 127). The artillery timetable had been introduced at Neuve Chapelle in March 1915, followed by the stepped barrage at Loos in September and the creeping barrage by the time of the Somme offensive in 1916 (Becke, 1931; Marble 2008, Chap. 6). The Tactical Note from Fourth Army HQ in May 1916 explained the principle:

> The ideal is for the artillery to keep their fire immediately in front of the infantry as the latter advances, battering down all opposition with a hurricane of projectiles. The difficulties of observation, especially in view of dust and smoke . . . the probable interruption of telephone communications between infantry and artillery . . . renders this idea very difficult to obtain.
>
> Experience has shown that the only safe method of artillery support during an advance, is *a fixed timetable of lifts to which both the infantry and artillery must rigidly conform.*
>
> This timetable must be regulated by the rate at which it is calculated the infantry can reach their successive objectives. (Macdonald, 1983, p. 46)

The Plan of Operations issued by XXI Corps repeated the same injunction:

> The advance of the infantry will be covered by a heavy barrage from all natures of guns and mortars. The heavy artillery barrage will lift direct from one line onto the next. The field artillery barrage will creep back by short lifts. Both will work *strictly according to time-table*. The lifts have been timed so as to allow the infantry plenty of time for the advance from one objective to the next . . . (Becke, 1931, Appendix 40)

These strictures were superimposed over model landscapes derived from air photographs. Some of them were scale models, a sort of topographical bas-relief. Blunden (1928/2000) described "an enormous model of the German systems" being "open for inspection, whether from the ground or from step-ladders raised beside, and this was popular, though whether from its charm as a model or value as a military aid is uncertain." (p. 150) (Fig. 4.4). Others were 1:1 simulacra—"we dug the trenches exactly as they were in the photographs" (Private William Holbrook,

---

the day of the attack] and afterwards will synchronise watches throughout the Brigade Group by a 'rated watch.'" Edmund Blunden (1928/2000) describes the practice: "Watches were synchronized and reconsigned to the officers" (p. 91); and again: "A runner came round distributing our watches, which had been synchronized at Bilge Street ['battle headquarters']" (p. 254). Wristwatches were originally worn by women and pocket watches carried by men, but wristwatches became favored by soldiers and airmen because they required a "hands-free" way of telling the time.

[24] That is surely something of an overstatement: just as the "optical war" was supplemented, subverted, and even resisted by quite other, intimately sensuous geographies so, too, must the impositions and regimentations of Walter Benjamin's (1940/2006) "homogeneous, empty time" have been registered and on occasion even refused in the persistence of other, more intimate temporalities.

Fig. 4.4 Trench model of Messines Ridge (Retrieved from http://www.expressandstar.com/wpmvc/wp/wp-content/uploads/2013/09/32047939.jpg. Copyright unknown)

4th Battalion, Royal Fusiliers, in Levine, 2009, p. 87)—built at considerable effort so that troops could practice their drills:

> Three weeks before the Big Push of July 1st [1916]—as the Battle of the Somme has been called—started, exact duplicates of the German trenches were dug about 30 kilos behind our lines. The layout of the trenches were [sic] taken from aeroplane photographs submitted by the Royal Flying Corps. The trenches were correct to the foot; they showed dugouts, saps, barbed wire defences, and danger spots.
>
> Battalions that were to go over in the first waves were sent back for three days to study these trenches, engage in practice attacks, and have night maneuvers. Each man was required to make a map of the trenches and familiarize himself with the names and location of the parts his battalion was to attack.[25] (Empey, 1917, p. 236)

---

[25] The models that were derived from aerial reconnaissance were also vulnerable to aerial reconnaissance: "These imitation trenches, or trench models, were well guarded from observation by numerous allied planes which constantly circled above them. No German aeroplane could approach within observing distance. A restricted area was maintained and no civilian was allowed within three miles . . ." But, Empey adds, "When we took over the front line we received an awful shock. The Germans displayed signboards over the top of their trench showing the names that we had called their trenches. The signs read 'Fair,' 'Fact,' 'Fate,' and 'Fancy' and so on, according to the code names on our map. Then to rub it in, they hoisted some more signs which read, 'When are you coming over?' or 'Come on, we are ready, stupid English'" (Empey, 1917, pp. 237–238).

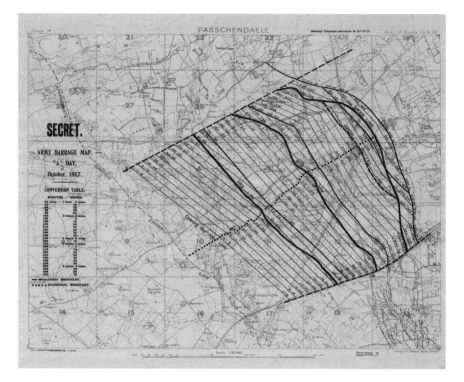

Fig. 4.5 Army barrage map, Passchendaele (Retrieved from http://upload.wikimedia.org/wikipedia/commons/f/f3/First_Battle_of_Passchendaele_-_barrage_map_%28colour_balance%29.jpg. Copyright by Wikipedia Commons)

Staff officers watched the rehearsals and re-calibrated the details of each paper offensive. "The plan for the attack has now come out," one artillery officer recorded in his diary, "about 100 pages of typed foolscap which had to be read through, digested and from which the battery programme had to be extracted and the calculations made" (Major Roderick Macleod, Royal Field Artillery, in Steel & Hart, 2000, p. 87). The (re)calibrations were projected onto a map whose timelines marched across the geometricized space in perfect military order (Fig. 4.5). Troops were to move in the same linear progression, their columns animated by the imperative future—and never conditional—tense of the typed orders:

> The left column will cross trenches 5, 6 and 7 by gangways; it will seize trenches D and E, drive out the defenders and occupy the communication trenches. . . . A detachment previously detailed for the purpose will face west; another similarly detailed will face east, and will enfilade trench B with a machine gun. As soon as the left column has reached the hostile trenches, the right column will debouch by trenches 8 and 9, and advance through the interval between them against trenches M and W. (*Trench Warfare*, 1915)

Nowhere was that attempt to order the future plainer than in the official British decree that the disordered space of "No Man's Land" did not exist: Allied territory extended all the way to the German front line (Deer, 2009, p. 23).

It did not take a Borges or a Korzybski to unpick the sutures between map, model, and territory: as one young private explained, once you were over the top and advancing behind the artillery curtain "there was barbed wire and artillery fire, and it wasn't like the practices" (Private Tom Bracey, 9th Battalion, Royal Fusiliers, in Levine, 2009, p. 87). Neither was it like the map. But from the air—the perspective from which the maps and models had been made—there was a disconcerting sense that the map had *preceded* the territory:

> The waves of attacking infantry as they came out of their trenches and trudged forward behind the curtain of shells laid down by the artillery had been an amazing sight. The men seemed to wander across No Mans Land and into the enemy trenches, as if the battle was a great bore to them. From the air it looked as though they did not realise they were at war and were taking it all entirely too easy. That is the way with clock-work warfare. These troops had been drilled to move forward at a given pace. They had been timed over and over again in marching a certain distance and from this timing the "creeping" or rolling barrage had been mathematically worked out. . . .
>
> I could not get the idea out of my head that it was just a game they were playing at; it all seemed so unreal. Nor could I believe that the little brown figures moving about below me were really men going to the glory of victory or the glory of death. I could not make myself realise the full truth or meaning of it all. It seemed that I was in an entirely different world, looking down from another sphere on this strange, uncanny puppet-show. (Bishop, 1918, pp. 97–98, 99)[26]

Yet those who had set these marionettes in motion were unable to watch the show. A dense web of telephone and telegraph lines ran from GHQ through division, brigade, and battalion headquarters to the front-line trenches but, as Keegan (2004) noted, it had "one disabling shortcoming: it stopped at the edge of no-man's-land. Once the troops left their trenches . . . they passed beyond the carry of their signals system into the unknown" (p. 260).[27] One subaltern saw troops running across a field towards Gommecourt Wood:

> Then they vanished into the smoke. And then there was nothing left but noise. And after this we saw nothing and we knew nothing. And we lived in a world of noise, simply noise.[28]

I want to follow those troops into the smoke and the noise, but before I do I want to pause to stake two claims. First, I do not mean to repeat the conventional (and casual) critique of GHQ and its staff officers. In March 1916 they had moved their departments from St. Omer to Montreuil, a small town even more distant from the

---

[26] Bishop had started his military career as a cavalry officer, and claimed that "It was the mud, I think, that made me take to flying" (1918, p. 17). Yet even those down in the mud used the same imagery. In Fredric Manning's (1929) semi-autobiographical novel *The Middle Parts of Fortune: Somme and Ancre, 1916* the troops are seen "moving forward in a way that seemed commonplace, mechanical, as though at some moment of ordinary routine . . . They had seemed so toy-like . . . they had moved forward mechanically" (p. 10).

[27] Keegan (2004, p. 260) continues: "The army had provided them with some makeshifts to indicate their position: rockets, tin triangles sewn to the backs of their packs as air recognition symbols, lamps and flags, and some one-way signaling expedients, Morse shutters, semaphore flags and carrier pigeons . . ."

[28] Captain Charles Carrington, in Arthur (2002, pp. 157–158).

Front, and there was an *experiential* break between the two worlds.[29] Keegan (2004) suggests that from the middle of the nineteenth century it had been accepted that "the main work of the general . . . had now to be done in his office and before the battle began," (p. 261) and, as Strachan (2006, pp. 171–172) reminds us, the combination of mass armies and massive firepower ensured that modern warfare would become ever more managerial. Staff officers were tied to their desks because, logistically and strategically, war on such a scale could only be administered through the telephone, the telegraph, and the wireless (see also Hall, 2009, 2012). Second, both contemporaries and critics have railed against the experiential detachment of the general staff from the front lines—I have no doubt this was true: none of them visited the Front in 1916 or 1917—but it was also an *epistemic* rupture. They were creatures as well as creators of an administrative apparatus that dictated the terms through which they apprehended the battle space. "If the work of a general occurs in the space of an office," Booth (1996) explains, "the space of a battlefield— physically expansive, perceptually elusive—must necessarily be shrunken and flattened to the plane of a map" (p. 88). The space of the map was supplemented by the space of the photograph, and together these were the optical-visual devices of a supremely abstract order. "If the emblematic figure for the collapse of vision was No Man's Land," so Deer (2009) argues, "it was the strategist's map that came to represent the struggle to recapture oversight, to survey and order the mud, chaos and horror of battle" (p. 24; see also Brantz, 2009).[30] This was, of course, precisely what Boyd's Gabriel had realized—and rejected—thousands of miles away on the coast of East Africa. In this struggle to reassert a cartographic order the battle *space* was mathematized and the simultaneous equations of clockwork war were solved, at least on paper, by bracketing the messiness and materiality of the battle *field*: it was as though "only mathematical space emptied of human experience but structured in abstract detail [could] provide the smooth sphere for the "pure" war of technology" (Hüppauf, 1993, p. 74). And yet, if this transformed time and space into what Hüppauf calls "predictable, calculable operations" at several removes from another, radically "impure" space—"the space of experience"—"constituted by fighting, suffering and dying soldiers," the fact remained that each co-produced the other (pp. 74–75).[31] The maps and the photographs, which were themselves a

---

[29] The same was true for the German High Command, and Jünger (1920) wryly describes "episodes [that] prove the futility of the system of higher command with its headquarters far in the rear" (p. 243) and operations that "had been ordered from the rear and by the map, for it could not have occurred to anyone who had seen the lay of the land to give such orders" (p. 261)—and of the occasional runner "who carried the paper war even into this secluded spot" (p. 254)—but quickly adds "though of course I do not question the necessity" (p. 243).

[30] "To many commanders, battlefields continued to be transposed onto maps" so that military strategies became "increasingly abstract" (Brantz, 2009, p. 74). Vismann (1997) draws a distinction between "the homogeneous space of geography" and "the specific space of the soil" (p. 47).

[31] It is not difficult to hear echoes of Lefebvre (1974/1991) in these formulations, who identifies the aggressive production of an abstract space with the violent triumph of a visual-geometric-phallocentric space that "entails a series of substitutions and displacements by means of which it overwhelms the whole body and usurps its role."

materialization not only of techno-scientific Reason but also of corporeal investment, were instrumental in the formation of what the soldier-poet Wilfred Owen
described as "the topography of Golgotha."[32] It is now time to descend into that
inferno.

The Corpography of the Slimescape

In this clockwork war, Erich Remarque (1929/2013) wrote in *All quiet on the
Western Front*, "the earth is the background of this restless, gloomy world of
automatons" (p. 87). But it was surely more than that: the earth was also the medium
in which and through which the war was conducted.[33] For many soldiers, the earth
was transformed into a mud of such cloying stickiness that it threatened to bring the
war to a juddering halt (Fig. 4.6).

Marc Bloch (1980) famously described his experience on the Aisne in 1914–
1915 as "the age of mud" (p. 152), and Arthur Empey (1917) complained that "the
men slept in mud, washed in mud, ate mud and dreamed mud" (p. 60). "At present,"
wrote one artillery major from Passchendaele in the summer of 1917, "I am more
likely to die from drowning than hostile fire. It has rained solidly for 3 days and the
place is knee deep in mud."[34] The weather was extraordinary for August—another
artillery officer there confirmed that "it rained absolutely continuously, one was as
afraid of getting drowned as of getting hit by shells"[35]—but, ironically, the quagmire
was also produced by artillery shells piercing the clay layer and forcing water to the
surface under pressure. In any event, the fear of drowning was real enough. "Deep
devouring mud spread deadly traps in all directions," recalled one British
guardsman: "We splashed and slithered, and dragged our feet from the pull of an
invisible enemy determined to suck us into its depth. Every few steps someone
would slide and stumble and, weighed down by rifle and equipment, rapidly sink
into the squelching mess."[36] Those who fell into one of the myriad waterlogged
shell-holes found themselves up to their waist in liquid, cloying mud and often had
to wait for hours, even days before they were rescued; many never made it out. One
subaltern described laying the wounded at Passchendaele on duckboards because
they had run out of stretchers and then, during a lull in the shelling, "we heard this
terrible kind of gurgling noise. It was the wounded, lying there sinking, and this

---

[32] "For 14 hours yesterday, I was at work—teaching Christ to lift his cross by numbers. . . and with
maps I make him familiar with the topography of Golgotha": Wilfred Owen, letter to Osbert
Sitwell, 4 July 1918. (The Topography of Golgotha, 1918). http://pw20c.mcmaster.ca/case-study/
topography-golgotha-mapping-trenches-first-world-war

[33] This is capable of generalization; I have explored the mud of the Western Front in the First World
War, the Western Desert in World War II, and the jungles of the Vietnam War in "The Natures of
War", *Antipode* (in press).

[34] Major Roderick Macleod, in Steel and Hart (2000), p. 138.

[35] Major Richard Talbot Kelly, in Arthur (2002, p. 218).

[36] Private Norman Cliff, in Hart (2013, p. 365).

Fig. 4.6 Mud at the Western front. Pilckem Ridge 1917 (Retrieved from http://upload.wikimedia. org/wikipedia/commons/6/6f/Q_005935PilckemRidge1August1917StretcherBearersBoesinghe. jpg. See also Brooke J W (Lt) © IWM (Q 5935). http://www.iwmprints.org.uk/image/743595/ brooke-j-w-lt-a-team-of-stretcher-bearers-struggle-through-deep-mud-to-carry-a-wounded-man-to-safety-near-boesinghe-on-1-august-1917-during-the-third-battle-of-ypres)

liquid mud burying them alive, running over their faces, into their mouth and nose."[37] "We live in a world of Somme mud," reported Edward Lynch (2008):

> We sleep in it, work in it, fight in it, wade in it and many of us die in it. We see it, feel it, eat it and curse it, but we can't escape it, not even by dying. (p. 147)

Not surprisingly perhaps, some began to see the mud as possessing a diabolical agency through which it possessed them:

> At night, crouching in a shell-hole and filling it, the mud watches, like an enormous octopus. The victim arrives. It throws its poisonous slobber out at him, blinds him, closes round him, buries him. One more *disparu*, one more gone. . . . For men die of mud, as they do from bullets, but more horribly."[38]

It was, still more horrifically, much more than mud: military operations commingled with the earth and the water to produce a cyborg nature in which mud

---

[37] Lt. James Annan, 1st/9th Bn Royal Scots Regiment, in Macdonald (1993, p. 126).

[38] *Le Bochofage: organe anticafardeux, Kaisericide et embuscophobe*, 26 March 1917, in Audoin-Rouzeau (1992, p. 38). *Le Bochofage* was a French trench journal.

mixed with barbed wire, shells and iron scraps, and with organic wastes, dead animals, and decomposing bodies, to form what Ernst Jünger described as "a garden full of strange plants" (see Huyssen, 1993, p. 15).[39] This "slimescape," as Das (2008, p. 37) calls it, had two effects on the neat and ordered lines of the battle space envisioned on the staff officers' maps and plans.

First, the slimescape multiplied scepticism at the order and reasonableness of the map a thousand times or more. The paper war was confounded at every turn. One of the artillery officers at Passchendaele watched through his binoculars as the infantry struggled to keep pace with the creeping barrage, which had been slowed down in an attempt to compensate for the terrain: "They were up to their knees in mud, and by the time they got half-way across it was virtually impossible for them to move either forward or back" (Macleod, quoted in Macdonald, 1993, p. 149). Even his own ordnance made little headway; his fellow artillery officer said that "the extraordinary quagmire nature of the Passchendaele battle masked much of the effect of the shells, which sank so deeply into the mud that the splinter and blast effect was to a large extent nullified" (Major Richard Talbot Kelly, in Arthur, 2002, p. 218). Horses, mules, artillery limbers strained to make it through the mud (Fig. 4.7), and it became desperately difficult to rescue the wounded:

> In normal conditions, even under fire, two men could carry a casualty from the line to the dressing-station. Now it took four, even six, men to haul a stretcher case to safety, and a journey of as little as 200 yards could take 2 hours of struggle through the lashing rain and the sucking mud. (Macdonald, 1993, p. 123)

Modern warfare seemed to be waged against the very earth itself. "Its new technology generated a capacity for destruction that no longer focused just on the killing of individual soldiers," Brantz (2009) suggests: "Now warfare also included the obliteration of entire landscapes" (p. 74). Hynes (1977) says much the same. In his view, the war "turns landscape into *anti-landscape*, and everything in that landscape into grotesque, broken, useless rubbish" (p. 8). Landscape is above all a visual construction—even a visual ideology[40]—and the power and significance of Hynes's insight resides in its implication that through the production of this anti-landscape the privileges accorded to vision in the constitution of "optical war" were challenged and even withdrawn by the soldiers most intimately involved in its execution.

Second, surviving the slimescape required a "re-mapping," what I call a corpography, in which other senses had to be heightened in order to apprehend and navigate the field of battle. Sight was no longer the master sense for those on the front line, especially the infantry, because the terrain had been pulverized—a European rural landscape that was so familiar to so many (but by no means all) of those who fought over it had been made strange—and its contours were successively reworked by each barrage and offensive that it became ever more unrecognizable. In a vivid anticipation of

---

[39] Huyssen (1993) sees Jünger directing his "entomological gaze" on this "garden" through an "armored eye".

[40] This *aperçu* was developed with most acuity by Cosgrove (1985).

AUSTRALIAN WAR MEMORIAL   E00963

Fig. 4.7 Ypres, 1917. Australian War Memorial, photograph E00963 (Retrieved from http://www.awm.gov.au/collection/E00963/ In the public domain)

Gabriel's despair at the orderliness of the map, one subaltern explained that "though we had studied the map so thoroughly beforehand, it was impossible to recognize anything in this chaos . . ."[41] His experience was a common one; here is another lieutenant:

> We sent out four runners to get to Battalion Headquarters at Minty's Farm, and every time, after an hour, the Adjutant rang up—because somehow or other we got a line laid—to ask, "When is your runner coming up to take the relief out?" That happened four times, and still he was on the blower, kicking up hell and asking where the runners were. Well, we weren't too happy about it either! So I said, "Well, the only thing I can do is have a go myself and see if I can get there." I walked right to it and there were no landmarks at all. You couldn't say, well, I know that tree, or I can see half a house there, or anything like that. There was nothing. Just one morass of mud as far as the horizon. The runners had simply got lost, and I didn't blame them at all.[42]

The battlefield was constantly shifting, not only as each advance swept forward and back, as trench lines were taken, lost and taken again, but as each wave of destruction broke over the land so its shapes and elements became ever more transitory. This meant that it was not only maps that became unreliable as the terrain became unreadable; memory became all but useless too. "I had to go round my sector once a night with the sergeant-major," another subaltern remarked. "And

[41] 2nd Lt. Thomas Hope Floyd, 2/5 Lancashire Fusiliers, 31 July 1917 in Barton (2007, p. 166).
[42] Lt. J. Annan, 1/9 Bn., Royal Scots Regiment, in Macdonald (1993, p. 133).

when we left one shell-hole we'd have to ask which way to go next, because each night the ground would have absolutely shifted."[43] Soldiers had to look for new markers—material or corporeal did not matter very much: "Left by the coil of wire, right by the French legs" (Brantz, 2009, p. 77)—but they were all increasingly impermanent. One runner returning to Brigade Headquarters across the Ypres Salient "by a quicker but more exposed route" looked for objects to help guide him. "I see a foot and it keeps me for the next time but it is not there long."[44]

The sense of radical instability is vital. Weir (2007) is right to insist that "wrinkles in the texture of destruction [became] coordinates which allow[ed] the striation of smooth space," that the destruction of the battlefield required and became "the starting-point for a new re-gridding" (p. 45): but there was nothing permanent about those makeshift griddings, which were fluid, improvisational processes rather than fixed cartographies. The stream of maps and photographs could not keep pace with these intimately local changes, and the gap between their representations—which remained crucial for the general staff and the artillery—and the stocks of local knowledge developed and mobilized by the infantry grew wider. Jünger (2003) describes being criticized by a staff officer, jabbing his finger at a map after the failure of a trench raid to take prisoners late in 1917: "I realised that the kind of confusion where notions like right and left just go out of the window was quite outside his experience. For him the whole thing had been a plan; for us an intensely experiential reality" (p. 189). It certainly was a matter of experience but it was also a matter of epistemology: of what counted as useful knowledge. In his classic account of *No Man's Land*, Leed (1981) explained the gulf between the infantry and the staff officers like this:

> Trench war is an environment that can never be known abstractly or from the outside. Onlookers could never understand a reality that must be crawled through and lived in. This life, in turn, equips the inhabitant with a knowledge that is difficult to generalize or explain. (p. 79)

The reason for that, Leed (1981, p. 74) argued, was that what he called "the knowledge gained in war"—he meant not the intelligence used by planners to ordain a future anterior through "the safe distance of the gaze" but an intensely practical, densely particular local knowledge used by, even inhabited by the infantry—resided in and derived from the "clumsy immediacy" of the combatant's body.

This is what Das (2008) variously calls a "phenomenological geography" through which the trenches and No Man's Land were known not in terms of the abstract, cognitive apparatus of "maps, places and names" but apprehended—*re*-cognized—as "sensuous states of experience," and also a "haptic geography" (p. 73):

> [T]he visual topography of the everyday world . . . was replaced by the haptic geography of the trenches and mud was a prime agent in this change. In an atmosphere of darkness, danger and uncertainty, sights, sounds and even smells are encountered as material presences against the flesh. (p. 23)

---

[43] Lt. Ulrich Burke, 2 Bn., Devonshire Regiment, in Arthur (2002, p. 241).
[44] Private Aston, in Weir (2007, p. 42).

These are both useful terms, but I prefer to call this a corpography: although it is a made-up word, it simultaneously speaks to cartography and undoes it through its own muffled corporeality, an almost subterranean acknowledgement of its implication in what Lynch (2008) called "the land of rotting men" (p. 357).[45]

There were three other senses that had to be heightened—three other sources of knowledge that had to be developed—if the soldiers were to survive. The first, of almost overwhelming importance, was sound. During an offensive the soldiers were thrust into a world of noise: not the sound detected by tunnelers as they listened through their stethoscopes and microphones for traces of the enemy digging towards them nor the arcs traced by the sound rangers on their oscilloscopes and filmstrips but a 'flat, unceasing noise' that was intensely corporeal: "You could feel the vibrations coming up through the earth, through your limbs, through your body. You were all of a tremor, just by artillery fire only." Or again: "We lie on the shuddering ground, rocking to the vibrations, under a shower of solid noise we feel we could reach out and touch."[46] Because the link between sight, space and danger was broken all along the Front, Das (2008) suggests there was an "exaggerated investment in sound" (p. 81). To capitalize on this, it became essential to learn to detect signals in the noise, to order the roaring soundscape, and A. M. Burrage (1930) captures this as well as anyone:

> We know by the singing of a shell when it is going to drop near us, when it is politic to duck and when one may treat the sound with contempt. We are becoming soldiers. We know the calibres of the shells which are sent over in search of us. The brute that explodes with a crash like that of much crockery being broken, and afterwards makes a "cheering" noise like the distant echoes of a football match, is a five-point-nine. The very sudden brute that you don't hear until it has passed you, and rushes with the hiss of escaping steam, is a whizz-bang. . . . The funny little chap who goes tonk-phew-bong is a little high-velocity shell which doesn't do much harm. . . . The thing which, without warning, suddenly utters a hissing sneeze behind us is one of our own trench-mortars. The dull bump which follows, and comes from the middle distance out in front, tells us that the ammunition is "dud." The German shell which arrives with the sound of a woman with a hare-lip trying to whistle, and makes very little sound when it bursts, almost certainly contains gas.
>
> We know when to ignore machine-gun and rifle bullets and when to take an interest in them. A steady phew-phew-phew means that they are not dangerously near. When on the other hand we get a sensation of whips being slashed in our ears we know that it is time to seek the embrace of Mother Earth. (pp. 78–79)[47]

And here is Edward Lynch (2008):

> Talk gets on to the sounds made by shells, and the *minenwerfers* that we can run from if our luck's in, and about the spiteful little whizz-bang that it's generally too late to run from

---

[45] Booth (1996, p. 50) writes of the "corpsescapes" of trench warfare, which also evokes Blunden's (1928/2000) description: "The whole zone was a corpse, and the mud itself mortified" (p. 98).

[46] Henry Holdstock, in Levine (2009, p. 94); Lynch (2008, p. 144).

[47] As that last sentence suggests, this fostered a sort of geo-intimacy. "Sometimes you wish the earth would shrink," one private said, "so as to let you in" (Private Thomas McIndoe, in Levine, 2009, p. 38). And here is Remarque (1928/2013): "To no man does the earth mean so much as to the soldier. When he presses himself down upon her, long and powerfully, when he buries his face and his limbs deep in her from the fear of death by shell-fire, then she is his only friend, his brother, his mother; he stifles his terror and his cries in her silence and her security . . ."(p. 41).

when it's heard. . . . More digging and the [machine-]gun fires again. Jacko makes to get down, but has a nasty shock when he sees that none of us has even bobbed. We explain that we knew by the sound of the gun that it was not firing in our direction. . . . Gas shells are sometimes hard to distinguish from duds. They land with a little putt-tt sort of sound. Just enough explosive in them to burst the case and release the gas without scattering it. (p. 95)

It was, in effect, a way of "seeing by listening" so that, as Brantz (2009) suggests, "trench life was, in many ways, a synesthetic experience" (p. 76).

The soldiers also inhabited an aggressive and intrusive smellscape compounded, as Ellis (1976) records, of a score of things: "the chloride of lime that was liberally scattered to minimise the risk of infection, the creosote that was sprayed around to get rid of the flies, the contents of the latrines, the smoke from the braziers and the sweat of the men" (pp. 58–59). Above all, it was the fetid odor of death, which Jünger (2003) described as "a persistent smell of carrion", or "Eau d'offensive" (p. 258). All smells are particulate, and there was something intensely, intimately physical about this apprehension of the killing fields. "I have not seen any dead," Wilfred Owen wrote after three weeks at the front, "I have done worse. In the dank air I have *perceived* it, and in the darkness *felt*" (Das, 2008, p. 7). It was commonplace yet never became a commonplace. "I never grew accustomed to the all-pervading stench of decayed and decaying flesh," one artillery officer said, "mingled with that of high explosive fumes that hung over miles and miles of what had been sweet countryside and now was one vast muck heap of murder."[48] But there were other smells that, if you knew them, could save your life. At Passchendaele, one corporal recalled, "the smells were very marked and very sweet. Very sweet indeed. The first smell one got going up the track was a very sweet smell which you only later found out was the smell of decaying bodies—men and mules." But then, he added,

You got the smell of chlorine gas, which was like the sort of pear drops you'd known as a child. In fact the stronger and more attractive the pear-drop smell became, the more gas there was and the more dangerous it was. When you were walking up the track a shell dropping into the mud and stirring it all up would release a great burst of these smells.[49]

The third sense was touch. Trench diaries, journals and memoirs are saturated with the predatory touch of the slimescape, the mud that invaded the body, "clogged the fingers, filled the nails, smeared the face, ringed the mouth and clung to the stubbly beard and hair," and which could all too silently infect wounds and kill soldiers.[50] But they could also be saved by their sense of touch, and those same sources are no less full of men subsisting in dugouts and crawling through the trenches, emerging to worm their way through the barbed wire and the mud. "Creep, crawl, worm, burrow," Das (2008) reminds us, "were the usual modes of movement during a night patrol in no man's land or while rescuing war-wounded in order to avoid being detected" (p. 43) and each of them—there are others too: plunge,

---

[48] Lt. R. G. Dixon, Royal Garrison Artillery, in Steel and Hart (2000, p. 198).

[49] Corporal Jack Dillon, Second Bn, Tank Corps, in Arthur (2002, p. 233).

[50] Private N. M. Ingram, in Barton (2007, p. 309).

immerse, scrape—registers a shift from the visual to the tactile.[51] Sight in those circumstances was of limited purchase, but where it was invoked it too became haptic, a facility described by Frederic Manning (1929) in *The Middle Parts of Fortune*, a novel based on his own experience in the Somme:

> [E]very nerve was stretched to the limit of apprehension. Staring into the darkness, behind which menace lurked, equally vigilant and furtive, his consciousness had pushed out through it, to take possession, gradually, and foot by foot, of some forty or fifty yards of territory within which nothing moved or breathed without his knowledge of it. Beyond this was a more dubious obscurity, into which he could only grope without certainty. The effort of mere sense to exceed its normal function had ended for the moment . . . (p. 224)

Stretching, pushing out, taking possession, groping: these are the probing moments of a profoundly haptic apprehension of the battlefield.

Conclusion

Paul Virilio's (1989) account of *War and cinema*, and particularly his rendering of the logistics of perception during World War I, remains a landmark analysis. He made much of the connections between aviation and cinema, and his arguments have informed the opening sections of my own essay. In his eyes, aerial reconnaissance—which stood in the closest of associations to the cartographic—became successively "chronophotographic" and then cinematographic, as these new methods struggled both to keep pace with and to produce the new motility of a war that merely appeared to be static and fixed in place. But Virilio also advanced another, more problematic claim: "As sight lost its direct quality and reeled out of phase, the soldier had the feeling of being not so much destroyed as de-realized or de-materialized, any sensory point of reference suddenly vanishing in a surfeit of optical targets" (pp. 14–15). Here he continues to privilege the visual-optical register of cartography and fails to register the bodily habitus that, as I have shown in the closing sections, was profoundly implicated in the actions and affects of the ordinary infantryman. Virilio was not alone. A. M. Burrage (1930) wrote that

> [W]e are slowly realising that the job of the infantry isn't to kill. It is the artillery and the machine-gun corps who do the killing. We are merely there to be killed. We are the little flags which the General sticks on the war-map to show the position of the front line. (p. 82)

In sketching the outlines of a countervailing corpography established by those on that front line, I do not wish to privilege one mode of knowing over the other: each sutures knowledge to power in vital, significant but none the less different ways, and

---

[51] Das (2008, p. 86) cites Merleau-Ponty to sharpen the contrast between ocular vision and touch: "It is through my body that I go to the world, and tactile experience occurs 'ahead' of me." There were of course other registers in which touch was central, and Das also beautifully illuminates the homo-sociality of this subterranean world in which forms of intimacy with other men—not just "mother earth"—were no less vital in rendering this stunted life endurable and meaningful.

each both advances and repels military violence. But I do sympathize with Edmund Blunden's (1928/2000) agonized question:

> Was it nearer the soul of war to adjust armies in coloured inks on vast maps at Montreuil or Whitehall, to hear of or to project colossal shocks in a sort of mathematical symbol, than to rub knees with some poor jaw-dropping resting sentry, under the dripping rubber sheet, balancing on the greasy fire-step . . . ? (p. 141)

Of course, "a map is a weapon," as Lt.-Col. E. M. Jack ("Maps GHQ") insisted, and those "vast maps," together with the panoply of trench maps, sketch maps, and all the rest, were some of the deadliest weapons in the staff officers' armory; but they were hardly sufficient sources of knowledge. And so I understand, too, why Blunden (1928/2000) concluded that venturing into the killing fields armed with its pure, abstract, mathematical knowledge alone was sheer folly:

> [T]he new Colonel . . . sent forward from C Camp an officer fresh from England, and one or two men with him, to patrol the land over which our assault was intended, . . . This officer took with him his set of the maps, panoramas, photographs and assault programmes which had been served round with such generosity for this battle. He never returned . . . (pp. 151–152)

Coda

In this essay I have been concerned with World War I but, as we approach its centenary, it is worth reflecting on the ways in which modern warfare has changed—and those in which it has not. Through the constant circulation of military imagery and its ghosting in video games, many of us have come to think of contemporary warfare as optical war hypostatized: a war fought on screens and through digital images, in which full motion video feeds from Predators and Reapers allow for an unprecedented degree of remoteness from the killing fields. In consequence, perhaps, many of us are tempted to think of the wars waged by advanced militaries, in contrast to World War I, as "surgical," even body-less. These are wars without fronts, whose complex geometries have required new investments in cartography and satellite imagery, and there have been major advances in political technologies of vision and in the development of a host of other sensors that have dramatically increased the volume of geo-spatial intelligence on which the administration of later modern military violence relies. All of this has transformed but not replaced the cartographic imaginary.

And yet, for all of their liquid violence, these wars are still shaped and even confounded by the multiple, acutely material environments through which they are fought. In Sebastian Junger's (2011) remarkable dispatch from Afghanistan, he notes that for the United States and its allies "the war diverged from the textbooks because it was fought in such axle-breaking, helicopter-crashing, spirit-killing, mind-bending terrain that few military plans survive intact for even an hour" (p. 47). If that sounds familiar, then so too will MacLeish's (2013) cautionary observations about soldiers as both vectors and victims of military violence:

The body's unruly matter is war's most necessary and most necessarily expendable raw material. While many analyses of US war violence have emphasized the technologically facilitated withdrawal of American bodies from combat zones in favour of air strikes, smart bombs, remotely piloted drones, and privately contracted fighting forces, the wars in Iraq and Afghanistan could not carry on without the physical presence of tens of thousands of such bodies. (p. 11)

In consequence, the troops have had to cultivate an intrinsically practical knowledge that, while its operating environment and technical armature are obviously different, still owes much to the tacit bodily awareness of the Tommy or the Poilu:

In the combat zone there is a balance to be struck, a cultivated operational knowledge, that comes in large part from first-hand experience about what can hurt you and what can't . . . So you need not only knowledge of what the weapons and armor can do for you and to you but a kind of bodily habitus as well—an ability to take in the sensory indications of danger and act on them without having to think too hard about it first. When you hear a shot, is it passing close by? Is it accurate or random? Is it of sufficient caliber to penetrate your vest, the window of your Humvee or the side of your tank? (MacLeish, 2013, p. 76)

In the intricate nexus formed by knowledge, space, and military power, later modern war still relies on cartographic vision—and its agents still produce their own corpographies.

References

Amad, P. (2012). From God's-eye to camera-eye: Aerial photography's post-humanist and neo-humanist visions of the world. *History of Photography, 36*, 66–86.

Anderson, R. (2001). The Battle of Tanga, 2–5 November 1914. *War in History, 8*, 294–332.

Arthur, M. (2002). *Forgotten voices of the Great War: A new history*. London: Imperial War Museum.

Audoin-Rouzeau, S. (1992). *Men at war 1914–1918: National sentiment and trench journalism in France during the First World War*. Oxford, UK: Berg.

Barton, P. (2007). *Passchendaele*. London: Constable and Robinson.

Barton, P., Doyle, P., & Vandewalle, J. (2010). *Beneath flanders fields: The tunnellers' war, 1914–1918*. Stroud, UK: The History Press.

Becke, A. F. (1931). The coming of the creeping barrage. *Journal of the Royal Artillery, 58*(1), 19–31.

Benjamin, W. (2006). On the concept of history. *Selected writings: Vol. 4. 1938–1940* (pp. 389–400). Cambridge, MA: Harvard University Press. (Original work published 1940).

Bennett, A. (1915). *Over there: War scenes on the Western Front*. London: Methuen.

Bishop, W. A. (1918). *Winged warfare*. New York: George Doran.

Bloch, M. (1980). *Memoirs of war 1914–1915* (C. Fink, Trans.). Ithaca, NY: Cornell University Press.

Blunden, E. (2000). *Undertones of war*. London: Penguin (Original work published 1928).

Booth, A. (1996). *Postcards from the trenches: Negotiating the space between modernism and the First World War*. New York: Oxford University Press.

Boyd, W. (2007, January 21). A bizarre and surreal conflict. *Sunday Times*.

Boyd, W. (1982). *An ice-cream war*. London: Hamish Hamilton.

Brantz, D. (2009). Environments of death: Trench warfare on the Western Front, 1914–1918. In C. Closmann (Ed.), *War and the environment: Military destruction in the modern age* (pp. 68–91). College Station, TX: A & M University Press.

Brodrick, W. (2008). *A whispered name*. London: Little, Brown.

Burrage, A. M. (1930). *War is war*. London: Victor Gollancz.

Chasseaud, P. (1999). *Artillery's astrologers: A history of British survey and mapping on the Western Front, 1914–1918*. Lewes: Mapbooks.

Chasseaud, P. (2002). British, French and German mapping and survey on the Western Front in the First World War. In P. Doyle & M. Bennett (Eds.), *Fields of battle: Terrain in military history* (pp. 171–204). Dordrecht, The Netherlands: Kluwer.

Chasseaud, P. (2013). *Mapping the First World War: The Great War through maps*. London: Collins/Imperial War Museum.

Cosgrove, D. (1985). Prospect, perspective and the evolution of the landscape idea. *Transactions of the Institute of British Geographers, 10*, 45–62.

Das, S. (2008). *Touch and intimacy in First World War literature*. Cambridge, UK: Cambridge University Press.

Deer, P. (2009). *Culture in camouflage: War, empire and modern British literature*. Oxford, UK: Oxford University Press.

Doyle, P., & Bennett, M. (1997). Military geography: Terrain evaluation and the British Western Front, 1914–1918. *Geographical Journal, 163*, 1–24.

Eksteins, M. (1989). *The rites of spring: The Great War and the birth of the modern age*. New York: Houghton Mifflin.

Ellis, J. (1976). *Eye-deep in hell: Trench warfare in World War I*. Baltimore: Johns Hopkins.

Empey, A. (1917). *Over the top*. New York: G. P. Putnam.

Finnegan, T. J. (2011). *Shooting the front: Allied aerial reconnaissance in the First World War*. Stroud, UK: Spellmount, The History Press.

Forsyth, I. (2013). Subversive patterning: The surficial qualities of camouflage. *Environment and Planning A, 45*, 1037–1052.

Forty, S. (2013). *Mapping the First World War: Battlefields of the great conflict from above*. London: Conway.

Godefroy, M. (2000). The Battle of Tanga Bay. *The Army Doctrine and Training Bulletin, 3*(3), 35–42.

Gough, P. (1998). Dead lines: Codified drawing and scopic vision in a hostile space. *POINT, Art and Design Research Journal, 6*(Autumn/Winter), 34–41.

Gough, P. (2009). "Calculating the future": Panoramic sketching, reconnaissance drawing and the material trace of war. In N. Saunders & P. Cornish (Eds.), *Contested objects: Material memories of the Great War* (pp. 237–251). London: Routledge.

Gregory, D. (in press). The natures of war. *Antipode*.

Gregory, D. (1994). *Geographical imaginations*. Oxford, UK: Blackwell.

Hall, B. (2009). *The British Expeditionary Force and communications on the Western Front, 1914–1918* (Doctoral dissertation). University of Salford, UK.

Hall, B. (2012). The British Army and wireless communication. *War in History, 19*, 290–312.

Hart, P. (2012). *Somme success: The Royal Flying Corps and the battle of the Somme, 1916*. Barnsley, UK: Pen and Sword.

Hart, P. (2013). *The Great War: A combat history of the First World War*. New York: Oxford University Press.

Hüppauf, B. (1993). Experiences of modern warfare and the crisis of representation. *New German Critique, 59*, 41–76.

Huyssen, A. (1993). Fortifying the heart—totally: Ernst Jünger's armored texts. *New German Critique, 59*, 3–23.

Hynes, S. (1977). *The soldiers' tale: Bearing witness to modern war*. New York: Allen Lane.

Innes, J. (1935). *Flash spotters and sound rangers: How they lived, worked and fought in the Great War*. London: George Allen & Unwin.

Jay, M. (1994). *Downcast eyes: The denigration of vision in twentieth-century French thought*. Berkeley: University of California Press.

Jones, H. A. (1928). *The war in the air* (Vol. 2). Oxford, UK: Clarendon.

Jones, S. (2010). *Underground warfare, 1914–1918*. Barnsley, UK: Pen & Sword.

Jünger, E. (2003). *Storm of steel* (M. Hofmann, Trans.). New York/London: Doubleday/Penguin. (Original work published 1920)

Junger, S. (2011). *War*. New York: Hachette.

Keegan, J. (2004). *The face of battle*. London: Pimlico.

Kennett, L. (1991). *The first air war, 1914–1918*. New York: Simon and Schuster.

Kern, S. (1983). *The culture of time and space, 1880–1918*. Cambridge, MA: Harvard University Press.

Kipling, R. (1910). *The brushwood boy*. London: Macmillan.

Leed, E. (1981). *No man's land: Combat and identity in World War I*. Cambridge, UK: Cambridge University Press.

Lefebvre, H. (1991). *The production of space* (D. Nicholson-Smith, Trans.). Oxford, UK: Blackwell. (Original work published 1974)

Levine, J. (2009). *Forgotten voices of the Somme*. London: Ebury Press.

Liddle, P. (1998). *Passchendaele in perspective: The third Battle of Ypres*. Barnsley, UK: Pen & Sword.

Lohman, A. (2012). East Africa in World War I: A geographic analysis. *Journal of Military Geography, 1*, 15–33.

Lynch, E. (2008). *Somme Mud: The experiences of an infantryman in France, 1916–1919*. London: Doubleday.

Macdonald, L. (1983). *Somme*. London: M. Joseph.

Macdonald, L. (1993). *They called it Passchendaele: The story of the Battle of Ypres and of the men who fought in it*. London: Penguin.

MacLeish, K. (2013). *Making war at Fort Hood: Life and uncertainty in a military community*. Princeton, NJ: Princeton University Press.

MacLeod, R. (2000). Sight and sound on the Western Front: Surveyors, scientists and the "battlefield laboratory", 1915–1918. *War and Society, 1*, 23–46. doi:10.1179/072924700791201405.

Manning, F. (1929). *The middle parts of fortune: The Somme and Ancre, 1916*. London: Piazza Press and Peter Davies.

Marble, S. (2008). *The infantry cannot do with a gun less: The place of the artillery in the British Expeditionary Force, 1914–1918*. New York: Columbia University Press.

Mattison, J. (n.d.). *Representations of war as autobiographical media*. Retrieved from http://www.walterdraycot.com

McCarthy, T. (2010). *C*. New York: Knopf.

Murray, J. (1988). British-Canadian military cartography on the Western Front, 1914–1918. *Archivaria, 26*, 52–65.

Paice, E. (2007). *Tip and run: The untold tragedy of the Great War in Africa*. London: Weidenfeld and Nicolson.

Pugliese, J. (2013). *State violence and the execution of law*. New York: Routledge.

Raleigh, W. (1922). *The war in the air* (Vol. 1). Oxford, UK: Clarendon.

Remarque, E. M. (2013). *All quiet on the Western Front* (B. Murdoch, Trans.). London: Vintage. (Original edition published 1928)

Rose, E., & Rosenbaum, M. S. (1993). British military geologists: The formative years to the end of the First World War. *Proceedings of the Geological Association, 104*, 41–49.

Ryan, R. (2013). *Dead man's land*. London: Simon & Schuster.

Saint-Amour, P. (2003). Modernist reconnaissance. *Modernism/Modernity, 10*, 349–380.

Saint-Amour, P. (2011). Applied modernism: Military and civilian uses of the aerial photomosaic. *Theory, culture and society, 28*(7–8), 241–269.

Sekula, A. (1975). The instrumental image: Steichen at war. *Artforum, 14*(4), 26–35.

Shell, H. R. (2012). *Hide and seek: Camouflage, photography, and the media of reconnaissance*. New York: Zone Books.

Slater, T. (n.d.). *British aerial photography and photographic interpretation on the Western Front*. Retrieved from http://tim-slater.blogspot.ca

Steel, N., & Hart, P. (2000). *Passchendaele: The sacrificial ground*. London: Cassell.

Strachan, H. (2006). *The First World War: A new illustrated history*. London: Simon and Schuster.

Strong, P., & Marble, S. (2011). *Artillery in the Great War*. Barnsley, UK: Pen and Sword.

Sykes, F. (1922). *Aviation in peace and war*. London: Edward Arnold.

Sykes, F. (1942). *From many angles: An autobiography*. London: Harrap.

The Camera at the front. (1917). *Scientific American, 117,* 380–381. doi:10.1038/scientificamerican11241917-380.

The Wrist watch speaks. (1918, February 15). *Stars and Stripes*, p. 2.

Trench warfare: Notes on attack and defence [Pamphlet]. (1915). Garvin Papers. London: British Library.

Virilio, P. (1989). *War and cinema: The logistics of perception* (P. Camiller, Trans.). London: Verso. (Original work published 1984)

Vismann, C. (1997). Starting from scratch: Concepts of order in no man's land. In B. Hüppauf (Ed.), *War, violence and the modern condition* (pp. 46–64). Berlin: Walter de Gruyter.

Weir, B. (2007). "Degrees in nothingness": Battlefield topography in the First World War. *Critical Quarterly, 49*(4), 40–55.

Telling the Future: Reflections on the Status of Divination in Ancient Near Eastern Politics

5

Stefan M. Maul

When there were political decisions to be made in the ancient Near East,[1] cuneiform sources from two millennia show us that kings and their counselors did not rely exclusively on their own professional expertise. They held off, rather, on putting a plan into action until its feasibility had been examined and confirmed by an independent "expert advisory board." The authority attributed to this examination can hardly be overestimated. This is proven by the mere fact that rulers submitted to it without dissent, in spite of the risk that their plan might be judged untenable. An assessment by these experts, on the other hand, had the benefit of guaranteeing reliable predictions as to the success of a given undertaking. For the experts had at their fingertips the knowledge and procedures to be able to look back, in a manner of speaking, from the vantage point of the future and see the consequences of an intended action, and thereby identify those plans and purposes that would lead to undesired outcomes. Naturally the prospect of such knowledge was of inestimable worth to political decision makers, because to those who sought such advice and received a positive verdict, it delivered the certainty of having chosen a path that was oriented toward the future and assured of success.

As insightful and rational as it may sound to test the viability of a given scheme before putting it into practice, the means by which such evaluations were made in the ancient Near East seem just as wrongheaded and downright absurd—at least from the perspective of our current worldview. Namely, the future prospects of a plan were regularly determined in royal palaces over the course of centuries from the color and shape of the liver of a sheep that had been slaughtered for this very purpose (Jeyes, 1993; Leiderer, 1990; Meyer, 1987; Starr, 1983, 1990).

---

[1] For a general introduction into the history and culture of the ancient Near East see Oppenheim (1996) and Sasson (1995).

S.M. Maul (✉)
Department of Languages and Cultures of the Near East – Assyriology, Heidelberg
University, Hauptstraße 126, 69117 Heidelberg, Germany
e-mail: stefan.maul@ori.uni-heidelberg.de

© Springer Netherlands 2015
P. Meusburger et al. (eds.), *Geographies of Knowledge and Power*,
Knowledge and Space 7, DOI 10.1007/978-94-017-9960-7_5

This procedure had developed into a proper "science" that correlated the appearance of a sheep's liver with future events. By the application of a system of rules, which in themselves seem quite systematic and logical, certain features on the surface of the liver were interpreted as favorable or unfavorable signs (Maul, 2003, pp. 69–82, 2013).

The experts checked systematically—going counter-clockwise—for the presence and undamaged condition of about a dozen anatomically constitutive elements on the surface of the liver (Koch-Westenholz, 2000, 2005), inspecting not only the gallbladder, but also looking for furrow-like markings and notches, distinctively textured surfaces, conspicuous protrusions, and the remains of ligaments that had been attached to the liver (Fig. 5.1).

The undamaged condition of the individual parts of the liver was seen as favorable. Furthermore, the location of certain features that could occur anywhere in the twelve regions of the liver played a fundamental role in the evaluation process. Among these were protruding lymph nodes, membranes, bubbles, warts, and holes in the liver tissue (Leiderer, 1990). The latter were open, occasionally calcified cavities in the surface of the liver caused by liver flukes, bladder worms, and other common parasites. Some of these characteristics, such as holes, were regarded as harbingers of evil. Others, however, such as slight bubbles caused by bladder worms, were positively construed (Fig. 5.2).

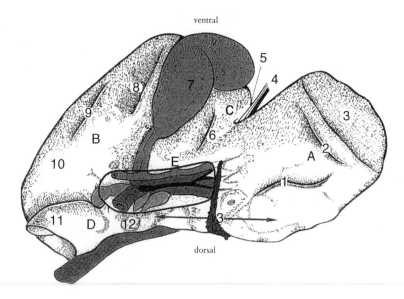

Fig. 5.1 Schematic drawing of a sheep's liver. The numbers 1–13 indicate the regions of the liver being checked, including the gallbladder (7) and several furrow like markings (1, 2, 6, 8, 9) (From *Anatomie der Schafsleber im babylonischen Leberorakel: Eine makroskopisch-analytische Studie* (p. 158, Fig. 2), by R. Leiderer, 1990, Munich: Zuckschwerdt. Copyright by Zuckschwerdt. Reprinted with permission)

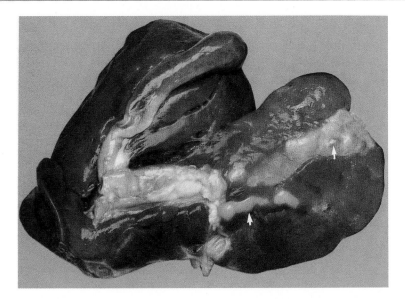

Fig. 5.2 A fresh sheep's liver infested by parasites. The arrows indicate bubbles caused by bladder worms. (From *Anatomie der Schafsleber im babylonischen Leberorakel: Eine makroskopisch-analytische Studie* (p. 161, Fig. 7), by R. Leiderer, 1990, Munich: Zuckschwerdt. Copyright by Zuckschwerdt. Reprinted with permission)

Although no two healthy livers are ever exactly alike, pathological phenomena ranging from inflammation to parasite infestation and necrosis lead to shockingly different findings. To facilitate the evaluation of certain characteristics as positive or negative, the inspection of a liver was approached with downright mathematical precision. A grid pattern was projected over each of the twelve constitutive parts of the liver, for instance over the gallbladder (Meyer, 1987 passim). The middle area of this grid was assigned to a fundamentally benign power of fate, the area to the right to an endorsement of the plan being evaluated, and the area to the left to the powers that opposed the plan. A characteristic construed as positive observed in the middle section was a positive finding, because fate was indeed revealing itself as benign. If the same sign appeared in the section to the right, this was also evaluated as a positive sign. But a sign of this type in the left section, which represented the powers opposing the plan, had the effect of strengthening those powers and thus became an unfavorable sign. A characteristic construed as negative, on the other hand, operated in the middle and right sections according to the mathematical formula $+ \times - = -$ (a positive times a negative equals a negative). In the left section of the grid, however, a weakening of the opposing powers amounted to strength, and hence a characteristic considered to be negative, if found in this section, was evaluated as a favorable sign. Admittedly, this procedure was in practice much more complicated. Not only did the system of analysis divide the gallbladder still further into subdivisions, which in turn were evaluated, but the entire liver was conceived of as a network of positively and negatively charged sections (Nougayrol, 1968). An

Fig. 5.3 An inscribed sheep's liver model from the Old Babylonian period (seventeenth century B.C.) (From British Museum Images, inventory number 00032437001. Copyright: The Trustees of the British Museum. Reprinted with permission)

Old Babylonian liver model from the seventeenth century B.C., made for teaching purposes, documented this layout for the ancient student using the example of a negatively construed cavity in the organ surface, as is seen in the photograph (Fig. 5.3). The respective meaning of the hole in each individual parcel is noted in cuneiform (Nougayrol, 1941, pp. 77–79).

Without going into further detail, this much can be said: professional haruspices claimed the ability, at least in the first millennium B.C., to calculate the validity of their predictions with a mathematical formula, in which certain numerical values must have been assigned on the basis of certain liver characteristics (Koch-Westenholz, 2005, pp. 63–66, 459–479).

The verdict on a given plan's prospects for success resulted from the simple addition of positive and negative signs. If there were more positive signs, the project was judged as desirable and cleared for implementation. If the negative signs were in the majority, then the evaluation was negative.

Decisions were made in this way in ancient Near Eastern courts about personnel issues, building projects, and even on the question of whether and when to go to war (Starr, 1983). Furthermore, in the eighteenth century B.C. it was routine in Old

Babylonian Mari[2] to inquire each month by means of extispicy[3] into the security of king, city and country, with the goal of warding off heretofore unrecognized dangers (Durand, 1988, pp. 57–58). The cause of the potential danger was established by means of carefully crafted questions, formulated in conjunction with the liver examination process (Lambert, 2007).

Such an examination procedure stands in opposition to modern conventional science, above all because it is blatantly unconcerned with the background and purpose of the undertaking in question. Nonetheless, over the course of more than two millennia, the Mesopotamians—and their neighbors as well—saw the mastery of such divinatory procedures as a decisive reason for the lasting cultural and geopolitical success of Babylonia and Assyria. Indeed, extispicy's reputation for effectiveness was so great that it outlived ancient Near Eastern civilization, being regarded as an indispensable means of political decision making in Greece, Etruria, and Rome (Collins, 2008; Pfiffig, 1975; Thulin, 1968).

The certainty that the process outlined here could afford a glimpse into the future rested on the belief, also current today, that the observable world contains traces of an unfolding future, which need to be recognized and interpreted. In the ancient Near East, every form of movement and change in all realms of experience, on earth as well as in the heavens, was understood as part of the vastly complex process of the world's development through time. All perceptible phenomena, however mundane they might be individually, were considered interrelated, for they are all part of the movement of the whole toward the future (Maul, 2003). Therefore, viewed individually or, even better, in conjunction, they allowed the ancient Near Eastern observer to project future events. The movement of the whole toward the future—as observed in the growth and development in nature, the alternation of day and night, the course of the year, and the progress of the heavens—was orderly and harmonious, and perceived as such. Every anomaly in nature, on the other hand, was thought to have been provoked by human beings. More precisely, it was considered to be a reaction to human deeds and probably desires as well. Deviations from regularity, such as abnormalities among plants and animals (Freedman, 1998; Moren, 1978), in the night sky (Rochberg-Halton, 2004), or on the surface of a sheep's liver (Koch-Westenholz, 2000, 2005), were perceived as messages to mankind that required their recipients to take stock of their situation and put things to rights, so that disorder could be eliminated and harmony restored (Maul, 1994).

The current experience with human-induced climate change may afford a perspective that gives an inkling of the rationale behind such notions. Be that as it may, the fundamental conviction that the entire cosmos is interactively centered on human beings subsided in the ancient oriental world in the seemingly compliant idea of gods mercifully using portent to guide humans down the right path, even though they ultimately had to bow to divine will anyway. Yet this conviction also fired an unquenchable spirit of exploration aimed at disclosing the inherent laws of

---

[2] Mari (modern Tell Hariri, Syria) was an ancient Sumerian and Amorite city.
[3] Extispicy is the inspection of the entrails of sacrificed animals, especially the livers of sheep and poultry.

the world's semiotic character and recognizing them in vastly different systems. Along with extispicy, the "science" of interpreting celestial omens had advanced far enough by the early first millennium B.C. that Babylonian and Assyrian kings regularly used it to make political decisions (Brown, 2000; Hunger, 1992; Koch-Westenholz, 1995; Maul, 2003, pp. 51–57; Rochberg-Halton, 2004). For the night sky—unlike extispicy—yielded unsolicited signs and offered an unceasing flow of information about the future. In the Neo-Assyrian period all of Mesopotamia was thus crisscrossed with a network of observation posts, which sent reports independently of one another to the king's palace in Nineveh, where these were compared and evaluated (Hunger, 1992; Oppenheim, 1969). Unlike the liver, however, the heavens—as the reflected image of the whole earth—were consulted for information not about the individual, but about matters of a national and even global nature. As such they even yielded predictions on the destiny of neighboring hostile lands. For this reason, astrology in the ancient Near East was of the greatest political interest. For, through the evaluation of the apparently irregular movements of celestial bodies, it seemed to offer the possibility of discerning opportunities as well as threats, of both avoiding disaster and taking advantage of particularly auspicious moments.

Numerous scholarly writings from the first millennium B.C. show that Mesopotamian diviners wanted to bring together insights from the two most important fields of divination, extispicy and astrology (Heeßel, 2008; Koch-Westenholz, 2005, pp. 30–31). Their reflections, which are still only partially understood, did not merely result in the liver's being conceived of as an emanation of the heavens, in a certain sense, and divided up into exactly twelve segments, like the zodiac. Mesopotamian scribes also thought it was possible to correlate the signs of the liver with equivalent astral signs (Reiner, 1995, p. 78; von Weiher, 1993, p. 159) and thus to trace the laws governing the dynamics of world events in various media.

Toward this end they collected signs not just for the purpose of telling the future. They also examined the present—that is, yesterday's future—in order to check for corresponding signs in the past that they might have overlooked. The *astronomical diaries* (Hunger & Sachs, 1988–2006) represent an ambitious attempt carried out over the course of centuries (with gaps from the seventh to first centuries B.C.) to shed more light on the interaction of causal events. In the form of yearly reports, the diaries record not only astral signs and the weather, but also water levels, price fluctuations, and historical events. The goal was to identify the laws that governed the world, in order to utilize them in the realm of politics. First millennium Babylonians thus developed mathematical astronomy, a branch of Babylonian learning that has survived until today (Hunger & Pingree, 1999; Neugebauer, 1975).

These various ancient Near Eastern divinatory procedures were intended to ensure that decisions and actions by those responsible for the common good remained in harmony with the all-encompassing flow of world events, which humans ultimately cannot resist. Mesopotamia's political and cultural dominance, a centuries-old tradition that had never been seriously questioned, and the embeddedness of divination in a type of scholarly system—reinforced by the considerable expenditure required by divinatory procedures—made the success of the Mesopotamian "science of telling the future" incontrovertible in the eyes of Mesopotamians and the surrounding peoples. They were convinced that divination guaranteed a high measure of stability and prosperity, and prevented major errors of judgment, thus offering a considerable and lasting advantage over others.

As a matter of course, Assyrian and Babylonian kings attempted to monopolize the knowledge and techniques of looking into the future and to bind its best practitioners to their courts (Pongratz-Leisten, 1999). This knowledge was so highly valued, that in times of war, tablets with divinatory content were plundered by explicit royal command (Lambert, 1957/1958, p. 44; see Parpola, 1983). The knowledge that dynasties of diviners had developed and transmitted from father to son was collected, systematized, and compiled in extensive text editions in the late second and early first millennium B.C.. The impetus for this was probably the ever increasing royal demand for divinatory counsel, which kept pace with the growing complexity of Mesopotamian ruling institutions. These new editions then formed the royally authorized and authoritative lexicon of classified divinatory expertise, which specialists in the service of the king were obliged to consult. Divinatory expertise was henceforth almost solely under royal control. Diviners in the king's service were involved in highly confidential matters and were required to take an oath of silence regarding any politically charged information they might have access to through their activities (Durand, 1988, pp. 13–15; see Parpola, p. 7) The profession of divination was so well organized in the Neo-Assyrian empire of the first millennium B.C. that reports on ominous events, mainly celestial but also terrestrial, came in regularly from the entire domain (Koch-Westenholz, 1995, pp. 180–185; Oppenheim, 1969). These mutually complimentary reports were directed to a commission that one could somewhat anachronistically call "the Ministry for the Future." Here they were harmonized, checked for internal consistency, and evaluated before any resulting political measures were taken.

That this was an effective form of political decision-making is proven by Mesopotamia's 3,000-year-long political and cultural domination of the entire Near East. And yet, from a modern perspective, the basis of the divinatory evaluation process is completely obsolete. So it is troubling, even scandalous, to us that such a thoroughly nonsensical procedure (by current standards) should have afforded such lasting success. The following reflections will be devoted to this contradiction. I outline my thoughts in five points:

1. First of all I must state that the success of any prognostic procedure is little affected by the question of whether it could actually provide a glimpse into the future, as long as (1) most people believed that the applied prognostic procedure worked and (2) it did not, at least ultimately or often, interfere with sensible decision making.[4] From the early second millennium B.C. until the end of cuneiform culture both conditions seem to have held true.
2. Granted, the divinatory evaluation process entailed the potential disadvantage of not being able to carry out a sensible plan because the relevant signs speak

---

[4] This, by the way, is also true for any modern prognostic procedure.

against it. But the advantages that accompanied a plan's approval by divinatory expertise should not be underestimated. As long as the procedure was accepted as plausible, a divinatory evaluation could convincingly justify political goals and actions by showing them to be in harmony with the cosmos and the will of the gods. Divine benevolence and resultant success were thus made concretely attainable. Once consensus and a widespread sense of "Gott mit uns" (God with us) had been achieved, this led on all important levels to optimism and self-assurance, strength of purpose and readiness for action, which in turn formed a sustainable basis for a stouthearted engagement with problems when they did occur.

3. If the signs delivered a negative verdict, then it was necessary to reconsider the plan in question. This required the political decision-making committee to re-open a discussion of pros and cons. It was not unlikely that issues on which previously no consensus could be reached were once more subject to debate, to then become the object of a renewed oracular inquiry. The inquiries that have been preserved for us in connection with extispicy are true works of art, which enumerate a plan with a detailed list of the individual steps (Lambert, 2007). If a plan was decided against, it did not have to be abandoned entirely, but could be resubmitted for examination in a slightly modified form. If the proposal was then positively evaluated, it meant that the detail that had been revised in the second inquiry was responsible for the original rejection. As a consequence, it was primarily those segments of a plan causing controversy in the original draft that were reexamined.

   Divinatory evaluation then, which looks to an authority higher than any human being, opens up space for substantive discussions that are relatively free of the constraints of hierarchy. Surprisingly, at first glance, divination actually promotes compromise between competing interests. The cherished Western myth of the origin of democracy in the Greek polis impedes the insight that a culture of negotiation is not necessarily bound to the *agora* or to the institutions with which we are familiar.

4. The theistic worldview of Mesopotamia understood divinatory evaluation as a deed-consequence relationship, in which a plan's approval or rejection was inter-preted as evidence of a reward or punishment from the gods. Those in positions of leadership had to justify themselves both in the eyes of the people and of the gods. For this reason it was important for leaders to comply with demands for social justice from the religious sphere. If divination revealed the presence of a threat, and hence underlying divine wrath, then those closest to the king had to address the possibility that he had alienated the gods through ritual, personal, or some other kind of misconduct. Numerous texts show us that it was not uncom-mon for the king himself to be personally confronted with this verdict. Although detailed sources for this are—for obvious reasons—mostly lacking, this shows that the discussion of an unfavorable prediction created space in which a small circle of people could question the legitimacy of a king's action or plan.

5. Astrology, which continually generates unsolicited signs, requires that predictions continually be harmonized with the present situation, and that the present be measured against that which has been predicted. By requiring continual reflection

on political actions, astrology thus produces an atmosphere of political vigilance. The astrological predictions commissioned by the Neo-Assyrian palace concerned the internal and external security of the land, the state of provisions and the outlook for the harvest (see e.g., van Soldt, 1995 passim). It was inevitable that omens concerning national security could not be discussed or even thought about without being connected to the current situation, for the future would surely unfold from the present. Forecasts of failure and defeat thus forced the review of internal and external security, of military and security force readiness, of advisor and allies trustworthiness, of the country's provision stockpiles, and many other areas. In this way, the continual astrological analysis of the expected really was, as the texts say, "the king's watch" (e.g., Parpola, 1993, p. 111, text no. 143). By identifying negative trends even before they were noticeable or of any consequence, it fulfilled the function of a political and social early warning system.

On second thought we must concede that it would be simply unwise to dismiss ancient Near Eastern divination as mere superstition or aberration. In its day, by giving shape to the future, creating space for negotiation, and helping to build consensus, it was a decisive means of reaching political goals.

References

Brown, D. (2000). *Mesopotamian planetary astronomy astrology* (Cuneiform monographs, Vol. 18). Groningen, The Netherlands: STYX Publications.

Collins, D. (2008). Mapping the entrails: The practice of Greek hepatoscopy. *American Journal of Philology, 129*, 319–345. doi:10.1353/ajp.0.0016.

Durand, J.-M. (1988). Archives épistolaires de Mari, I/1 [Epistolary archives of Mari, I/1]. In J.-M. Durand (Ed.), *Archives royales de Mari: Vol. 26/1*. Paris: Édition Recherche sur les Civilisations.

Freedman, S. M. (1998). *If a city is set on a height: The Akkadian omen series Šumma Alu ina Mēlê Šakin* (Vol. 1, Tablets 1–21) (Occasional publications of the Samuel Noah Kramer Fund: Vol. 17). Philadelphia: Samuel Noah Kramer Fund.

Heeßel, N. P. (2008). Astrological medicine in Babylonia. In A. Akasoy, C. Burnett, & R. Yoeli-Tlalim (Eds.), *Astro-medicine: Astrology and medicine, East and West* (pp. 1–16). Florence, Italy: SISMEL, Editione del Galluzzo.

Hunger, H. (Ed.). (1992). *Astrological reports to Assyrian kings* (State archives of Assyria, Vol. 18). Helsinki, Finland: Helsinki University Press.

Hunger, H., & Pingree, D. (1999). *HdO: Astral sciences in Mesopotamia.* Handbuch der Orientalistik, Abteilung 1, 44. Leiden, The Netherlands: Brill.

Hunger, H., & Sachs, A. (1988–2006). *Astronomical diaries and related texts from Babylonia, Vol. I: Diaries from 652 B.C. to 262 B.C.* [1988]; *Vol. II: Diaries from 261 B.C. to 165 B.C.* [1989]; *Vol. III: Diaries from 164 B.C. to 61 B.C.* [1996]; *Vol. V: Lunar and planetary texts* [2001]; *Vol. VI: Goal year texts* [2006]. Vienna, Austria: Verlag der Österreichischen Akademie der Wissenschaften.

Jeyes, U. (1993). Divination as a science in ancient Mesopotamia. *Jaarbericht van het Vooraziatisch-Egyptisch Genootschap Ex Oriente Lux, 32*, 23–41.

Koch-Westenholz, U. (2005). *Secrets of extispicy: The chapter Multābiltu of the Babylonian extispicy series and Nisirti bārûti texts mainly from Aššurbanipal's library* (Alter Orient und Altes Testament, Vol. 326). Münster, Germany: Ugarit-Verlag.

Koch-Westenholz, U. (1995). *Mesopotamian astrology: An introduction to Babylonian and Assyrian celestial divination* (CNI publications, Vol. 19). Copenhagen, Denmark: Museum Tusculanum Press.

Koch-Westenholz, U. (2000). *Babylonian liver omens: The chapters Manzāzu, Padānu and Pān tākalti of the Babylonian extispicy series mainly from Aššurbanipal's library* (CNI Publications, Vol. 25). Copenhagen, Denmark: Museum Tusculanum Press.

Lambert, W. G. (1957–1958). Three unpublished fragments of the Tukulti-Ninurta epic. *Archiv für Orientforschung, 18*, 38–51.

Lambert, W. G. (2007). *Babylonian oracle questions*. Winona Lake, IN: Eisenbrauns.

Leiderer, R. (1990). *Anatomie der Schafsleber im babylonischen Leberorakel: Eine makroskopisch-analytische Studie* [Anatomy of the sheep liver in the liver oracle of Babylonia: A macroscopic analytic study]. Munich, Germany: Zuckschwerdt.

Maul, S. M. (1994). *Zukunftsbewältigung: Eine Untersuchung altorientalischen Denkens anhand der babylonisch-assyrischen Löserituale (Namburbi)* [Coping with the future: A study of ancient oriental thinking on the basis of Babylonian-Assyrian solving rituals (Namburbi)] (Baghdader Forschungen: Vol. 18). Mainz, Germany: Philipp von Zabern.

Maul, S. M. (2003). Omina und Orakel. A. Mesopotamien [Omina and oracle. A. Mesopotamia]. In D. O. Edzard & M. P. Streck (Eds.), *Reallexikon der Assyriologie und Vorderasiatischen Archäologie,* (Vol. 10, pp. 45–88). Berlin, Germany: De Gruyter.

Maul, S. M. (2013). *Die Wahrsagekunst im Alten Orient. Zeichen des Himmels und der Erde,* München: Beck.

Meyer, J.-W. (1987). *Untersuchungen zu den Tonlebermodellen aus dem Alten Orient* [Studies of the clay liver models from the Old Orient] (Alter Orient und Altes Testament: Vol. 39). Kevelaer, Germany: Butzon & Bercker.

Moren, S. M. (1978). *The omen series "Shumma alu": A preliminary investigation* (Doctoral dissertation). Philadelphia: University of Pennsylvania.

Neugebauer, O. (1975). *A history of ancient mathematical astronomy*. Berlin, Germany: Springer.

Nougayrol, J. (1941). Textes hépatoscopiques d'époque ancienne conservés au musée du Louvre [Hepatoscopic texts from ancient times in the Musée du Louvre]. *Revue d'assyriologie et d'archéologie orientale, 38,* 67–88.

Nougayrol, J. (1968). Le foie "d'orientation" BM 50494 [The liver "guide" BM 50494]. *Revue d'assyriologie et d'archéologie orientale, 62,* 31–50.

Oppenheim, A. L. (1996). *Ancient Mesopotamia: Portrait of a dead civilization* (Revised ed. completed by Erica Reiner). Chicago: University of Chicago Press.

Oppenheim, A. L. (1969). Divination and celestial observation in the last Assyrian empire. *Centaurus, 14*, 97–135.

Parpola, S. (1983). Assyrian library records. *Journal of Near Eastern Studies, 42,* 1–29.

Parpola, S. (Ed.). (1993). *Letters from Babylonian and Assyrian scholars* (State archives of Assyria, Vol. 10). Helsinki, Finland: Helsinki University Press.

Pfiffig, A. J. (1975). *Religio etrusca* [The Etruscan religion]. Graz, Austria: Akademische Druck- und Verlagsanstalt.

Pongratz-Leisten, B. (1999). *Herrschaftswissen in Mesopotamien: Formen der Kommunikation zwischen Gott und König im 2. und 1. Jahrtausend v. Chr.* [The knowledge of rule in Mesopotamia: Forms of communication between god and king in the 2nd and 1st millennia B.C.] (State Archives of Assyria Studies: Vol. 10). Helsinki, Finland: Neo-Assyrian Text Corpus Project.

Reiner, E. (1995). Astral magic in Babylonia. *Transactions of the American Philosophical Society, 85*(4), 1–150.

Rochberg-Halton, F. (2004). *The heavenly writing: Divination, horoscopy, and astronomy in Mesopotamian culture*. Cambridge, UK: Cambridge University Press.

Sasson, J. M. (Ed.). (1995). *Civilizations of the ancient Near East*. New York: Scribner.

Starr, I. (1990). *Queries to the Sungod: Divination and politics in Sargonid Assyria* (State archives of Assyria, Vol. 4). Helsinki, Finland: Helsinki University Press.

Starr, I. (1983). *The rituals of the diviner* (Bibliotheca mesopotamica, Vol. 12). Malibu, CA: Undena Publications.

Thulin, C. O. (1968). *Die etruskische Disziplin* [The Etruscan discipline]. Darmstadt, Germany: Wissenschaftliche Buchgesellschaft.

van Soldt, W. H. (1995). *Solar omens of Enuma Anu Enlil: Tablets 23 (24)–29 (30)* (Uitgaven van het Nederlands Historisch-Archeologisch Instituut te Istanbul, Vol. 73). Istanbul, Turkey: Nederlands Historisch-Archaeologisch Instituut.

von Weiher, E. (1993). *Uruk, Spätbabylonische Texte aus dem Planquadrat U 18, 4: Ausgrabungen in Uruk Warka*. Endberichte herausgegeben von Rainer Michael Boehmer, 12. [Uruk, late Babylonian texts from plan grid square U 18, Part 4: Excavations in Uruk Warka (Final reports, R. M. Boehmer, Ed.)], Vol. 12. Mainz, Germany: Philipp von Zabern.

Who Gets the Past? The Changing Face of Islamic Authority and Religious Knowledge

6

Dale F. Eickelman

Historians and sociologists often take at face value the ideological claim in Islam of the fixed nature of religious knowledge. Consequently, they give less attention to how such a system of knowledge is affected by changing modes of transmission, who takes part in the increasingly widespread debates over what is valued knowledge, and how these debates have shifted over time.

Competing Claims to Authoritative Religious Knowledge

The problem of defining what knowledge is valued and how it relates to faith and authority is increasingly a subject of intense debate in Muslim societies. Innovation—even when denied outright—can emerge from surprising quarters. In March 2009, for example, conservative religious scholars in Saudi Arabia argued in the local Arabic press that the secluding and covering of women was an innovation (Arabic, *bid'a*) that was not practiced in the time of the Prophet Muhammad and therefore could not be considered "Islamic." A Kuwaiti scholar (Alatiqi, 2009), entrusted as a government official with enforcing gender separation at private universities in his country, offered in his private capacity a powerful public version of the same argument. Trained not in the religious sciences but rather as a civil engineer, Alatiqi based his argument on a consideration of the recognized sources of "authentic" Islamic tradition—including the Qur'an, the sayings (*hadith*) of the Prophet Muhammad, and accounts of the Prophet's life.

Alatiqi, like members of the Kuwaiti parliament who enacted university gender separation regulations in the first place, bases his argument on claims about what happened in the past, particularly in the time of the Prophet Muhammad (d. 632).

D.F. Eickelman (✉)
Department of Anthropology, Dartmouth College,
Silsby Hall 6047, Hanover, NH 03755, USA
e-mail: dale.f.eickelman@dartmouth.edu

© Springer Netherlands 2015
P. Meusburger et al. (eds.), *Geographies of Knowledge and Power*,
Knowledge and Space 7, DOI 10.1007/978-94-017-9960-7_6

The fact that one can separate official duties from private opinions in Kuwaiti and Saudi public space also indicates the new settings in which beliefs and practices can be argued in public.

To use a term made popular by Oxford philosopher W. B. Gallie (1968), "innovation" in the Islamic tradition is an "essentially contested concept." Innovation concerns not only the content and context of ideas and practices, but also who takes part in the discussion about them and who is influenced by those discussions. In the Islamic tradition, the easiest way to claim legitimacy for innovation is to deny that it has taken place (see, e.g., Kamarava, 2011). Like concepts of "good governance," "duty," and "social justice," innovation in Islamic thought and practice is impossible to define once and for all. People can justify why they hold one interpretation over others, and authorities can attempt to block public debate, but the "proper" meaning of an essentially contested concept cannot by definition be settled once and for all.

The clarification of such claims to fixed religious knowledge involves considering how differing parties have used the concept throughout its history. The uncontested experts once were the *'ulama*, or men of learning, the generally recognized authorities of prior generations. Yet, as the Sorbonne-educated Sudanese lawyer and politician al-Turabi (1983, p. 245) has argued, all knowledge is "divine and religious," so that all those who possess knowledge (*'ilm*) are the equals of those who possess specialist religious knowledge.

This view is still strongly contested. For example, Sa'id Ramadan al-Buti (d. 2013), a Syrian religious scholar and television preacher, argued that just as one goes to an architect for a building and a medical doctor for illness, one goes to a properly trained specialist for religious questions (personal communication, Damascus, August 12, 1999). The addition of women to these debates further shapes the field of what is no longer taken for granted.

Struggles for control of the mantle of religious and political authority in Muslim-majority societies are often phrased in opaque interpretations, blurring lines between tradition and modernity and concealing the vigor of the underlying debates. This opacity is quickly becoming transparent through new media, which enable key religious leaders to be regularly seen on satellite television and in streaming video. Disciples and coworkers regularly post catechism-like documents and Web links, answers to religious questions, and simplifications of complex arguments in multiple languages to expand the reach of their *shaykh* (see, e.g., http://naseemalsham.com).

In the prescient words of Castells (1996, p. 373), the new media have increasingly become a "real virtuality"—not just a channel through which the appearance of reality is communicated, but experience itself. Since at least the mid-twentieth century, the increased availability of mass education, especially mass higher education, the greater ease of travel, and new communication technologies have reshaped struggles over religious and political authority in South and Southeast Asia, the Middle East, Turkey, and North Africa even as the protagonists in these struggles claim to sustain old ideas and practices.

In any challenge to political and religious authority, incumbents decidedly have the advantage. Nonetheless, in the hotly contested Iranian elections of June 12, 2009, and in the "Arab Spring" demonstrations from 2011 onward, opposition effectively mobilized and reacted to government actions via mobile telephones, the Internet, Twitter, YouTube, Facebook, and text messages, as well as older forms of communication. State authorities try to block subversive communications, but the ensuing cat-and-mouse games between those authorities and their opposition have become increasingly fragmented and multidimensional.

Public Islam and the Common Good

The notion of "public Islam" refers to the highly diverse invocations of Islam as ideas and practices that religious scholars, self-ascribed religious authorities, secular intellectuals, members of Sufi orders, mothers, students, workers, engineers, and many others make in public life. These debates make a difference in configuring the politics and social life of large parts of the globe. They make a difference not only as a template for ideas and practices but also as a way of envisioning alternative political realities and, increasingly, in acting on both global and local stages, thus reconfiguring established boundaries of civil and social life.

Advancing levels of education, greater ease of travel, and the rise of new communications media throughout the Muslim-majority world have contributed to the emergence of a public sphere in which large numbers of people, and not just an educated, political, and economic elite, want a say in political and religious issues. The result has been to challenge authoritarianism, fragment religious and political authority, and increasingly open discussion of issues related to the "common good" (*al-maslaha al-'amma*), an essentially contested concept that is at the core of public life in Muslim-majority countries. The trend toward this greater openness and inclusion has, however, been uneven and often contradictory.

Not all of these trends are unique to the modern world. Cook's (2000) majestic study of "commanding right and forbidding wrong" in Islamic thought from the early Islamic centuries to the present depicts how issues of the common good and community responsibilities have engaged both Muslim jurists and a wider Muslim public well before the last two centuries. As in the present, some fundamentalists seek solace in literal attempts to imitate the life of the Prophet Muhammad. Others emphasize the necessity of interpreting the Qur'an as if it were revealed in the present and in interpreting the life and sayings of the Prophet metaphorically and not literally, engaging critical reason. This approach underlies that of the Andalusian jurist Abu Ishaq al-Shatibi (d. 1388; Masud, 1995) as much as it does the writings of the Syrian engineer Shahrur (2009), whose published work since 1990 in Arabic has gained an increasingly significant audience in the Arab world and, in translation, elsewhere.

Many of the emerging new voices and the leaders of movements within the proliferating public space of the contemporary Muslim world—a social location which is simultaneously physical and virtual—claim authoritatively to interpret basic religious texts and ideas, and work in local or transnational contexts. These new interpreters of how religion shapes, or should shape, societies and politics, like their counterparts in Poland's Solidarity movement and the liberation theology movements in Latin America in the 1980s, often lack the technical textual sophistication of the religious scholars of earlier eras who previously led such discussions. Such new leaders and spokespeople have nonetheless succeeded in capturing the imagination of large numbers of people. These trends often intensify the ties that bind Muslim communities in the Muslim-majority world with Muslims in Europe, North America, and elsewhere in the world.

The issues and themes in Muslim politics increasingly transcend the specifics of region or place. Thus the contemporary "publicization" of Islam is more commonly rooted in communicative practice than in formal ideology (Adelkhah, 2002). It has created new social spaces, a trend significantly accelerated since the mid-twentieth century, and facilitated modern and distinctively open senses of political and religious identity.

Such practices involve both emotional and intellectual engagement among participants in overlapping circles of communication, solidarity, and the building of bonds of identity and trust. Some of these circles are based on local communities. Others are geographically diffuse yet targeted to receptive audiences. One example is the use of e-mail among the Indonesian university students who coordinated the nationwide campus protests that contributed to the downfall of President Suharto in 1998, a use of technology that seems archaic in light of the use of newer media in Iran, Jordan, Pakistan, and Morocco since then. These modern practices and new communication technologies create new and effective bases for effective mobilization that are not dependent on geographical propinquity. At the same time, they can threaten tolerance and civil society by facilitating publicity and calls to action by extremist groups (Hefner, 2003).

Social practices that are based on ideas of the common good and that contribute to shaping public Islam include collective rituals, such as popular festivals and religious and secular commemorations. They also encompass disciplining and performance practices as diverse as Sufi rituals, regional pilgrimages, the informal economy, the routines of modern schooling, and the use of the press and modern communications technologies.

Public Islam and Modernity

Mid-twentieth century theories of modernity and modernization assumed that religious movements, identities, and practices had become increasingly marginal and that only religious intellectuals and leaders who attached themselves to the nation-state would continue to play a significant role in public life. Assertions about the eclipse of religion in the public life of North America and Europe were exaggerated. Casanova (1994) was one of the first to remind us of several major developments in the 1970s that challenged the idea of the eclipse of religion in public life: the Iranian revolution, the rise of the Solidarity movement in Poland, the role of liberation theology in political movements throughout Latin America, and the return of Christian fundamentalism as a force in American politics.

In the Muslim-majority world, however, the role of religion in society and community life never receded, though it did change and develop in ways often underemphasized by Western observers and by Muslims themselves (Zaman, 2002). Only since the mid-1990s has the idea of an "Islamic public sphere"—*Islamische Öffentlichkeit* in German—come to the fore. Schulze (1995/2000), responding to the work of Jürgen Habermas, discerned this phenomenon as forming the infrastructure of communication and discourse of a new intellectual class that had emerged from the classic era of Islamic reform in the late nineteenth century through the structural transformations of the 1960s and 1970s.

A trope in the Muslim-majority world is to claim that these ideas of the common good are a return to an immutable heritage of religious or normative traditions fixed by Muhammad in seventh-century Arabia. They are not. They are defined by ethical notions and social values contested and redefined through interaction, practice, and transmission over generations.

In a parallel way, sectarianism in Christian Europe provided the habitus and congregational form for developing ideas of the public. It is possible to see in the Sufi tradition and other Muslim religious practices a similar contribution to learning how to participate in the public sphere. Like the Christian sects, the more orthodox forms of Sufism and other styles of public piety have contributed to shaping reasoning selves and to reconfiguring the relationship between legitimate authority and independent pursuit of truth. Public reasoning has a long tradition in Islamic jurisprudence. However, both Sunni and Shia awareness of this tradition is deflected by claims that anything new actually originated in the valued past of the time of the Prophet Muhammad.

As Casanova (1994) argues, various sectarian movements in Europe played a major role in developing the idea of the modular self, empowered with a moral conscience and confronting the authority both of established religion and of the state. According to this European trajectory, only when the freedom of individual conscience is recognized and tolerated can a public sphere develop. Nonetheless, religious ideas and practices can similarly foster the emergence of a public sphere.

Ideas of the public are historically situated and have strong links with culturally shared senses of self and community. They are located at the strategic intersection of practice and discourse. A recent book in France, *Penser le Coran* (or "Thinking the Qur'an"; Hussein, 2009), persuasively indicates how Qur'anic revelation is situationally linked to the understanding of revelation in seventh-century Arabia. In the context of the contemporary state, techniques of authority, persuasion, and control are also historically situated. Modern techniques often promote a secular outlook of citizenship and social membership, but these ideas exist alongside religious traditions and the emergence of new socioreligious discourses and leaderships that intersect with and challenge nation-state projects. In Morocco, for example, there is resurgent interest among the middle classes in collective Qur'anic chanting and the recitation of Sufi poetry, often composed by "pious ones" (*salihun*, or saints) known equally for their piety and their religious knowledge. The popularity of such piety pervades all social classes, and rural as well as urban milieus. Visitors to the royal compound in Rabat quickly note that only two ministries are situated within it—the Ministry of Defense and the Ministry of Pious Endowments and Religious Affairs.

Religious and Secular Identities

How does this continued pervasiveness of religious ideas and practice match views of the public sphere that are premised on the existence of religiously neutral or "secular" access to public debate? Some ideas of the "secular" divest participants in public exchanges of their religious and cultural identities, or at least marginalize these identities. However, the creation of a public culture promoting exchange and discussion can also build on traditions of religious faith and practice. Such traditions can also encourage the gradual emergence of ever more abstract patterns of membership and citizenship that rest on obligations and rights which increasingly fit a legal vocabulary and a contractual view of society.

Such developments including the discontinuities between tradition and modernity created by the emergence of a "culture of publicness," have been the focus of interest of political philosophers, social scientists, and historians alike. It suffices here to mention such diverse authors as Giambattista Vico, Adam Smith, Immanuel Kant, Alexis de Tocqueville, Ferdinand Tönnies, and John Dewey. These thinkers have concentrated on developments in Europe and North America, developments that are specifically Western but regarded as exemplary of universal trends. As John Agnew argues in this volume, the unexamined assumption that European and North American views are universal is all too common.

In spite of the growing recognition that religion plays an important role in public life and can contribute to the common good, it remains necessary to challenge the common assumption that secularism and secularly oriented practical rationality constitute the exclusive normative base for "modern" public life (Eickelman, 2000; Salvatore, 1997, 2001). Religious thought and practice in the Muslim world can inspire rational-practical orientations as much as do secular approaches to social action.

For both the nineteenth century and the contemporary era, it is possible to identify the norms of exchange and discourse that are the product of these interactions and clashes, and also the emergence of explicit and implicit Muslim forms of civility and publicness. Identifying these norms requires an effort to discern the social history, or genealogy, of the emergence of a sense and structure of public communication and participation in societies shaped by Muslim cultural, religious, and political traditions.

The present period differs from earlier ones in the speed, intensity, and large numbers of people involved in shaping the contours of tradition, but the publics of an earlier era were equally engaged in doing so. The reshaping of religious identity and forms of communication and publicness in the nineteenth-century Ottoman Empire is especially salient in this respect. Consider, for example, Istanbul, a city inhabited by a religiously, ethnically, and linguistically diverse population that outnumbered the Muslims for a good part of the Ottoman era. The most commonly held assumption is that the confessional communities of the empire lived separately, with minimal interaction, and developed social bonds and allegiances exclusively within their own communities.

This assumption fails to appreciate the mobile and relational aspect of community relations in Ottoman Istanbul, and it says little about the people's sense of identity and of collective allegiance. Re-examining the ongoing transformations of the Ottoman Empire from the nineteenth century to the present facilitates a better grasp of the possibilities for change in the contemporary Muslim-majority world (see, e.g., Çinar, 2001; Frierson, 2004; Meeker, 2002).

The collective historical experience of coexistence among Muslims and non-Muslims in the Ottoman Empire can be analyzed on the basis of their common interests as members of a vibrant society. In India, the relation between Hindus and Muslims is crucial to the development of ideas of secularism and religiosity in relation to the public sphere (Ahmad, 2009). In such an historical and interreligious perspective, forms of public Islam in the twentieth century appear as contingent crystallizations of much more complex historical processes that were present in earlier periods. For example, imperial encounters have been of great importance in the historical development of public debate in the metropole as well as the colony—a circumstance that the U.S. occupation of Iraq in 2003 and increased involvement in Afghanistan from the Soviet invasion of 1979 onward bring once again to light.

Notwithstanding their diversity of historical experience, most Muslims share inherited conceptions of the common good, and these ideas from the past shape contemporary understandings of publicness in Muslim societies (see Eickelman & Salvatore, 2004, pp. 15–20). For example, Islamic religious scholars, the 'ulama, claim that God reveals ideas of the common good to humankind. Yet these scholars also regard themselves alone as capable of discerning these ideas through their expertise in the science of scriptural hermeneutics. However, their agreement about the common good and how to understand the past still lead to vigorous debate. Moreover, Muslims increasingly are disinclined to allow conventionally trained religious scholars the final word in interpreting such vital questions as "What is Islam?" "How is it important to my life?" and "How do I interpret the past?" Participants in these debates may assert universal scope, but all such claims are locally situated, such as Tarek Fatah's vigorous attacks on adherents of the ideal of an Islamic state both in the present and since the death of the Prophet Muhammad in AD 632 (Fatah, 2008).

As the writings of Fatah—a self-described left-wing student leader and later a journalist in Pakistan who is now a Canadian—and many others make clear, interpreting the Islamic past as a means to legitimate the present is too important a task to be left to conventional Islamic scholars or to received wisdom. The authority of conventional religious scholars remains strong in the modern world but is increasingly challenged by alternative religious authorities who often lack formal training in the traditional religious sciences. Even the constitution of the Islamic Republic of Iran is based on two conflicting principles, the absolute sovereignty of God (Principles 2 and 56) and the people's right to determine their own destiny (Principle 3:8) (Islamic Republic of Iran, 1980), thus opening the door to wide debate over issues of government and society. Within Sunni Islam, it is also becoming increasingly common for lay personalities to lead the Friday prayers at mosques. Thus, like the state, the 'ulama rarely maintain a monopoly over the implicit understandings and formal ethical pronouncements guiding the Muslim community. Morocco's Minister of Pious Endowments and Religious Affairs since 2002 was trained as an historian, not as a religious scholar, and his writings include novels, not religious treatises.

The increasing accessibility of new media, including satellite television and the Internet, and new uses of older media such as video- and audiocassettes and CDs contribute to the fragmentation of the traditional structures of religious authority. It also facilitates innovative ideas on religious authority and representing Islam in public in unexpected ways (Gonzalez-Quijano, 2003; Gonzalez-Quijano & Guaaybess, 2009; Hefner, 2003). There are numerous combinations of fragmented and sustained old and new forms of religious authority and influence in the public sphere, making debates about what constitutes "good" or authentic Islam much more contentious than has been the case in the past.

One paradox of modern Muslim publics is that despite the discursive expansion in many Muslim-majority states and communities, which includes respect and tolerance for non-Muslim "others," the public good is increasingly defined within the parameters of Islam. Some states, such as republican Turkey, vigorously sought to domesticate and neutralize Islamic institutions and ideas in the first half of the twentieth century, yet mutual accommodation and tacit bargaining among proponents of the different alternatives define the main approaches to current Turkish politics. The guardians of secularism and those who participate in Turkey's public sphere and civic life learn mutual accommodation through public debate and practice (White, 2002). As Adelkhah (2004) suggests for Iran, the most powerful achievement of the women's movement is not formal and recognized organizations, all monitored and repressed by the state, but women's activities in the informal economy and in shaping religious practices. As in the French Revolution, Adelkhah argues that such "informal" activities can be at least as powerful a vehicle for changing gender roles and ideas of Islam as explicit ideological statements and formal organizations. In all cases, Islamic ideas of the common good shift in content and elaboration over time and, despite explicit denials, may often converge with Western understandings of such major issues as democracy and tolerance for religious diversity (Hefner, 2000; Sulaiman, 1998). Thus the role of Islam in shaping understandings of the common good is unlikely to recede in importance in the years to come.

Muslims participate in crafting the idea of the common good in a variety of ways, and they also contribute to shaping the definitions of wider and more inclusive publics in societies where they are not a majority, as in Europe (Kepel, 1994/1997; Khosrokhavar, 1997; Schiffauer, 2001); or, as in Syria and Turkey, where they are confronted with a profoundly secular elite; or, as in Iran, with an increasingly unpopular, although powerful, clerical elite (Adelkhah, 2004). In India, Muslims live in a secular state strongly buffeted by religious extremism (van der Veer, 1994). Such historically situated and contemporary discourses speak against efforts to find a single, overarching idea of the common good shared by all Muslim societies, even if some ideologues—both those claiming to represent Islam and those attacking it—make such essentializing claims. It is often the case that such discussions or conflicts about what "good" or "true" Islam entails disrupt implicit conceptions of the public sphere, as in many communities throughout the Muslim world. These debates and the contexts in which they occur throw into relief competing claims to speak in public, revealing threads of consensus and points of divergence or rupture.

Authorities and Audiences

The participation of religious authorities in public religious debate cannot be understood without an analysis of the audiences to which their discourses are directed and the elements that connect the followers of religious leaders to their persona. New media, including sermons on tape, popular journals, and local radio broadcasts, may combine with more conventional media (including gossip, published fatwas, and religious interpretations) to broaden spheres of participation and make them more complex. The degree to which the participation or influence of these new audiences alters conceptions and implementation of the common good, however, is a question that must always be asked rather than assumed (Eickelman & Salvatore, 2004, pp. 15–20). New authorities or speakers emerge in the space between the state and more traditional religious authorities, and thus come to represent alternative sites of power.

Religious authorities can be an essential part of the construction of public religious discourse. For example, the participation of Sufis in public religious debate combines modern forms of conceptualizing and presenting religious arguments with membership in a hierarchical and intensely personalized religious framework. Public articulation of the common good does not require the equality of all participants in order to raise a claim to truth and justice. The relationship between religious authority—whether claimed by traditional religious scholars or by "new" religious intellectuals (Roy, 1992/1994)—and the public sphere is profoundly ambiguous and more complex than conventional Habermasian theories would have us believe. Even in places where there is a state-sponsored Islamic ideology, as in Pakistan and Iran, individuals, groups, and communities often appropriate this ideology—or strive to disregard it—in order to reinforce their position in public religious debate by claiming Islamic credentials rooted in the historical past for defining the common good, or by furthering particular interests in the guise of shared ones, a strategy prevalent in public spheres everywhere.

Well before September 2001, the growing number of Muslims in Europe and North America began to foreground questions about national identity, citizenship, and multiple loyalties, as Muslims in France and Germany did before them. Events since then have further illuminated the vulnerability of, and misconceptions about, Muslims living in Europe and North America. This situation has at times led to efforts to organize for more effective participation in the political life of the societies in question; at other times it has led to waves of self-estrangement, exposing the fragility of multicultural discourse. Even in such a predicament, however, a positive outcome of double estrangement within the home and the receiving societies is to encourage engagement with transnational Muslim causes, especially where Muslims are the victims of human rights abuses.

In short, there is no singular public Islam, but rather a multiplicity of overlapping forms of practice, discourse, and invocations based on readings of the past. The competing claims represent the varied historical and political trajectories of Muslim communities and their links and influences with societies elsewhere. Debates about the common good encompass both words and actions. In spite of competing claims to represent the past authentically, these representations are profoundly shaped by new practices, new forms of publication and communication, and new ways of thinking about religious and political authority.

References

Adelkhah, F. (2002). *Being modern in Iran*. New York: Columbia University Press.

Adelkhah, F. (2004). Framing the public sphere: Women in the Islamic Republic. In A. Salvatore & D. F. Eickelman (Eds.), *Public Islam and the common good* (pp. 227–241). Leiden, The Netherlands: Brill.

Ahmad, I. (2009). *Islamism and democracy in India: The transformation of the Jamaat-e-Islami*. Princeton, NJ: Princeton University Press.

Alatiqi, I. (2009, March). *Male-female encounters in early Islamic society: Examples and a case study*. Presented at the conference "University development and critical thinking: Education in the Arabian Peninsula for a global future," Kuwait.

al-Turabi, H. (1983). The Islamic state. In J. L. Esposito (Ed.), *Voices of resurgent Islam* (pp. 241–251). New York: Oxford University Press.

Casanova, J. (1994). *Public religions in the modern world*. Chicago: University of Chicago Press.

Castells, M. (1996). *The rise of the network society*. Boston: Blackwell.

Çinar, A. (2001). National history as a contested site: The conquest of Istanbul and Islamist negotiations of the nation. *Comparative Studies in Society and History, 43*, 364–391. doi:10.1017/S0010417501003528.

Cook, M. (2000). *Commanding right and forbidding wrong in Islamic thought*. Cambridge, UK: Cambridge University Press.

Eickelman, D. F. (2000). Islam and the languages of modernity. *Daedalus, 129*(1), 119–133.

Eickelman, D. F., & Salvatore, A. (2004). Muslim publics. In A. Salvatore & D. F. Eickelman (Eds.), *Public Islam and the common good* (pp. 3–27). Leiden, The Netherlands: Brill.

Fatah, T. (2008). *Chasing a mirage: The tragic illusion of an Islamic state*. Mississauga, Canada: Wiley.

Frierson, E. (2004). Gender, consumption, and patriotism: The emergence of an Ottoman public sphere. In A. Salvatore & D. F. Eickelman (Eds.), *Public Islam and the common good* (pp. 99–125). Leiden, The Netherlands: Brill.

Gallie, W. B. (1968). *Philosophy and the historical understanding*. New York: Shocken.

Gonzalez-Quijano, Y. (2003). The birth of a media ecosystem: Lebanon in the internet age. In D. F. Eickelman & J. W. Anderson (Eds.), *New media in the Muslim world: The emerging public sphere* (2nd ed., pp. 61–79). Bloomington, IN: Indiana University Press.

Gonzalez-Quijano, Y., & Guaaybess, T. (Eds.). (2009). *Les Arabes parlent aux arabes* [Arabs speaking to Arabs]. Paris: Sindbad.

Hefner, R. W. (2000). *Civil Islam: Muslims and democratization in Indonesia*. Princeton, NJ: Princeton University Press.

Hefner, R. W. (2003). Civic pluralism denied? The new media and *Jihadi* violence in Indonesia. In D. F. Eickelman & J. W. Anderson (Eds.), *New media in the Muslim world: The emerging public sphere* (2nd ed., pp. 158–179). Bloomington, IN: Indiana University Press.

Hussein, M. (2009). *Penser le Coran* [Thinking the Qur'an]. Paris: Grasset.

Islamic Republic of Iran. (1980). Constitution of the Islamic Republic of Iran (ratified December 2–3, 1979). *Middle East Journal, 34*, 181–204.

Kamarava, M. (Ed.). (2011). *Innovation in Islam: Traditions and contributions*. Berkeley, CA: University of California Press.

Kepel, G. (1997). *Allah in the West: Islamic movements in America and Europe* (S. Milner, Trans.). Palo Alto, CA: Stanford University Press. (Original work in French published 1994)

Khosrokhavar, F. (1997). *L'Islam des jeunes* [The Islam of the young]. Paris: Flammarion.

Masud, M. K. (1995). *Shatibi's philosophy of Islamic law*. Islamabad, Pakistan: Islamic Research Institute.

Meeker, M. E. (2002). *A nation of empire: The Ottoman legacy of Turkish modernity*. Berkeley, CA: University of California Press. doi:10.1525/california/9780520225268.001.0001.

Roy, O. (1994). *The failure of political Islam* (C. Volk, Trans.). Cambridge MA: Harvard University Press. (Original work in French published 1992)

Salvatore, A. (1997). *Islam and the political discourse of modernity*. Reading, UK: Ithaca Press.

Salvatore, A. (2001). Introduction: The problem of the ingraining of civilizing traditions into social governance. In A. Salvatore (Ed.), *Muslim traditions and modern techniques of power, Yearbook of the sociology of Islam 3* (pp. 9–42). Hamburg, Germany: Lit Verlag.

Schiffauer, W. (2001). Production of fundamentalism: On the dynamics of producing the radically different. In H. de Vries & S. Weber (Eds.), *Religion and media* (pp. 435–455). Palo Alto, CA: Stanford University Press.

Schulze, R. (2000). *A modern history of the Islamic world*. New York: New York University Press. (Original work published 1995)

Shahrur, M. (2009). *The Qur'an, morality, and critical reason: The essential Muhammad Shahrur* (A. Christmann, Trans., Ed.). Leiden, The Netherlands: Brill.

Sulaiman, S. J. (1998). Democracy and *shura*. In C. Kurzman (Ed.), *Liberal Islam: A sourcebook* (pp. 96–98). New York: Oxford University Press.

Van der Veer, P. (1994). *Religious nationalism*. Berkeley, CA: University of California Press.

White, J. B. (2002). *Islamist mobilization in Turkey: A study in vernacular politics*. Seattle, WA: University of Washington Press.

Zaman, M. Q. (2002). *The ulama in contemporary Islam: Custodians of change*. Princeton, NJ: Princeton University Press.

"An Heavenly Kingdom Shall Descend": How Millennialism Spread from New England to the United States of America

7

Robert Jewett

Introduction

This volume on knowledge and power takes account of cultural factors in the assessment of geography. This chapter explains how millennialism—a theory about a 1,000-year kingdom and its relation to the end of time—influenced early American colonists. Timothy Dwight's poem "America" includes the line quoted in the chapter's title and presents the puzzle of its millennial vision of America as the "heavenly kingdom," as derived from Revelation 20.[1] By the eve of the American Revolution, this sense of being the nation destined to usher in the millennial age was clearly developed (Ahlstrom, 1972, p. 52). Timothy Dwight's poem "America," published in 1771 (Dwight, 1969; quoted by Tuveson, 1968, pp. 105–106), described the hopeless state of the world before the discovery of the new promised land and set forth the promise of the millennial peace that would soon be administered by the saints in America.[2] Dwight was a Congregational minister and poet (1752–1817) who became president of Yale University (see Dowling, 1999, pp. 192–194). His poem (see Dwight, pp. 11–12) was widely cited, which suggests that its breathtaking geographic extension was considered self-evident.

> With Freedom's fire their gen'rous bosoms glow'd,
> Warm for the Truth, and zealous for their God....
> By these inspired, their zeal unshaken stood,

---

[1] This essay draws on material from my study *Mission and Menace: Four Centuries of American Religious Zeal* (Minneapolis: Fortress Press, 2008); German translation published by Vandenhoeck & Ruprecht.

[2] The material in this section is adapted from the book *Captain America and the Crusade Against Evil: The Dilemma of Zealous Nationalism* (Jewett & Lawrence, 2003, pp. 55, 57–58).

R. Jewett (✉)
Department of Scientific Theology, University of Heidelberg,
Kisselgasse 1, 69117 Heidelberg, Germany
e-mail: robert.jewett@wts.uni-heidelberg.de

© Springer Netherlands 2015

P. Meusburger et al. (eds.), *Geographies of Knowledge and Power*,
Knowledge and Space 7, DOI 10.1007/978-94-017-9960-7_7

And bravely dar'd each danger—to be good. [...]
Hail Land of light and joy! Thy power shall grow
Far as the seas, which round thy regions flow;
Through earth's wide realms thy glory shall extend,
And savage nations at thy scepter bend ...
No more shall War her fearful horrors found,
Nor strew her thousands on th' embattled ground ...
Then, then an heavenly kingdom shall descend,
And Light and Glory through the world extend.
And every region smile in endless peace;
Till the last trump the slumbering dead inspire,
Shake the wide heavens, and set the world on fire.

The idea of the heavenly kingdom descending to earth at the conclusion of the battle of Armageddon comes from the book of Revelation. The glory and power will extend as "far as the seas," and "savage nations" will submit to the rule of the saints. Such peace, of course, could come only through violence that sets "the world on fire." Dwight pictures the American troops as joining with the heavenly host in the manner of the ancient Israelite ideology (see Dwight, 1969, p. 10). This idea spread from New England to the other colonies, which resulted in the establishment of the United States as the "new order of the ages," to use the wording of the national seal. How this came about is the subject of this essay.

The colonies outside New England as of 1740 were mostly non-millennial, and the sober religious orientation of their populations was generally consistent with the anti-millennial nature of European religion. New York, New Jersey, the Carolinas, and Delaware were settled for commercial purposes; Virginia's planter culture was Anglican and as uninterested in millennial politics as were the Lutheran and Mennonite immigrants in Pennsylvania. How could this European legacy have been replaced by Dwight's millennialism? How can we account for its geographic extension from New England to the rest of the colonies that formed the United States of America?

From 1789 to the present day, this millennialism remains evident in the peculiar orientation of Americans toward the future and peculiar belief in their alleged innocence and power, including their widespread conviction that they are in some sense a chosen people, destined to exercise global leadership. I begin my investigation at the point of origin, the Puritan colonies of New England.

New England as a Millennial Seedbed

Like other Puritans, the Massachusetts Bay colonists had an apocalyptic view of history, which held that their colony would play a central role in the final drama of world history (see Spillmann, 1984, pp. 55–73). New England saw itself as the millennial Protestant realm that fulfilled the ideal of the heavenly kingdom descending to earth, as in Revelation 20. A distinctive sense of mission to redeem the entire world marked the first generation of emigrants in New England. The Puritans derived from the book of Revelation and portions of the Old Testament their dualistic worldview and their belief that violence would inaugurate God's kingdom. They thought of themselves as standing in the succession of Christian warriors and martyrs from the Bible down to seventeenth-century England. Between 1630 and 1640, when their cause was in decline in England, thousands of Puritans emigrated to New England with this mission in mind. It was the call to battle that quickened their spirits, and they were fully convinced that such warfare had to be waged in the civil realm against the forms of corruption they felt were afflicting England. John Fiske said they were animated with "the desire to lead godly lives and to drive out sin from the community" (Fiske, 1889, p. 147). Their hope was that with the successful completion of such a purifying campaign, the millennial kingdom promised in the book of Revelation would surely arrive.

The idea of a 1,000-year kingdom that would follow a great battle between the forces of God and the forces of the demonic realm originated in Jewish writings (Massyngbaerde Ford, 1992, pp. 832–834; for a helpful discussion of classic texts, see also Aune, 1999, pp. 136–137); it is mentioned only once in the Bible (Revelation 20:1–7), in one of the most influential passages in Holy Writ as far as American religion is concerned. In this grandiose vision of the future, after the great battle that destroys his forces, Satan would be disabled for a 1,000 years while the saints ruled the earth. The Puritans combined this idea of a millennial kingdom with their campaign against monarchy and episcopacy in England. Michael Walzer (1965) pointed out the decisive role of such ideas in the creation of the Puritan radicals:

> What finally made men revolutionaries, however, was … an increasingly secure feeling that the saints did know the purposes of God…. Beginning at some point before 1640, a group of writers, including Joseph Meade of Cambridge University, began the work of integrating the spiritual warfare of the preachers with the apocalyptic history of Daniel and Revelation. The religious wars on the continent and then the struggle against the English king were seen by these men as parts of the ancient warfare of Satan and the elect, which had begun with Jews and Philistines and would continue until Armageddon. (p. 291)

The zealous leaders whom Walzer described had shifted the 1,000-year kingdom of Revelation 20 from the past to the immediate future and had reinterpreted the role of the saints in martial categories. Thus, when the revolution came in England, preachers rose in Parliament to proclaim that the final battle with Satan was at hand. As one of them declared in 1643, "When the kings of the earth have given their power to the beast, these choice-soldiers … will be so faithful to the King of kings, as to oppose the beast, though armed with kinglike power" (Walzer, 1965, p. 294). Stephen Marshall exhorted the troops in Parliament in 1644: "Go now and fight the battles of the Lord…. Do now see that the question in England is whether Christ or Anti-Christ shall be lord or king." Henry Wilkenson wrote that Parliament's "business lies professedly against the apocalyptical beast and all his complices" (p. 294). The battle was directed, of course, not only against the Cavaliers but also against moral corruption everywhere. The purge of heretics, worldlings, and adulterers was viewed as part of the same battle by which "the whore of Babylon shall be destroyed with fire and sword" (p. 295). The terminology of this discourse derives almost exclusively from the book of Revelation.

When the revolution was overthrown in England in 1660, there was a sense among the Puritans that the American colonies had become the new bearers of Protestant destiny to usher in this millennium (Walzer, 1965, p. 296). Increase Mather returned to Boston the following year with this idea in mind, "believing it was the last stronghold of Protestantism," as Perry Miller (1967b, p. 72) described it. With such convictions, the New England colonists resisted the efforts of the Restoration regime to topple the rule of the saints. They evaded Charles II's letter of complaints in 1662, frustrated the royal commissioners in 1664, and evaded compliance with the Navigation Acts for the next 10 years.[3] Even after their charter was revoked in 1684, they resisted the efforts of Governor Andros and had the nerve to imprison him the moment they heard of the Glorious Revolution of 1688.

Ernest Lee Tuveson traces the development of this theocratic millennialism in *Redeemer Nation*, noting the preachers' retention of the "fanatic notion" of overturning evil by the forceful rule of the saints (Tuveson, 1968, pp. 97–99). He notes the impact of Jonathan Edwards's idea that with the religious revival of the eighteenth century, "divine providence is preparing the way for the future glorious times of the church, where Satan's kingdom shall be overthrown throughout the whole habitable globe" (Edwards, 1989; cited by Tuveson, p. 100; see also Baumgartner, 1999, pp. 127–130). As J. F. Maclear (1971) shows, the idea that America was the millennial nation "gave to all succeeding American events a continuing cosmic importance" (p. 190). This orientation encouraged militant resistance against Anglicanism and other elements of British authority as the "Great Beast" of Daniel and Revelation. Although this millennial idea was not initially shared by other colonies, its impact was augmented by New England's superior educational system and the intellectual vigor of its clergy.

The Impact of the Great Awakening

Whereas earlier revivals in the colonies began in established churches and generally remained local in their effect, the so-called Great Awakening, which began in the late 1730s and came to a climax two decades thereafter, influenced the entire colonial culture (see Andersen, 2006, pp. 26–29). It was associated above all with the itinerant preaching of George Whitefield, whose techniques developed in the years prior to his first visit to the colonies in 1739. His visit is usually credited with inaugurating the Great Awakening, the first broadly based revival in American history (for a discussion of the terminology of "revival" and "awakening," see Cottret, 2000; Gäbler, 1989; Richey, 1993). In his well-organized tours of 1739–1741, 1744–1748, 1763–1765, and 1769–1770, George Whitefield preached to huge crowds in all of the American colonies. It is estimated that in his lifetime Whitefield preached more than 7,500 sermons, reaching literally millions of people.

---

[3] The English Navigation Acts were a series of laws between 1651 and 1847 designed primarily to expand English trade (after 1707, British trade) and limit trade by British colonies with countries that were rivals of Great Britain (e.g., the Netherlands, France, and other European countries).

The Great Awakening was the first experience shared by all of the colonies, providing a new sense of continental identity as God's New Israel, the millennial nation promised by Revelation 20. The Great Awakening provided a new form that Nathan Hatch identified as "civic millennialism" (Hatch, 1977, pp. 28–31). In 1743, some seventy New England clergy signed a manifesto stating that with the Great Awakening the 1,000-year kingdom had come. John Moorehead, a Boston preacher, proclaimed, "The Millennium is begun…. Christ dwells with men on earth" (cited by Boyer, 1992, p. 70, from Stein, 1984, p. 358; see also Ziff, 1973, pp. 303–311). Jonathan Edwards interpreted the remarkable revival as the coming of a spiritualized 1,000-year kingdom. In place of wars to overcome evil, the world was to be converted by the Gospel. Evident here is the emergence of the belief that not simply "New England" but all of the American colonies comprised in some sense a chosen people whose task was to usher in the new age.

The Great Awakening resulted in a new majority of Baptists and Presbyterians who were prepared to accept a millennial theology and had particular reasons to oppose established churches supported by public taxation. The groups that benefited the most from the Great Awakening were those that stressed the need for religious experience and allowed for the rise of charismatic lay preachers who were prepared to become itinerants (Finke & Stark, 2002, p. 50). These preachers went from house to house, revival to revival, camp meeting to camp meeting in the frontier regions, founding new churches in settlement after settlement. The various Baptist groups gained the most from the revival, growing from around 90 congregations in 1740 to more than 370 in 1776. Their loose, congregational structure and emphasis on believer's baptism as well as their Calvinist theology suited them well in following up on Whitefield's successes. Some of the more evangelistic forms of Presbyterianism also benefited, but their higher educational standards for the ministry tended to suit them less well to the frontier situation. The new religious majority was inclined to favor churches that were independent of government control. In Virginia, for example, which had been largely unchurched prior to the Great Awakening, despite an Anglican establishment supported by taxes, Thomas Jefferson estimated that two thirds of the inhabitants were associated with dissenting churches (Lambert, 2003, p. 226) that were the products of the Great Awakening.

One of the results of the rise of independent churches was to challenge the tax system supporting the established clergy in various colonies. By the end of the Great Awakening, Baptist congregations temporarily gained the right to be exempt from such taxes in Massachusetts. There were similar struggles in the other colonies against the Anglican tax system. Since the time of Roger Williams, the Baptists had argued for freedom of religion and the noninterference of the state in religious affairs. Isaac Backus (1724–1806), a Baptist itinerant preacher in Connecticut and Massachusetts (see Dunn, 1999, pp. 836–838; McLoughlin, 1967, pp. 110–192; Miller, 1988, pp. 210–216), wrote a book entitled *A Fish Caught in His Own Net* (1768), which argued that Congregationalists had violated their own principles in giving clergy associations the right to determine eligibility for ministry and to employ public coercion in order to enforce such decisions to disallow Baptist clergy. He argued for complete freedom of conscience and a complete separation of church and state because religion should rely on "persuasion alone" rather than on coercion (Lambert, 2003, p. 201; McLoughlin, p. 127). This line of argument was taken up in Backus's appeal to the Massachusetts legislature in 1774, protesting the tax on Baptist congregations to support Congregational churches (see Gaustad & Noll, 2003, pp. 225–227). Baptists refused to pay such taxes because that would be "implicitly allowing to men that authority which we believe in our consciences belongs only to God. Here, therefore, we demand charter rights, liberty of conscience" (p. 227). Backus and his Baptist colleagues were demanding not tolerance in the Enlightenment sense but intrinsic rights of religious freedom. This demand reveals the link between the Great Awakening, which gave such growth to the Baptists and other independent churches, and the cause of political liberty. Leaders like Backus ended up supporting the revolution against England on religious grounds, in defense of their view of religious freedom that they felt the Anglicans and other establishment figures as well as the British government were threatening (McLoughlin, pp. 136–137). They fought this battle with the apocalyptic rhetoric of the book of Revelation, following earlier New England Puritans in identifying the British church and government as the Whore of Babylon and the Antichrist. The American colonies, on the other hand, began to be pictured as the nation that pioneered in religious freedom, a theme that surfaces in Timothy Dwight's poetry.

The Apocalyptic Interpretation of American Wars

The new form of civil millennialism surfaced in many colonists' interpretation of the so-called French and Indian War from 1754 to 1763 (Hatch, 1977, pp. 36–44). Because the French Catholics in Canada had allied themselves with Native American tribes to fight against the American colonists and their native allies, it was natural to employ the rhetoric that had been used in the Puritan revolution against England. The Whore of Babylon mentioned as the enemy of the church in Revelation 13 had long been identified with Rome and its alleged allies in the Anglican Church. Congregational minister and historian Thomas Prince (1687–1758) saw the French and Indian War as "opening the way to enlighten the utmost regions of America preparatory to the millennial reign" (cited by Maclear, 1971, p. 190). In *Longing for the End*, Frederic Baumgartner (1999) confirms that

> for the Puritans, the French and Indian War in North America also served as a millennial event.... The French and their native allies served Antichrist by waging war on the people of God, and their early victories were signs that the great tribulation was beginning. The British victory in turn confirmed the deeply held belief among the English colonists that they were a chosen people building the New Kingdom in America. (p. 131)

At first glance, it seems contradictory to view Britain as fighting against the Antichrist, but that appeared to make no difference. In a variety of ways through American religious history down to the present moment, victory against God's alleged enemies has assumed a high priority, whoever those enemies happen to be. The triumph was viewed as a confirmation of providential destiny, shared now by all of the colonies. It was not long before this apocalyptic orientation turned against Great Britain itself.

While some churches were neutral or opposed the American Revolution, it was advocacy on the part of Protestants shaped by the Great Awakening that turned the conflict into a veritable "religious war." The Calvinist strain of North American religion stressed the sovereignty of God over human affairs and was thus inclined to believe that God takes sides in military conflicts. The Presbyterian minister Samuel Davies (1723–1761) blessed the Continental Army with the language of Old Testament holy war: "May the Lord of hosts, the God of the armies of Israel, go forth along with you! May he teach your hands to war, and gird you with strength in battle!" (Albanese, 1976, p. 85). When Governor Hutchinson's house was burned by a revolutionary mob in Boston, one of the leaders explained his behavior to the court by claiming that he was excited by a sermon by the Congregational minister Jonathan Mayhew (1720–1766) and "thought he was doing God's business" (Lambert, 2003, p. 216). When the Continental Army was defeated, the preachers interpreted the defeat as punishment for the colonists' sins, while its victories were due to divine providence. In the Battle of Long Island, an unusual fog at 2 o'clock in the morning allowed the Continental Army to retreat in safety, and this "providential shifting of the wind" was viewed as a sign that God supported the revolution (Albanese, p. 86).

A popular song written by William Billings (1746–1800) included the stanza, "Let tyrants shake their iron rod/And slavery clank her galling chains;/We fear them not; we trust in God–/New England's God for ever reigns" (cited by Albanese, 1976, p. 25; see also De Jong, 1985; Morin, 1941). Here the God of New England's churches becomes the God of all thirteen colonies. In virtually claiming that the Deity was on the side of a political entity, Billings was expressing the "zealous nationalist" form of civil religion (Jewett & Lawrence, 2003, pp. 55–106). In the issue of slavery versus freedom, God takes sides. One sees this tendency to place the revolution in a context of holy war in various ways. One observer wrote,

> The clergy of New England were a numerous, learned and respectable body who had a great ascendancy over the minds of their hearers. They connected religion and patriotism, and in their sermons and prayers, represented the cause of America as the cause of Heaven. (David Ramsay quoted in Albanese, 1976, p. 37; see also Hatch, 1977, pp. 85–91)

Part of this fervor derived from the religious resentments held by free-church Protestants. The Congregationalists remembered former leaders burned at the stake by Catholic and Anglican authorities and resented Anglican efforts to reestablish their dominance in colonies that had enjoyed a form of democratic self-government for more than 140 years. The Baptists opposed all efforts to impose government control over churches (see Clark, 1994, pp. 372–381). The Scottish Presbyterians harbored centuries of resentment against English domination, and the Scotch-Irish had experienced forcible relocation to Northern Ireland followed by legislation that violated their rights to Presbyterian activities. They emigrated to the colonies with these resentments still fresh and unforgotten (Clark, p. 362).

The historian Catherine Albanese (1976) draws some of these themes together under the banner of the millennial context, in which the colonists were not just defending their rights but also advancing the cause of freedom for the entire world and thereby ushering in the 1,000-year kingdom:Oratory echoed with the millennial theme of the land of plenty where, as in Israel's dreams of a future of bliss, each man would dwell under his own fig tree in the shade of his vines, while his wife would be a fruitful olive branch surrounded by her joyful children. (p. 29) Further, she writes, "one hears" in these sermons and speeches, "the echoes of revivalist ardor and millennial zeal, for Armageddon was surely close at hand when divine enthusiasm was unleashed by righteous patriots against demonic British soldiers" (p. 42).

Hugh Henry Brackenridge based his *Six Political Discourses Founded on the Scripture* (1778) on the same set of premises observed with Timothy Dwight. He argued that King George was inspired by Satan and that Providence sided with the Americans in the great revolution. "Heaven hath taken an active part, and waged war for us.... Heaven knows nothing of neutrality.... There is not one Tory to be found amongst the order of the seraphim" (cited in Miller, 1967a, p. 95). In the revolutionary period, of course, the Tories supported the British Empire while the Whigs supported the revolution. In Brackenridge's argumentation, these political parties were based on eternal realities, and heaven was claimed to side with the Whig revolution. Historian Perry Miller has described "how effective were generations of Protestant preaching in evoking patriotic enthusiasm" during the revolution (Miller, p. 97; see also Butler, Wacker, & Balmer, 2003, p. 149). In particular he traces the precedents and implications of the "day of publick humiliation, fasting, and prayer" called by the Continental Congress in 1775. All over the colonies the belief was that God would respond to such repentance, bless the impending revolution, and usher in an era of peace for the saints. This belief provided a powerful motivation for carrying out a rebellion against the greatest power on earth. A widely distributed oration by John Allen, a Baptist preacher in Boston, proclaimed, "Liberty ... is the native right of the Americans" because "they were never in bondage to any man" (Gaustad & Noll, 2003, pp. 221–222). He went on to argue that no institution of royal government had a right to tax the Americans without their consent, and that the Americans should stand "upon their own strength" in resisting such efforts. The Congregational minister Samuel Sherwood placed the revolution for the sake of liberty in the millennial context of the final battle of world history promised in the book of Revelation (see Gaustad & Noll):

> Liberty has been planted here; and the more it is attacked, the more it grows and flourishes. The time is coming and hastening on, when Babylon the great shall fall to rise no more; when all wicked tyrants and oppressors shall be destroyed forever.... These commotions and convulsions in the British Empire may be leading to the fulfillment of such prophecies as relate to his [i.e., Satan's] downfall and overthrow, and to the future glory and prosperity of Christ's church. (p. 228)

A noteworthy feature was that this religious interpretation of the national destiny correlated closely with the views of citizens who were more secular in outlook. In the eighteenth century, Deism developed a secular form of millennialism. It was a religion that "assumes a correspondence between the rational structure of the physical universe and the rational capacity of the human mind, so that by discovering the universe one may come to know its creator" (Van Til, 1990, p. 347; see also Gestrich, 1981, pp. 392–406). Deists rejected Christian doctrines that are not supported by reason, such as the Trinity, the virgin birth, and the divinity of Christ, while at the same time maintaining the structure of divine sovereignty that guides history toward progress and freedom. Some accounts of Deism do not take the providence issue into account, but it played a major role among American Deists (see, e.g., Gaustad & Noll, 2003, p. 266). God was less personal than in classical Christian doctrine, but divine providence still guided history and struggled on behalf of justice and freedom. While preferring terms such as "providence" to the term "God," the Deists held a millennial view that enlightenment would change the entire world and usher in the golden age. It is well known that many revolutionary leaders, such as Benjamin Franklin, were inclined to Deism, but it is also clear that this inclination did not place them at odds with the more orthodox Calvinists with respect to issues of the revolution. Deists and evangelicals were on parallel tracks. They all agreed that government should be based on the consent of the governed— an idea derived from covenantal theology and developed in John Locke's theory of government. They all agreed that government must rest on freely chosen covenants in which the governed give consent to their governors, and that individuals must be free to make religious and political choices.

A belief in the possible attainment of human perfection linked the Deists with the revivalists, both of whom were inclined to millennialism by the latter part of the eighteenth century. For example, Benjamin Franklin wrote,

> It is impossible to imagine the Height to which may be carried, in a 1,000 years, the Power of Man over Matter.... Agriculture may diminish its Labor and double its Produce; all Diseases may by sure means be prevented or cured.... Men would cease to be wolves to one another. (cited by Lambert, 2003, p. 171)

This belief was a secular form of the biblical visions of the millennial age found in Isaiah and Revelation. After the revolution, the correspondence between John Adams and Thomas Jefferson addressed the Deist topics, on which they largely agreed (see Gaustad & Noll, 2003, pp. 269–271). Adams's letter of November 13, 1813, is particularly revealing. In it, the former president Adams claimed nothing less than millennial sainthood for Americans: "Many 100 years must roll away before we shall be corrupted. Our pure, virtuous, public spirited, federative republic will last forever, govern the globe and introduce the perfection of man" (cited by Kohn, 1957, p. 13; also in Jewett & Lawrence, 2003, p. 211).[4] The breathtaking optimism of this otherwise sober politician reveals the fusion of enlightenment enthusiasm, democratic ideology, and biblical millennialism that has shaped American civil religion since its beginnings in the early colonies (see Moorhead,

---

[4] See also John Adams's entry in his diary in February 1765: "I always consider the settlement of America with Reverence and Wonder—as the Opening of a grand scene and Design in Providence, for the Illumination of the Ignorant and the Emancipation of the slavish Part of Mankind over all the Earth" (Adams, 1961, p. 257).

1999, pp. xii–xv, 2–16). Strengthened by the constitutional establishment after the Revolutionary War, this civic millennialism made its way from New England to all thirteen colonies.

The United States as the Apocalyptic "New Order of the Ages"

The celebrations from city to city following the ratification of the Constitution focused on the inauguration of a millennial republic. In these celebrations, popular clergymen from Congregational, Presbyterian, and Baptist churches were typically asked to deliver the main patriotic addresses. They employed the word "miracle," which had been used by George Washington and James Madison in letters to their friends concerning the Constitutional Convention (Bowen, 1966, p. ix). That Providence had guided the process of debate, compromise, and public discussion was widely assumed and celebrated. The *Federalist Papers* noted the consensus that "Providence has in a particular manner" blessed the country, and a protégé of Washington wrote that America was the "theater for displaying the illustrious designs of Providence in its dispensations to the human race" (cited by McCartney, 2006, pp. 28–29).

This interpretation of the national destiny was anticipated in works such as Timothy Dwight's epic poem of 1785, "The Conquest of Canaan." The poem was dedicated to George Washington and celebrates the triumph of freedom over tyranny that Providence ensured. The poem fuses the biblical accounts of Israel's conquest of Canaan with the visions of the peaceable kingdom found in Isaiah and the book of Revelation, which Dwight believed were being fulfilled after the revolution. In contrast to earlier wars that had produced a legacy of destruction and tyranny, a new democratic world order is celebrated as spreading over an immense geographic realm (see Dwight, 1969, pp. 274–275):

> To nobler bliss yon western world shall rise.
> Unlike all former realms, by war that stood,
> And saw the guilty throne ascend in blood,
> Here union'd Choice shall form a rule divine;
> Here countless lands in one great system join;
> The sway of Law unbroke, unrivall'd grow,
> And bid her blessings every land o'erflow....
> Here Empire's last, and brightest throne shall rise;
> And Peace, and Right, and Freedom, greet the skies.

This fulfillment of the ancient visions resulted not from divine fiat but from a combination of divine providence and human "Choice," reflected in the union of the colonies. The idea of government by compact, which had animated the earliest colonists in New England and led to the doctrine of the consent of the governed, comes full circle to its perceived fulfillment, ushering in the peaceable republic of "Peace, and Right, and Freedom."

Two years after the publication of Dwight's poem, a fellow graduate of Yale, Joel Barlow (1754–1812), issued his epic poem "The Vision of Columbus," "to celebrate the United States as the most advanced embodiment of an enlightened Republican culture" (Schloss, 2003, p. 139). In the final section of the poem, an angel reveals the rise of "a world civilization, a league of states resembling the United Nations with the individual member states all modeled after the newly established American republic" (p. 143). Here the Isaianic vision of the impartial world court on "the mountain of the Lord" that would allow nations to "beat their swords into plowshares" (Isaiah 2:3–4) is fulfilled by the Republican triumph. It is seen now to spread its influence over the entire globe (Barlow, 1970, pp. 256–257):

> From all the bounds of space (their labours done),
> Shall wing their triumphs to the eternal throne;
> Each, from his far dim sky, illumines the road,
> And sails and centres tow'ard the mount of God....
> So, from all climes of earth, where nations rise,
> Or lands or oceans bound the incumbent skies,
> Wing'd with unwonted speed, the gathering throng
> In ships and chariots, shape their course along....
> There, hail the splendid seat by Heaven assign'd,
> To hear and give the counsels of mankind....
> To give each realm its limit and its laws;
> Bid the last breath of dire contention cease,
> And bind all regions in the leagues of peace.

Barlow intended to show that "on the basis of the republican principle" not only "good government" but also the "hopes of permanent peace must be founded" (cited by Schloss, 2003, p. 143). Dwight's poem "Greenfield Hill" celebrates the equality provided by this republican system, and once again its geographic extension is described (cited by Schloss, 1999, p. 28, from Dwight, 1969, p. 511; capitalization of "heaven" in original):

> See the wide realm in equal shares possess'd!
> How few the rich, or poor! how many bless'd!
> O happy state! the state by HEAVEN design'd
> To rein, protect, employ, and bless mankind.

These themes reappear in the remarkable celebrations of the ratification of the Constitution. The most elaborate of these celebrations, analyzed by Dietmar Schloss (2001, pp. 44–62), was the Grand Federal Procession, which occurred in Philadelphia on July 4, 1788. The parade was divided into three parts, beginning with twenty-five groups illustrating American history, followed by fifty groups of farmers, workers, and artisans, and concluded by walking professionals. In place of marching soldiers, a triumphal arch, and a hierarchical social order, which were typical of European victory parades, some five thousand citizens presented themselves as supporting and being supported by the Constitution. The order of the professions was determined by lot, expressing the egalitarian ethos. The most elaborate float was the "Grand Federal Edifice," with its roof held up by thirteen columns and with a "cupola crowned by a figure of the goddess of plenty carrying a cornucopia" (Schloss, p. 53), symbolizing the golden age inaugurated by the Constitution. The biblical theme of Isaiah's peaceable kingdom was represented by the master blacksmith hammering plowshares and pruning hooks out of old swords (p. 57). For contemporary witnesses it was particularly striking to see clergymen of different denominations, including a Jewish rabbi, "all walking arm in arm" (Schloss, p. 48), which conveyed the sense of divine providence blessing the tolerant enterprise. The *Pennsylvania Gazette* observed that this sight of "almost every denomination, united in charity and brotherly love," was a "circumstance which probably never occurred in such extent" (cited by Bowen, 1966, p. 308). At the conclusion of the parade, participants and spectators were invited to tables in a broad circle, where they were addressed by James Wilson. He concluded with ten toasts, the first addressed to the "people of the United States" and the last to "the whole family of mankind" (Schloss, p. 49), conveying the millennial sense that the democratic new order of the ages would be a blessing to the entire human race.

When the time came to create the national seal, a millennial motto in Latin was selected: *Novus ordo saeclorum*, which usually was translated as "the new order of the ages." The inscription *Annuit Coeptis* ("He has ordered our way") stands over the classical pyramid (Butler et al., 2003, p. 174). One could understand these references as purely political claims, in that the constitutional system was a new form of government, which was true at the time. One could understand them in Deist or enlightenment terms as the beginning of a democratic age of progress. Or one could take these references as the announcement of the beginning of the 1,000-year kingdom ruled by the Christian saints in North America. These political artifacts convey the millennial idea of America as the promised land where messianic hopes were being fulfilled.

Conclusion

This chapter demonstrates the historical interaction of geography and ideology and the coalition between knowledge and power. At first only New England thought of itself as "God's New Israel," the arena in which the millennial destiny of Revelation 20 would be fulfilled. Within 120 years, this vision had spread to all of the colonies, as celebrated in the poetry of Timothy Dwight. The populations of the other colonies were initially uninterested in the political fulfillment of millennial ideals, and it seems highly unlikely that without the experience of the Great Awakening, the Indian wars, the conflicts over religious establishment, and dissatisfaction with British rule, this millennial viewpoint would have become dominant. It appears, moreover, that millennialism was closely linked with developing a national identity, as compared with the earlier colonial identities. Without this emerging identity as *Americans*, it seems unlikely that the Continental Army led by Washington could have prevailed after its many defeats. In this historical example, there is therefore a case to be made for a link between knowledge, in the form of millennial ideology, and power. In part because of the global dimensions of the biblical millennialism that inspired this development, this ideology also has a tendency toward geographic expansion. What started in New England spread throughout the other colonies. The descending "heavenly kingdom" envisioned by Timothy Dwight in his poem of

1771 extends itself over "countless lands" by 1785; Joel Barlow envisions the binding of "all regions in the leagues of peace," following the American example. Despite the disappointments, frustrations, and betrayals of later history, this idea of being a geographical region called to advance freedom, democracy, and peace around the world remains a characteristic feature of American civil religion.

References

Adams, J. (1961). Diary and autobiography of John Adams: Diary 1755–1770. In L. H. Butterfield, L. C. Faber, & W. D. Garrett (Eds.), *The Adams papers* (Vol. 1). Cambridge, MA: Belknap Press of Harvard University Press.

Ahlstrom, S. E. (1972). *A religious history of the American people*. New Haven, CT: Yale University Press.

Albanese, C. L. (1976). *Sons of the fathers: The civil religion of the American Revolution*. Philadelphia: Temple University Press.

Andersen, M. (2006). *Between secular and sacred: America's fourth Great Awakening and the return of religious rhetoric in American politics*. Ph.D. dissertation, University of Copenhagen, Denmark.

Aune, D. E. (1999). Chiliasmus [Chiliasm]. In H. D. Betz (Ed.), *Religion in Geschichte und Gegenwart: Handwörterbuch für Theologie und Religionswissenschaft* (4th ed., pp. 136–137). Tübingen, Germany: Mohr Siebeck.

Barlow, J. (1970). *The works of Joel Barlow: Vol. 2 Poetry*. Gainesville, FL: Scholars' Facsimiles & Reprints. (Original work published 1781–1825)

Baumgartner, F. J. (1999). *Longing for the end: A history of millennialism in Western civilization*. New York: St. Martin's Press.

Bowen, C. D. (1966). *Miracle at Philadelphia: The story of the constitutional convention May to September 1787*. Boston: Little Brown and Company.

Boyer, P. S. (1992). *When time shall be no more: Prophecy belief in modern American culture*. Cambridge, MA: Harvard University Press.

Butler, J., Wacker, G., & Balmer, R. (2003). *Religion in American life: A short history*. Oxford, UK: Oxford University Press.

Clark, J. C. D. (1994). *The language of liberty, 1600–1832: Political discourse and social dynamics in the Anglo-American world*. Cambridge, UK: Cambridge University Press.

Cottret, B. (2000). Le paradigme perdu: Le Great Awakening, entre la faute et l'innocence [The lost paradigm: The Great Awakening, between guilt and innocence]. *Études Theologiques et Religieuses, 75*, 211–228.

De Jong, M. G. (1985). "Both pleasure and profit": William Billings and the uses of music. *The William and Mary Quarterly, a Magazine of Early American History and Culture, 42*, 104–116.

Dowling, W. G. (1999). Dwight, Timothy. In J. A. Garraty & M. C. Carnes (Eds.), *American national biography* (Vol. 7, pp. 192–194). New York: Oxford University Press.

Dunn, E. E. (1999). Backus, Isaac. In J. A. Garraty & M. C. Carnes (Eds.), *American national biography* (Vol. 1, pp. 836–838). New York: Oxford University Press.

Dwight, T. (1969). *The major poems of Timothy Dwight (1752–1817) with a dissertation on the history, eloquence, and poetry of the Bible*. Gainesville, FL: Scholars' Facsimiles & Reprints.

Edwards, J. (1989). *History of the work of redemption*. Reprinted by New Haven, CT: Yale University Press.

Finke, R., & Stark, R. (2002). *The churching of America 1776–1990: Winners and losers in our religious economy*. New Brunswick, NJ: Rutgers University Press.

Fiske, J. (1889). *The beginnings of New England: Or, the Puritan theocracy in its relations to civil and religious liberty*. Boston: Houghton & Mifflin.

Gäbler, U. (1989). Erweckung im europäischen und amerikanischen Protestantismus [Awakening in European and American Protestantism]. *Pietismus und Neuzeit, 15,* 24–39.

Gaustad, E. S., & Noll, M. A. (2003). *A documentary history of religion in America: Since 1877* (3rd ed.). Grand Rapids, MI: Eerdmans.

Gestrich, C. (1981). Deismus [Deism]. In G. Müller, H. Balz, J. K. Cameron, C. Grethlein, S. G. Hall, B. L. Hebblethwaite, K. Hoheisel, J. Wolfgang, V. Leppin, K. Schäferdiek, G. Seebaß, H. Spieckermann, G. Stemberger, & K. Stock (Eds.), *Theologische Realenzyklopädie* (Vol. 8, pp. 392–406). Berlin, Germany: De Gruyter.

Hatch, N. O. (1977). *The sacred cause of liberty: Republican thought and the millennium in revolutionary New England.* New Haven, CT: Yale University Press.

Jewett, R. (2008). *Mission and menace: Four centuries of American religious zeal.* Minneapolis, MN: Fortress Press. German translation published by Vandenhoeck & Ruprecht.

Jewett, R., & Lawrence, J. S. (2003). *Captain America and the crusade against evil: The dilemma of zealous nationalism.* Grand Rapids, MI: Eerdmans.

Kohn, H. (1957). *American nationalism: An interpretive essay.* New York: Macmillan.

Lambert, F. (2003). *The founding fathers and the place of religion in America.* Princeton, NJ: Princeton University Press.

Maclear, J. F. (1971). The republic and the millennium. In E. A. Smith (Ed.), *The religion of the republic* (pp. 183–216). Philadelphia: Fortress Press.

Massyngbaerde Ford, J. (1992). Millennium. In D. N. Freedman (Ed.), *The anchor bible dictionary* (Vol. 4, pp. 832–834). New York: Doubleday.

McCartney, P. T. (2006). *Power and progress: American national identity, the war of 1898, and the rise of American imperialism.* Baton Rouge, LA: Louisiana State University Press.

McLoughlin, W. G. (1967). *Isaac Backus and the American pietistic tradition.* New York: Brown, Little, and Company.

Miller, P. (1967a). From the covenant to the revival. In P. Miller (Ed.), *Nature's nation* (pp. 90–120). Cambridge, MA: Belknap Press of Harvard University Press.

Miller, P. (1967b). Preparation for salvation in seventeenth-century New England. In P. Miller (Ed.), *Nature's nation* (pp. 50–77). Cambridge, MA: Belknap Press of Harvard University Press.

Miller, W. L. (1988). *The first liberty: Religion and the American republic.* New York: Paragon.

Moorhead, J. H. (1999). *World without end: Mainstream Protestant visions of the last things, 1880–1925.* Bloomington, IN: Indiana University Press.

Morin, R. (1941). William Billings: Pioneer in American music. *The New England Quarterly, 14,* 25–33. doi:10.2307/360095.

Richey, R. E. (1993). Revivalism: In search of a definition. *Wesleyan Theological Journal, 28,* 165–175.

Schloss, D. (1999). Politics and religion in Timothy Dwight's *Greenfield Hill* (1794). *Annales du Monde Anglophone* (Special issue: Religion et Culture aux États-Unis), *9,* 17–34.

Schloss, D. (2001). The nation as spectacle: The Grand Federal Procession in Philadelphia, 1788. In J. Heideking, G. Fabre, & K. Dreisbach (Eds.), *Celebrating ethnicity and nation: American festive culture from the revolution to the early twentieth century* (pp. 44–62). New York: Berghahn Books.

Schloss, D. (2003). Joel Barlow's vision of Columbus (1787) and the discovery of the ethnic. In H.-J. Grabbe (Ed.), *Colonial encounters: Essays in early American history and culture* (pp. 39–155). Heidelberg, Germany: Universitätsverlag Winter.

Spillmann, K. R. (1984). *Amerikas Ideologie des Friedens: Ursprünge, Formwandlungen und geschichtliche Auswirkungen des amerikanischen Glaubens an den Mythos von einer friedlichen Weltordnung* [America's ideology of peace: Origins, transformations, and historical impacts of the American belief in the myth of a peaceful world order]. Bern, Switzerland: Peter Lang.

Stein, S. J. (1984). Transatlantic extensions: Apocalyptic in early New England. In C. A. Patrides & J. Wittrich (Eds.), *Apocalypse in English renaissance thought and literature* (pp. 266–298). Manchester, UK: Manchester University Press.

Tuveson, E. L. (1968). *Redeemer nation: The idea of America's millennial role*. Chicago: University of Chicago Press.

Van Til, L. J. (1990). Deism. In D. G. Reid, R. D. Linder, B. L. Shelley, & H. S. Stout (Eds.), *Dictionary of Christianity in America* (p. 347). Downers Grove, IL: Inter Varsity Press.

Walzer, M. L. (1965). *The revolution of the saints: A study in the origins of radical politics*. Cambridge, MA: Harvard University Press.

Ziff, L. (1973). *Puritanism in America: New culture in a new world*. New York: Viking.

The Power of Words and the Tides of History: Reflections on *Man and Nature* and *Silent Spring*

8

Graeme Wynn

> *The silent power of books is a great power in the world....*
> *Silent, passive, and noiseless though they be, they may yet set in*
> *action countless multitudes, and change the order of nations*
>
> Giles (n. d.).

So wrote Henry Giles, an Irish-born, American clergyman, who gained a modest reputation, in the mid-nineteenth century, as a skilled orator, lecturer, and author (Rich, 1891). Although he is now largely forgotten, these words provide a forceful touchstone for this essay, which seeks to explore some of the interconnections among knowledge, power, and action by examining how two passive and noiseless artifacts—books published in the nineteenth and twentieth centuries—brought new understanding (knowledge) to a diverse if not necessarily countless body of readers, and worked, more or less efficaciously, to change the ways in which some of them (if not entire nations) thought about human-environment relations.

To frame this rumination, I begin with three propositions—two from the nineteenth-century Scottish philosopher, historian, and satirical essayist, Thomas Carlyle, and one of more recent and less distinct provenance—that point broadly toward what might be characterized, in more formal discourse, as the notions of agency, structure, and the immutable mobile.

1. "the History of the World is . . .the Biography of Great Men." (Carlyle, 1840, Project Gutenberg E-Text 1091)
2. Lives are pebbles dropped into the sea of history. They have an impact, but it is ephemeral. Spreading ripples chart their effect and draw the attention of people

G. Wynn (✉)
Department of Geography, University of British Columbia,
West Mall 1984, Vancouver, BC V6T 1Z2, Canada
e-mail: wynn@geog.ubc.ca

© Springer Netherlands 2015
P. Meusburger et al. (eds.), *Geographies of Knowledge and Power*,
Knowledge and Space 7, DOI 10.1007/978-94-017-9960-7_8

nearby. But the swell they create soon fades, to be enveloped by the deeper tide of social and historical forces. [1]

3. "The Writer of a Book, is not he a Preacher preaching not to this parish or that, on this day or that, but to all men in all times and places?" (Carlyle, 1840/2007, p. 101)

These claims provide a foundation for considering how one man and one woman—who were certainly preachers and people whom Carlyle might have considered "great"—made splashes that helped to move the tides of time. This man and woman lived and worked approximately a century apart. Both were Americans, one from Vermont, the other from Pennsylvania. The man, George Perkins Marsh (1801–1882), has been described as "the fountainhead of the conservation movement" and the woman, Rachel Carson (1907–1964), as someone who "altered the balance of power in the world" (Hynes, 1989, p. 3) by encouraging the emergence of the new environmental movement. They stand therefore as key players in the development of what the historian Samuel P. Hays (1959, 1987) described, and contrasted, as the production- and amenity-oriented attitudes toward the environment characteristic (respectively) of the early- and late-twentieth century. Both were prolific authors, but their reputations rest, largely, on single works: *Man and Nature; or, Physical Geography as Modified by Human Action*, in Marsh's case, and *Silent Spring* in Carson's (Carson, 1962; Marsh, 1864).

Given the iconic status of these works and their authors, and the massive influence attributed to them, I seek to understand the power of words and the knowledge they convey by asking where, when, how, and why the ideas in Marsh's and Carson's landmark books were so important? Doing so raises several other questions: Were the arguments unprecedented? Where did they come from? Where did they go? How did they work? Were they framed in particularly novel and/or compelling ways? What facilitated their dissemination? How did they gain purchase? Were they lightning bolts that ignited inert populations or merely winds that fanned already-glowing embers and flickering flames?

In short, I interrogate the contents of these widely-cited books and attempt to excavate something of the social, economic, political, environmental, and intellectual contexts into which first Marsh's and then Carson's ideas were released, to see not only what a book or two can do—but also why and how they exercise influence. My approach is thus two-pronged. First, I chart some of the links between the books and their consequences, between the knowledge they contained and the power they exercised. This is, so to speak, to explore their "public lives" and to suggest why and how their challenging arguments had the impacts that they did—it is therefore an effort to reveal something of the ways in which they spoke truth to power. In a second, related, vein my aim is to explore how (and to what extent) these books laid

---

[1] I have a long-standing interest in George Perkins Marsh—see Wynn (2004) and Wynn (2008)—but this paper reflects a special debt to a study by P. C Murphy (2005) considering Rachel Carson's *Silent Spring* from a "history of the book" perspective and a luminous essay by Adam Gopnik (2009), both of which prompted me to think anew about Marsh and *Man and Nature*. I draw the pebbles in the sea of history analogy from Gopnik.

the basis for what Jürgen Habermas (1998 and 1984, 1987) called communicative action, based on a "shared understanding that the goals [they articulated] are inherently reasonable or merit-worthy" (Bohman & Rehg, 2009). In other words, I hope that this little foray might encourage deeper appreciation of both these books, as contributions to a discourse of environmental concern that has exhibited sufficient staying power to influence government policies.

Man and Nature: A Book and Its Reception

Marsh began writing *Man and Nature* in his home town of Burlington, Vermont, in the spring of 1860 and completed it in Italy, where he was serving as his country's ambassador, in 1864. By his own account, he first imagined the book as "a little volume" intended to challenge prevailing ideas that "the earth made man" by demonstrating that "man in fact made the earth" (Lowenthal, 2000a, p. 267). In the end, the "burly volume" (p. 269) ran to 465 pages (and subsequent editions were even longer). The purpose of the book was set out, plainly, in its first few lines (Marsh, 1864):

> The object …is: to indicate the character and, approximately, the extent of the changes produced by human action in the physical conditions of the globe we inhabit; to point out the dangers of imprudence and the necessity of caution in all operations which, on a large scale, interfere with the spontaneous arrangements of the organic or the inorganic world; to suggest the possibility and the importance of the restoration of disturbed harmonies and the material improvement of waste and exhausted regions; and, incidentally, to illustrate the doctrine, that man is, in both kind and degree, a power of a higher order than any of the other forms of animated life, which, like him, are nourished at the table of bounteous nature (p. iii).

Six chapters follow, each of them intimidating in scope and erudition. Chapter 1 is essentially an essay on "the ravages committed by man." Here, the author laid out his thesis, dealing in broad brush strokes with "the general effects and the prospective consequences of human action upon the earth's surface and the life which peoples it." The chapter opens with a powerful, five-page rumination on the natural advantages and physical decay of the territory of the Roman Empire, and of other parts of the Old World, and ends some fifty pages later with one of the book's signature sentences:

> But we are, even now, breaking up the floor and wainscoting and doors and window frames of our [earthly] dwelling, for fuel to warm our bodies and seethe our pottage, and the world cannot afford to wait till the slow and sure progress of exact science has taught it a better economy (Marsh, 1864, p. 55).

Chapter 2 deals with the "Transfer, Modification and Extirpation of Vegetable and Animal Species." Chapter 3, 200 pages long, focuses on "The Woods" (Forests), Chap. 4 on "The Waters" and Chap. 5 on "The Sands." The book then moves to an end with a series of reflections on "Projected or Possible Geographical Changes by Man."

Page after page, Marsh offers up an astonishingly diverse array of sources: the book is a heady, and often difficult, brew of interpretations, clarifications, asides and quotations; quotations from classical texts, quotations from the works of engineers and foresters, quotations from newspapers and plays, quotations from dictionaries and personal letters, all of which are blended in what David Lowenthal called a "stylistic mélange" with data from censuses and accounts from life (Marsh, 1864/1965, p. xx). Marsh probably had a smile on his face when he informed a friend that *Man and Nature* was an effort "to tell everything I know & have not told" elsewhere. But the weary reader working through this complex long-winded work might justifiably conclude that he was not far off the mark. The result of Marsh's labors was clearly (as even his biographer David Lowenthal (2000a, pp. 268–269) conceded "a volume not fully digested nor easily digestible."

Yet Lowenthal (1958) describes this selfsame book as "the most important and original American geographical work of the nineteenth century," and argues that "*Man and Nature* ushered in a revolution in how people conceived their relations with the earth" (p. 246, see also 2000b). Others have been equally enthusiastic. The modern-day environmental historian William Cronon ranks it as one of the "three books by American authors that have had the greatest impact on environmental politics and on the struggle to build more responsible human relations with the natural world." (Cronon, 2000, p. ix). More than this, Lowenthal argues, *Man and Nature* stood second only to Charles Darwin's *On the Origin of Species* as "the most influential text of its time to link culture with nature, science with society, landscape with history" (Lowenthal, 2000a, p. xv).

In fact, *Man and Nature* had a mixed reception. Initial responses were perhaps surprisingly favorable, given—as one reviewer of the second edition had it—that "the matters of which Mr. Marsh treats were only of curious interest" in 1864 (Anon, 1875, p. 124). Although Marsh feared that it would ruin his publisher, more than a thousand copies of the book were sold within months of its release. Early in the 1870s, asserts Lowenthal (in Marsh, 1864/1965, p. xxii), the book "was a classic of international repute." A contributor to *The Nation* (1874, cited in Marsh, 1864/1965 p. xxii), reviewing the enlarged and rebranded (with *The Earth* replacing *Physical geography* in the subtitle), but otherwise not greatly changed second edition of 1874, described it as "one of the most useful and suggestive works ever published" and thought that it carried "the force of a revelation." On the strength of this work, observed John Bigelow, sometime owner of the *New York Evening Post* and American Minister to France, in a letter to Marsh, he would stand among geographers as Adam Smith did among political economists and the Comte de Buffon among natural historians (Lowenthal, 2000a, p. 302).

The book quickly caught the attention of those concerned about the destruction of eastern North American forests. Franklin B. Hough, who had noted the decline in standing timber between 1855 and 1870 charted by the New York State census (which he supervised), drew on Marsh's insights in a presentation to the 1873 meeting of the American Association for the Advancement of Science: *On the Duty of Governments in the Preservation of Forests* (Hough, 1873). Hough was subsequently appointed to the U.S. Department of Agriculture to assess the state of American forests and he became the first chief of the Division of Forestry in USDA in 1883. His successor credited Marsh with identifying "our destructive treatment of the forests and the necessity of adopting a different course" (Lowenthal, 1958, p. 269). Even Gifford Pinchot, widely regarded as the founder of the American forest conservation movement, described Marsh's book as "epoch-making", although (ever anxious to portray himself as "breaking new ground") he elsewhere insisted that few Americans had read it and that it had little impact upon popular opinion (Miller, 2001, pp. 55–56; Pinchot, 1947, pp. xvi–xvii).

Beyond the United States, *Man and Nature* similarly influenced scholars and foresters in the decade or two after its publication. In France, Élisée Reclus incorporated its insights into his *La Terre*, published in 1868; geologists Charles Lyell and Arnold Guyot, whose earlier views Marsh challenged, acknowledged its importance; and Italian legislators incorporated references to the book in forest laws approved in the 1870s and 1880s. The book shaped practice in the Imperial Forestry Department of India, one of the officers of which wrote Marsh in 1868 to say that he had "carried … [*Man and Nature*] with [him] along the slope of the Northern Himalaya and into Kashmir and Tibet" (Lowenthal, 2000a, p. 304, 2000b, p. 4). In the same year, New Zealand legislators quoted liberally from *Man and Nature* (although often without attribution) in their efforts to halt the deforestation and "barbarous improvidence" that threatened to turn their recently-colonized "land of milk and honey" into a "howling desolation" (Wynn, 1977, 1979).

For all that, Harvard professor and public intellectual Charles Eliot Norton lamented, a quarter century after the publication of *Man and Nature*, that Marsh's warnings had fallen "upon deaf ears." Although a third edition of the book was published in 1884, and reprinted as late as 1907, Charles S. Sargent, an eminent botanist and director of Harvard University's Arnold Arboretum, reflected, in 1908, that "the younger generation" seemed to know nothing of it (Lowenthal, 2000a, p. 305). Perhaps most books and ideas follow a similar trajectory: they enjoy a more or less bright and protracted period in the sun (and 25 years, praise to the heavens, and world-wide policy impact are no trifles) before they fade from public consciousness. But *Man and Nature's* day was not yet done—although its revitalization was prolonged and wavering.

In 1920, the American historian and thinker Lewis Mumford learned of Marsh's work in correspondence with the Scottish polymath Patrick Geddes. Four years later he referred to Marsh in *Sticks and Stones*, a study of American architecture and civilization. In *The Brown Decades*, early in the 1930s, Mumford coined the resonant description of Marsh as the fountainhead of the conservation movement (and later claimed that *Man and Nature* was "quite forgotten even by geographers" until this act of resurrection). A few years later, the geographer Carl O. Sauer (1938/1963) added credence to this claim by describing Marsh as a "forgotten scientist" (pp. 147–148; see also Lowenthal, 2000a, 2000b, Koelsch, 2012, and Lowenthal, 2013).[2] In 1954, in another attempt at rebirth, Sauer's student Andrew Hill Clark

---

[2] Patrick Geddes to Lewis Mumford, November 13, 1920, in Novak (1995), Mumford (1955, p. 201), Sauer (1963, pp. 147–148). Mumford's "quite forgotten" claim is in Mumford to Babette Deutsch, October 30, 1960, in Lewis Mumford (1979, p. 177). See also Lowenthal (2000a, 2000b), where these sources are noted.

(1954) described Marsh as "among the first, and . . . one of the greatest of, our historical geographers"—but added (in ironic and probably unknowing refutation of John Bigelow), that "all too few modern geographers" think of him "as one of their own" (p. 81). A year later, however, the cobwebs of neglect were more vigorously brushed away when Mumford and Sauer dedicated a symposium considering "man's role in changing the face of the earth" to Marsh (Thomas Jr., 1956). In 1963, a year after the publication of *Silent Spring*, U.S. Secretary of the Interior Stewart Udall hailed *Man and Nature* as "the beginning of land wisdom in this country"— although then as now Thoreau, Muir, and Leopold (and Rachel Carson, who was described that same year as the fountainhead of the new environmental movement by none other than Udall himself) almost certainly ranked well ahead of Marsh in public consciousness of these things (Udall, 1963, pp. 69–82).[3]

In recent decades several questions have been raised about the influence and oft-proclaimed primacy of *Man and Nature*. Some scholars have baulked at Lowenthal's claims that the first "realisation of human impact on Earth stems from Marsh's *Man and Nature*" and that "only the most scanty ecological awareness antedates Marsh's own writings" (Lowenthal, 2000a, pp. 419–422). Environmental historian Richard Grove has argued that "western environmental concern and concomitant attempts at conservationist intervention" long pre-dated the publication of *Man and Nature* (Grove, 1995; in related vein see Girard, 1990, pp. 63–80). In his view, "reasoned awareness of the wholesale vulnerability of the earth to man" as well as the idea of state-directed environmental (or resource) conservation emerged gradually from the experience of colonial encounters with tropical regions and island ecosystems well before 1864. Others have noted that Immanuel Kant (1802) included humankind among the natural phenomena producing environmental change (in his *Physische Geographie*), and that long before Marsh, the Comte de Buffon (1782) wrote that "the state in which we see nature today is as much our work as it is hers. We have learned to temper her, to modify her, to fit her to our needs and our desires" (see Glacken 1967, pp. 568–575, 658–659, 666, 698–702).

On a somewhat different tack, the American scholar Richard Judd (1997) has found much evidence that ordinary people working the land of early New England developed grassroots strategies of resource conservation as integral elements of their local cultures well before the middle of the nineteenth century, and he goes so far as to insist that ecological principles were "common currency in early American natural history." On this account, many of Marsh's most cogent claims were foreshadowed in the actions of ordinary early New Englanders. And they were certainly adumbrated in print as early as 1835 when Titus Smith (1835) of Nova Scotia drew examples from the once prosperous, then desiccated, landscapes of the eastern Mediterranean to argue "that man has, by mismanagement, impoverished some of the finest countries on earth."

Lowenthal (2000a) has pushed back against what he describes as these "Marsh put-down[s]" by people who would diminish the reputation of "the prophet of

---

[3] Lowenthal (2000b), footnote 50 includes the following: "From 1955 to 1987 the *Science Citation Index* had 413 references to Thoreau, 248 to Muir, and 68 to Marsh."

conservation" by elevating "unsung hoi polloi on the mainstream's margins" (p. 419) to unwarranted prominence. The reputation of the great Marsh, he contends, is being tarnished by modern "wilderness-bent" environmentalists who too readily associate him with the managerialist emphasis of the 1955 *Man's Role* symposium, "impose their own apartheid on the past" and dismiss him as a resource-conserving economist rather than a preservationist poet. It is being scanted by "populist revisionists" who celebrate rural virtues and indict Marsh for turning the folk wisdom of his neighbors into a coda that justified restrictions on resource use and disempowered ordinary citizens. And it is being undermined by claims that his insights were "largely mistaken . . . unoriginal or inconsequential," (p. 423) and that his influence trended toward "technocratic, elitist, socially regressive imperialist or anthropocentric" (p. 423) outcomes. All such criticism, says Lowenthal is "unfounded or irrelevant" (p. 423, Lowenthal, 2000b, passim).

To summarize a long story too starkly, *Man and Nature* was read (if not always cited) in the quarter century after its publication, and cited (but not much read) through the next 75 (or hundred) years. In recent times new and in some cases explicitly revisionist accounts of changing attitudes toward the environment have whittled away at the underpinnings of Marsh's reputation but they have not quite or yet dislodged the man and his book from their pedestal. Modern environmental texts, observes David Lowenthal (2000a, p. 415) "pay almost obligatory homage to *Man and Nature*, then mention it no more.

Man and Nature: The Fate and Power of Words

How then to explain the lasting reputation yet uneven influence of *Man and Nature*? Why, so to speak, has the size of its parish and the respect afforded its preachings varied so greatly over the last 150 years? One approach, which offers an approximation of an answer, is to map interest in the book against changing patterns of environmental concern. *Man and Nature* spoke most clearly, in the third quarter of the nineteenth century, to those concerned with the fate of forests because eastern North Americans were confronting (and documenting) the consequences of a prolonged assault on the resource, rising prices for fuelwood, and so on (Williams, 1989). In Europe, and especially in India, where the Imperial Forest Department was established in 1864, the book fell in timely fashion into the hands of an emerging cadre of professionals newly charged with managing and administering forest and woodlands (Rajan, 2006). And in recently-settled New Zealand, Marsh's stentorian warnings about the erosive consequences of deforestation seemed highly pertinent in a dynamic geological environment in which upland denudation was far more active than in the "old countries" from which most settlers came (Wynn, 2002). By the last decades of the nineteenth century however, the extension of the American railroad network, the opening to exploitation of enormous forest stands in the upper Great Lakes states and early engagements with the magnificent coastal forests of the west had allayed (at least temporarily) North American fears of timber famine (Williams). A few years later, Gifford Pinchot and Theodore Roosevelt began to construct their own stories about the rise of American conservation and these had little room for precursors (Miller, 2001).

William Greeley's article (1925) on "The Relation of Geography to Timber Supply," in the first issue of *Economic Geography*, both marked and gave graphic expression to, renewed concerns about resource depletion in the 1920s, but the impetus toward conservation moved resolutely along tracks laid down with reference to the gospel of efficiency (Hays, 1959). *Man and Nature* fell from public consciousness and its author disappeared from the intellectual horizons of all but a handful of interacting scholars left to lament that Marsh had been forgotten. The leaders of this small group dedicated their 1955 symposium on *Man's Role in Changing the Face of the Earth* to Marsh, and brought his contribution back into the limelight, albeit in the context of a meeting in which complacency and optimistic belief in the capacity of human ingenuity and technology to address environmental ills were more common than were Marsh's more apocalyptic concerns. For a few years thereafter, Marsh was a name to be conjured with, but the times they truly were a changing. New concerns—nuclear Armageddon, the bioaccumulation of toxic substances, over-population—seized imaginations and new clairvoyants—Carson, Commoner, Ehrlich—pondered futures and global predicaments beyond those ever imagined by Marsh (Egan, 2007; Ehrlich, 1968). Beauty, health, and permanence became the watchwords of the new environmental movement and the newly-identified threats to permanence seemed far more urgent than the specter of desiccation that formed the centerpiece of Marsh's argument and that had, in some metaphorical sense at least, been stared down during the Dust Bowl of the 1930s (Hays, 1987). Author and title lingered on library shelves, but *Man and Nature* was a compendium of words from and for another era. Its ripples were but faint traces on the larger tide of changed times. Lip service was all that most people afforded the great book—unless their purpose was to "correct the record" and tell another tale about the importance (or otherwise) of *Man and Nature*.

David Lowenthal offers a rather different story about the reputational trajectory of Marsh and *Man and Nature*, the essence of which has been the foil for recent critiques of both the author and his book. For Lowenthal (who has spent half a century in Marsh's literary presence and who is undoubtedly the most knowledgeable student of the man and his works), the impact of *Man and Nature* owed almost everything to its author's unique gifts. It was "the sweep of his data, the clarity of his synthesis, and the force of his conclusion [that] made *Man and Nature* an almost instant classic" (Lowenthal, 2000b, p. 4). Marsh was a visionary, a prophet, and his light continues to shine undimmed down the decades. He was "the first to show that human actions had unintended consequences of unforeseeable magnitude" (Lowenthal, 2000a, p. 430). Or as Lowenthal (2000a) has it in one of his most pointed phrases "[a]nyone with a hoe or an ax knows what he is doing, but before Marsh no one had seen the total effects of all axes and hoes" (p. xxvii). The "perceptive powers" that allowed Marsh to see the big picture so clearly, derived (again according to Lowenthal, 2000a), from "the creative coincidence" of Marsh's "own special skills and circumstances with a habit . . . of contrasting Old World and New World perspectives" (p. 430). In this account—and it has been elaborated repeatedly by Lowenthal —Marsh was a great man and his book a lightning bolt of white hot insight that forged "a truly modern way of looking at the world, of thinking about how people live in and react on the fabric of the landscape they inhabit" (Lowenthal, 2000a, pp. 429–430). By combining ecological insight with an appreciation of the need for social reform, says Lowenthal, Marsh framed arguments that retain "a lasting force four generations later." That they are not more widely appreciated in the twenty-first century has more to do with the myopia of the present than the power and prescience of those arguments. Marsh's words can yet help to "bridge the gulf between the environment we have and the environment we need" (Lowenthal, 2000b, p. 16).

Both of these accounts hold some water. For all their thumbnail-sketch brevity, however, neither seems capable of providing an entirely satisfying account of how *Man and Nature* had the impact it did and why it has been so little engaged in recent time. In an effort to address this conundrum, I turn now to look more closely at the form and content of Marsh's book and the context into which it was released. This, it seems to me, is an important thing to do. To modern eyes, *Man and Nature* is a very peculiar book indeed. The essence of Marsh's approach lies in the observation that "labor is life"; seeking to stimulate rather than to satisfy curiosity, he flatly denies any desire "to save my readers the labor of observation or of thought" (Marsh, 1864, p. 10). If this makes the book tough to read, then so be it. "Self is the schoolmaster whose lessons are best worth his wages," and those who harbor doubts would do well to recall that "Death lives where power lies unused" (p. 10).[4]

Man and Nature requires the reader to develop that "power most important to cultivate, and, at the same time, hardest to acquire," the power "of seeing what is before him" (Marsh, 1864, p. 10). There are "no more important practical lessons in this earthly life of ours," asserted Marsh, "than those relating to the employment of the sense of vision in the study of nature." But, he cautioned, "the eye sees only what it seeks"; like a mirror, "it does not necessarily perceive what it reflects" (p. 10). Sight, said Marsh "is a faculty; seeing, an art"–and then he elaborated on this seven-word claim (in a manner that is entirely typical) with a 700 word footnote (p. 10). This note begins with the observation that "skill in marksmanship . . . depends more upon the training of the eye than is generally supposed," (p. 11) and discusses the use of firearms and almost every other known projectile weapon. Then there is a comment on how the Indians of the Amazon shoot tortoises:

> As the arrow, if aimed directly at the floating tortoise, would strike it at a small angle, and glance from its flat and wet shell, the archers have a peculiar method of shooting. They are able to calculate exactly their own muscular effort, the velocity of the stream, the distance and size of the tortoise, and they shoot the arrow directly up into the air, so that it falls almost vertically upon the shell of the tortoise, and sticks in it. (p. 11)

---

[4] Marsh (1864, p. 10) renders this phrase as "Death lives where power lives unused," and attributes it to a verse addressed to Sir Walter Raleigh and quoted by Hakluyt. Christopher Hill (1997, p. 141) attributes the version used here to George Chapman, who prefixed a poem including this line to Lawrence Kemyis "Relation of the second voyage to Guiana (1596)".

This is followed by a riff on the etymology of the word *aim*, a discussion of how blind children are taught to write, and reflections on the visual acuity of Classical artists: "Glasses ground convex have been found at Pompeii," Marsh informs his readers, "but they are too rudely fashioned and too imperfectly polished to have been of any practical use for optical purposes" (p. 12).

Examples could be multiplied and multiplied again. There is a quality of considered judgment and a sort of magic realism evident on every page of the book. Tracing the impacts of human actions on the physical earth was no easy task. A couple of centuries back (i.e., before the middle of the seventeenth century) knowledge of meteorological conditions derived from imperfect sources, "from the vague statements of ancient historians and geographers in regard to the volume of rivers, and . . . from other almost purely casual sources of information" (Marsh, 1864, p. 16). Ancient dwelling sites, "memorials of races which have left no written records" (p. 16), have yielded animal and vegetable remains from which "ingenious inferences have been drawn as to the climates of Central and Northern Europe" (p. 17) in earlier times. But, a note of caution:

> Even if we suppose an identity of species, of race, and of habit to be established between a given ancient and modern plant, the negative fact that the latter will not grow now where it flourished 2,000 years ago does not in all cases prove a change of climate. The same result might follow from the exhaustion of the soil,—or from a change in the quantity of moisture it habitually contains. (p. 20)

More generally, it is important to remember that "There are . . . sources of error which have not always been sufficiently guarded against in making these estimates" (p. 17). If you are having a hard time figuring out quite how all this soil chemistry and capillary moisture retention works, then there is a footnote on the purported introduction of madder to southern France and the decline in the quality of the crop over a century (p. 20). But if this is too much, then consider this:

> When a boat, composed of several pieces of wood fastened together by pins of the same material, is dug out of a bog, it is inferred that the vessel, the skeletons, and the implements found with it, belong to an age when the use of iron was not known to the builders. But this conclusion is not warranted by the simple fact that metals were not employed in its construction; for the Nubians at this day build boats large enough to carry half a dozen persons across the Nile, out of small pieces of acacia wood pinned together entirely with wooden bolts. (p. 17)

And in similar vein,

> although it has been said that stone weapons are not found in Sicily, except in certain caves half filled with the skeletons of extinct animals. . . . I suspect . . . [this] is because eyes familiar with such objects have not sought for them. In January, 1854, I picked up an arrow head of quartz in a little ravine or furrow just washed out by a heavy rain, in a field near the Simeto. It is rudely fashioned, but its artificial character and its special purpose are quite unequivocal. (p. 18)

There are important rhetorical qualities at work and on display in this passage. *Man and Nature* offers a long argument for ordinary readers. Marsh was explicit in proclaiming that his book was addressed "not to professed physicists, but to the general intelligence of educated, observing, and thinking men; and that . . . [his] purpose is rather to make practical suggestions than to indulge in theoretical speculations" (Marsh, 1864, p. vi). It is remarkable for the sheer range of knowledge that it encompasses, and for (what is now called) its sympathetic summary of the ideas and interpretations of others. In this it bears comparison with the other great book of its age, *The Origin of Species* (Darwin, 1859). Both Marsh and Darwin were what Adam Gopnik (in his brilliant short study of Darwin and Lincoln) has called "nearsighted visionaries" (Gopnik, 2009, p. 19). They *"particularized* in everything" and their arguments, their "big ideas," emerged from the welter of detail that they laid out for their readers. "They built their inspiration from induction" and relied upon the "slow crawl of fact"—much of it "certified" by the author's personal observation—to give potency to their arguments (p. 20). Marsh, like Darwin, had (as Gopnik puts it)

> written a book whose tone of empirical exactitude, fair-minded summary, and above all sweeping argumentative force—so subtly orchestrated that it acted not as a straitjacket on the argument pressing it in, but as a tide behind it, driving it forward—was almost impossible to resist. (p. 147)

This was a tide for its times. Like *The Origin of Species*, and Abraham Lincoln's speech-making, *Man and Nature* is marked by a certain eloquence (now perhaps regarded as somewhat dated), by an expectation that curious and gentle readers would work hard to find their ways through great thickets of detail, and by an "insistent need to persuade and convince, argue and substantiate, talk and justify" (Gopnik, 2009, p. 183). This was a style much used in the Victorian era and perhaps especially common in natural history writing. In some ways it drew its inspiration from Alexander von Humboldt: think of his unrelenting efforts to catalogue minutiae, of his hauling a barometer across the spine of central America to chart variations in air pressure, of his conviction that large issues might be understood by detailed observations of small things and of his assertion of "mutual dependence and connection" in nature (Humboldt, 1845/1858, p. 8). But it was widely evident—from Gilbert White's observations of the miniscule in Selborne to John Ruskin's obsessive devotion to measurement in *Stones of Venice* (Ruskin, 1851–1853; White, 1789). There was, perhaps inevitably, a sort of helter-skelter quality to much of this prose. Yet these rhetorical commitments had consequences. According to Adam Gopnik (2009, pp. 73, 184), Lincoln ("who lived in a society of speaking") used "the narrow language of the law to arrive at a voice of liberalism still resonant and convincing today" and Darwin ("who lived in a society of seeing") used "the still more narrow language of natural observation . . . to change our ideas of life and time and history. Marsh, it seems to me, embodied both tendencies; he combined legal training with the habit of "close amateur looking" to drive home the message that "man has done much to mould the form of the earth's surface" (Marsh, 1864, p. 13). Yet the tide behind *Man and Nature* lost energy as that behind *On the Origin of Species* did not, and Marsh's memorial in Washington is the Smithsonian Institution (which he helped to found), not a statute on the National Mall. Why is this, one must ask?

One of the fundamental underpinnings of Marsh's argument, the very bedrock of *Man and Nature*'s rhetorical power, was the notion of "nature's harmony" and its disruption. Nature was durable—until its balance was destroyed by human action. This was not to say that nature was stable and unchanging. As Marsh saw it,

> every generation of trees leaves the soil in a different state from that in which it found it; every tree that springs up in a group of trees of another species than its own, grows under different influences of light and shade and atmosphere from its predecessors. (1864, p. 22)

Nature was dynamic, within limits, and the natural world had a certain resilience. The glacial pace of geological and astronomical changes allowed nature to respond without loss of equilibrium. Other small-scale changes could also be absorbed. When disturbed by natural forces, nature sought "at once to repair the superficial damage, and to restore . . . the former aspect" (p. 27). But human *ravages* destroyed nature's *balance*. They were hammer blows to the web of natural life. Wherever humans settled, there ensued "almost indiscriminate warfare" that "gradually eradicates or transforms every spontaneous product of the soil" (p. 41). Human actions—clearing forests first among them—amplified the intensity of erosion. Two or three generations of human action were capable of producing "effects as blasting as those generally ascribed to geological convulsions" and "laid waste the face of the earth more hopelessly than if it had been buried by a current of lava or a shower of volcanic sand" (p. 262). As societies armed with ever more powerful technologies exercised dominion over the earth, nature was despoiled beyond its capacity to heal itself. Marsh's book denounced these tendencies, so forcefully that large parts of it have an apocalyptic tone—although a second (less noticed) side of *Man and Nature* celebrates people's capacity to rebuild, restore, and reconstruct lands laid waste by the destructiveness of humankind, and urges societies to better stewardship of nature (Hall, 2005).

In assessing the impact of Darwin's great book, Adam Gopnik observes that "Scientific ideas become a whole climate of opinion when they can provide a set of metaphors for people who aren't doing science" (2009, pp. 152–153). Marsh (1864) certainly provided both powerful metaphors—"Breaking up the floor and wainscoting and doors and window frames of our [earthly] dwelling, for fuel to warm our bodies and seethe our pottage" is but the most well known—and lugubrious warnings (or moral injunctions)—". . . man is everywhere a disturbing agent. Wherever he plants his foot, the harmonies of nature are turned to discords" (p. 36)—to his readers in reminding them that "Man has too long forgotten that the earth was given to him for usufruct alone, not for consumption, still less for profligate waste" (p. 35). These echoing phrases struck a chord among his contemporaries confronting shortages of fuelwood, the erosion of hillslopes, the silting of millponds, and concerns about resource depletion.

But—Gopnik (2009) again—"for a new scientific theory to become . . . vastly influential" beyond its immediate sphere, it has to help "thinking people …interrogate the world in a new way." (p. 153). Despite Lowenthal's claims for Marsh's prescience, *Man and Nature* never quite achieved this level of probing insight. Marsh's embrace of the balance of nature was entirely orthodox. Although

Lowenthal (2000a) argues that "Marsh's vision of a self-regulating nature . . . became, in its essential vision the ecological paradigm of the early twentieth century," (p. 292) the idea is in fact a very old one. *Man and Nature* was at its most powerful in urging that humans had an impact upon nature. But this was no revelation in 1864. Marsh's phrases may have been more compelling than those of other writers, but they did not bring light to the world. Fully 60 years earlier, Alexander von Humboldt traveling through the Equinoctial regions of America, had come upon Lake Tacarigua high in the mountains of Venezuela. The level of the lake had been sinking for years. It was surrounded by desiccated landscapes, "vast tracts of land . . . formerly inundated, now dry" (Humboldt quoted in Sachs, 2006, p. 77). The locals thought that the lake was draining through some subterranean outlet. But Humboldt offered a different explanation—and this is only the baldest summary of it: the changes were attributable to "the destruction of forests, the clearing of plains, and the [irrigated] cultivation of indigo" (p. 77).

In retrospect it seems to me that Marsh's book was most effective in answering "what" questions, and in framing somewhat familiar arguments in powerful ways, and rather less successful at identifying "why" things happened as they did and revealing the world in truly new ways. People changed places, humankind modified the earth. There was little room to doubt this proposition in the mid-nineteenth century (indeed, rather like the idea of evolution, it was hardly revolutionary in 1859 (Stott, 2012)). To borrow Adam Gopnik's (2009) wonderful imagery, Darwin's great achievement was in taking "a poetic figure familiar to his grandfathers" and putting "an engine and a fan belt in it" (pp. 7–8). In other words, his triumph lay in finding the mechanism that drove evolution. Marsh never came as close to accounting for the transformations he documented; he told his readers what people did to the earth but rarely explained why.

There is a further point of intersection between *On the Origin of Species* and *Man and Nature* that warrants attention. The last lines of Marsh's book serve both to explain the dense assemblage of detail in the preceding pages and to identify the big question with which it was most concerned.

> The collection of phenomena must precede the analysis of them, and every new fact, illustrative of the action and, reaction between humanity and the material world around it, is another step toward the determination of the great question, whether man is of nature or above her. (Marsh, 1864, p. 549)

In the end, *Man and Nature* and *On the Origin of Species* address the same question—is man *of* nature or *above* her—and produce very different answers to it. Marsh concludes "above," Darwin "of." There is no evidence that Darwin knew of the arguments Marsh would articulate in *Man and Nature* when he wrote *The Origin*, but Marsh was certainly familiar with Darwin's writings. Indeed he housed a certain suspicion of Darwin's concept of evolution by natural selection, at least as it was brought to bear on cultural rather than natural history. To the philologist—and at least one obituary of Marsh celebrated his contributions to this realm above those represented by *Man and Nature*—the branching, ever more diverse, tree of life mapped a pattern that was utterly contrary to that revealed by the evolution of languages. "History teaches us, the further back we go the wider was the diversity of speech among men" (Lowenthal, 2000a, p 306). Marsh also took gentle issue with Darwin for arguments that underestimated the duration and extent of human modifications of the earth. Most fundamentally, however, where Darwin saw humans located within nature—most famously expressed in the final chapter of *The Descent of Man*: "We thus learn that man is descended from a hairy quadruped, furnished with a tail and pointed ears, probably arboreal in its habits, and an inhabitant of the Old World" (Darwin, 1871, p. 291)—Marsh believed "that man is, in both kind and degree a power of higher order than any of the other forms of animated life" (Marsh, 1864, p. iii). On this he was firm. Man was not "part of nature" nor was "his action . . . controlled by what are called the laws of nature". Indeed "a leading object" of *Man and Nature* was "to enforce the opposite opinion, and to illustrate that man . . . is a free moral agent working independently of nature" (Marsh to C. Scribner, 10 September 1863, cited by Lowenthal, 2000a, p. 291; see also more generally 2000b, p. 5).

These are the reasons why Marsh is now most generally acknowledged as a precursor rather than honored as a prophet, and why *Man and Nature* is known but hardly read these days. The two fundamental suppositions on which the book (and Marsh's reputation) rest—that there is a balance of nature and that humankind stands apart from nature—have been reconsidered in recent years. Charles Elton, Daniel Botkin, and chaos theory largely put paid to the former (at least in the sense used by Marsh), and Rachel Carson (among others) gave the lie to the latter (Botkin, 1990; Elton, 1942).

Silent Spring: The Fate and Power of Words

Rachel Carson's *Silent Spring*, a controversial best-selling book about the toxic side effects of widely-used chemical pesticides, herbicides, and fungicides, published by Houghton Mifflin in 1962, spawned immediate controversy in the media. Like *Man and Nature* (which opens, remember, by looking back to the desiccation of the Mediterranean littoral due to the improvidence of "man"), *Silent Spring* also opens with a bang, but Carson's "hook" is forward-looking and fictional. Her *Fable for Tomorrow* tells of a beautiful (albeit non-existent) town nestled in the heart of American plenitude where, suddenly, animals die, people succumb to "mysterious maladies," the bees disappear, and no birds sing. It was not witchcraft or enemy action that "silenced the rebirth of new life in this stricken world," but a white granular powder that fell from the skies. "The people had done it themselves" (Carson, 1962, pp. 1–3).

Marsh's argument was similar in some ways: man modified the earth, the consequences were deleterious, fields dried, soil blew, lakes shrank (until perhaps sedges withered at their former edges), and people brought all of this on themselves by their careless and profligate actions. But there were important differences between Marsh's and Carson's books. In pursuing her story, Carson developed a narrower, sharper identification of the villains of her piece than Marsh offered in his scolding volume. Yes 'the people' were ultimately responsible for their plight. But they had remained inert—and allowed the specter of a Silent Spring into being—because they had been kept in the dark. Awareness of the threats posed by chemical pesticides was very limited because "this is an era of specialists . . . [and] also an era dominated by industry, in which the right to make a dollar at whatever cost is seldom challenged" (Carson, 1962, p. 13). Chemical companies and their employees, supporters, and spokespersons (who could not be expected to bite the hand that feeds them), had failed to disclose or denied what they knew to be true and, worse, had sometimes fobbed off public anxiety with "little tranquilizing pills of half truth" (p. 13). Even the government, which people had a right to regard as the protector of its citizens, had failed in its imputed responsibility to "secure [individuals] against lethal poisons" (p. 12). Research costing a fraction of the sum spent on developing toxic sprays could "keep poisons out of our waterways," noted Carson in the final lines of a chapter entitled "Rivers of Death," before ending with the rhetorical question: "When will the public become sufficiently aware of the facts to demand . . . action" (p. 152)? All of this offered concerned citizens a clear set of targets, set up the possibility of an "us" against "them" struggle, and dressed the battle in Old Testament cloth as a confrontation between David and Goliath—in marked contrast to Marsh's Pogo-esque, and at some level debilitating, conclusion that "we have met the enemy and he is us".[5]

Like that offered by *Man and Nature*, the originality of the argument in *Silent Spring* has been exaggerated. Celebrated as the fountainhead of the new environmental movement, Carson has been described by former vice president of the United States, Al Gore as planting "the seeds of a new activism that has grown into one of the great popular forces of all time" and providing "a shaft of light that for the first time illuminated what is arguably the most important issue of our era" (Gore, n.d.). But Carson drew insight and evidence, as did Marsh, from the work of many other writers and scientists. Both books include lengthy bibliographies of works consulted. In the late 1940s, veterinarians had raised questions about the harmful effects of DDT on animals, and the Audubon Society did likewise with respect to birds several times during the 1950s (see also more generally Whorton, 1975). In 1957 residents of Long Island sued the USDA for their aerial spraying of several communities in an effort to eradicate gypsy moths, and 2 years later American Thanksgiving celebrations were thrown into turmoil by a report that cranberries had been contaminated by aminatraizole, a weed-killer known to cause cancer in rats.[6] Just as others had written of environmental decline in the eastern Mediterranean before 1864, so Carson's powerful image of the Silent Spring was adumbrated in

---

[5] The phrase "We Have Met The Enemy and He Is Us" was used by Walter Kelly creator of the Pogo comic strip on a poster for Earth Day in 1970. It then appeared as the title of a book: Kelly, W. (1972). *Pogo: We have met the enemy and he is us.* New York, NY: Simon and Schuster.

[6] Carson (1962, pp. 154–159) discusses the Long Island gypsy moth issue. The cranberry incident was well reported in Larry Gosnell's National Film Board of Canada documentary *Poisons, Pests and People* produced and aired on the Canadian Broadcasting Corporation (CBC) in Bairstow and Gosnell (1960), and available at: http://beta.nfb.ca/film/Poisons_Pests_People

earlier writing. In July 1946, for example, John Terres wrote about the spraying of DDT on Moscow, Pennsylvania (Terres, 1946). He told the story of a bright May morning, when birdsong ran through the oak woodlands, but the trees were losing their leaves to the voracious gypsy moth. An airplane droned overhead and released a fine mist. "The effect was instantaneous. The destructive caterpillars caught in the deadly rain, died by the thousands". But the next morning "the sun rose on a forest of great silence—the silence of total death. Not a bird call broke the ominous quiet." (see also Davis, 1971; Lear, 1992).

Just as *Man and Nature* found a broadly receptive audience in the decade or two after its first publication, so did *Silent Spring*. If anything, the latter book appeared at a more propitious moment than did its predecessor. Post-World War II economic prosperity had begun to reconnect Americans with nature. Automobiles facilitated visits to national parks, forests, and other places of natural beauty; suburban homes had lawns and gardens to tend and beautify; garden clubs grew; and the numbers of those hunting and fishing rose moderately (Charbonneau & Lyons, 1980, pp. 121–126; Rome, 2001; Sutter, 2002). The plight of some 10,000 children worldwide, born between 1957 and 1962 with physical deformities attributable to the prescription of thalidomide to their mothers as an inhibitor of morning sickness, was widely publicized and raised awareness of what chemicals could do to human bodies (Campaign Against Fraudulent Medical Research, 1996). The newly-elected Kennedy administration was more open (than many administrations before and since might have been) to such arguments as Carson presented. Not least, moreover, new media—including magazines—such as the *New Yorker*, which ran an abbreviated version of Carson's account in three parts before publication of the book—television and radio massively increased the reach of Carson's words. They were also important vehicles for dissemination of the vigorous and vicious critique of Carson and her book mounted by the very chemical companies she criticized. Debate was quickly polarized, but the specialist and financial resources of corporate America were unable to quell the groundswell of interest in and support for Carson's arguments. In challenging large chemical pesticide producers, she was perhaps tapping into the growing unease with corporate and bureaucratic America given early expression in the works of sociologists David Riesman, C. Wright Mills, and Vance Packard, and soon to flourish into the counter-culture movements of the 1960s (Mills, 1951; Packard, 1959; Riesman, 1950; Roszak, 1969).

Some have found in *Silent Spring* an argument for the balance of nature, but in my reading Carson's case rests on an acknowledgment of the interrelatedness of nature's parts rather than upon a belief (qua Marsh) in nature's somewhat mystical harmony. This is an important distinction. Carson understood nature as an intra-dependent system and she was concerned about the ways in which human actions affected the world around us and then stood to redound upon humans themselves. Carson was a scientist, a trained zoologist with a Master's degree and a long career in the U.S. Fish and Wildlife Service, much of it as editor in chief of its publications division.[7] In her prodigious research she combined the approaches of the professional scholar and the investigative journalist; "checking and digging and research" she wrote her agent, "are matters I would never turn over to another person" (Murphy, 2005, p. 26). Her dedication to inquiry and her quest for deeper understanding led her to a grave concern, echoed in the dedication of *Silent Spring*: "To Albert Schweitzer who said 'Man has lost the capacity to foresee and to forestall. He will end by destroying the Earth.'" Until her publisher suggested the John Keats-inspired title of her last book she planned to call it *Man Against Nature* (Murphy, p. 31).

There are echoes in this, and tangled ironies too. Recognizing their respective views of humankind's place in nature—as independent and above, and as intradependent and within—it is hardly a stretch to suggest that Marsh might more accurately have conveyed the message of his book by calling it *Man Against Nature* and that Carson might happily have used *Man and Nature* to describe her work, had that title not been in circulation already. In the end however, both Marsh and Carson were anxious about humankind's failure to exercise due stewardship of the earth, and their books were intended to change this. Clearly—no surprise here—they wrote in very different ways, and their works, published a century apart, engaged radically different contexts. As these pages have demonstrated, both *Man and Nature* and *Silent Spring* have been widely recognized as landmark contributions to ongoing debates about human environment relations, and each volume has been said by enthusiastic supporters to have changed the ways in which people thought, and continue to think, about the world.

But there are reasons to doubt such claims. So in conclusion I turn to consider two things: first, whether these books and their authors stand as bolts of pure genius that metamorphosed understanding and transformed the landscape or whether they are better seen as ripples reverberating across the tides of time, the marks of people and productions whose impact left surficial traces on the deeper ocean of discourse; and second, whether these radically different books offer any insights that may assist in dealing with the looming environmental challenges precipitated by what historian John McNeill has called twentieth-century humankind's propensity "to play dice with the planet, without knowing all the rules of the game" (McNeill, 2000, p. 4).

Words at Work

In *The Forbidden Best-sellers of Revolutionary France*, Robert Darnton (1995) wonders whether books cause revolutions. Although some might insist otherwise, these days the guarded, historian's, answer has to be *no*. By the lights of our time, events are seen as complex, contingent, and interdependent—and contrary to the

---

[7] The fullest account of Carson's life is Lear (1997), but see also Souder (2012). The following are also valuable in the larger context of this discussion: (Lytle, 2007; Waddell, 2000 and Dunlap, 1974; Dunlap, 2008).

lessons of many high school text books—most historical moments are not thought to be properly explained by a ranked list of *reasons why*. Nor are ideas unitary. They find expression in many forms, some of which reverberate more strongly in some circles than in others. Indeed, and because of this, it is extremely hard to trace the diffusion of ideas. It is even more difficult to ascertain the effects or results of reading any particular text. Reader response theory has emphasized that reading is not a passive act and insisted that the meaning any and every reader takes from a text is influenced by her personal circumstances and cultural setting. Thus, in some sense, much of the debate reviewed above about the primacy or otherwise of the ideas found in *Man and Nature* and *Silent Spring* is unfounded or irrelevant. Yet none of this negates Henry Giles sense that books might "set in action countless multitudes" and change the course of events. It is necessary to evaluate the effects and influence of these two books, and to ponder their method of working, if their importance to both past and present is to be understood, and their lessons turned into action.

Both *Man and Nature* and *Silent Spring* might be described—in a phrase more common in our day than in theirs—as works of nonfiction designed to raise public consciousness. Historians of the book have spent a good deal of effort charting the processes by which books are produced and their messages disseminated and received, and even the most cursory acquaintance with this work suggests that *Silent Spring* had a much greater chance of raising public consciousness than did *Man and Nature*. It was launched into a mass society. Broadcast media; publicity departments; book clubs—all helped to create a buzz around the book and to draw attention to its message. So too did Carson's powerful indictment of the academic scientists, the government agencies, and the chemical industry whom she held responsible for the hazardous use of pesticides. The industry, in particular, responded vigorously to rebut Carson's claims. Her science was challenged, and her credibility impugned, often with pointed comments about her gender (why, asked a former Secretary of Agriculture, would a "spinster with no children . . . [be] worried about genetics"? (Ezra Taft Benson, as quoted in Murphy, 2005, p. 106). The Monsanto corporation even responded with a parody of Carson's *Fable for Tomorrow*, called "The Desolate Year," which described in lurid prose the terrible effects that the tightening "garrotte of Nature rampant" (The Desolate Year, pp. 4–9 as quoted in Murphy, p. 100) would have in a world without pesticides. The particular con-juncture presented by the early 1960s, with its concerns about the appearance of Strontium in breast milk, nuclear build up (Jarvis, Brown, & Tiefenbach, 1963),[8] and so on was also highly conducive to creating a receptive audience for Carson's work. *Silent Spring* was a Book of the Month Club selection, a special edition was produced for distribution to members of the Consumers Union, and it was much

---

[8] This appeared a few months after the publication of *Silent Spring*, but its notes reference earlier work on the topic. For the nuclear build-up, see various items available in CBC archives under the heading "Cold War Culture: The Nuclear Fear of the 1950s and 1960s," available at: http://archives.cbc.ca/war_conflict/cold_war/topics/274/ and the important article by (Lutts, 1985).

serialized in periodicals; it was on best-seller list for weeks, and within 3 months of the book's publication half a million copies were in print.

All of this is in stark contrast with the reception of *Man and Nature*. It sold well in its day, to be sure. But 1,000 copies against half a million. This reveals a good deal about the reach of Marsh's book. Written for "intelligent observing, and thinking men," its audience was, it would seem, largely limited to a few of them. It spoke, as almost all sizeable and serious books must have done in the mid-nineteenth century, to an elite male readership. Identifying no villains—and avoiding the challenge to religious orthodoxy that led powerful persons to rise up against (and draw notice to) Darwin's ideas—*Man and Nature* provoked no heated opposition. Its influence—on a few—was profound, and that chain of effect ran, not inconsequentially, from Franklin Hough through the Imperial Forest Department in India to New Zealand parliamentarians W. T. L. Travers and Thomas Potts, and on to Carl O. Sauer and beyond, eventually providing a stimulus for my own work. But *Man and Nature* a bolt of pure genius, the touchpaper of an environmental revolution? I think not.

In the final analysis this is probably too much to claim even for Rachel Carson, though *Silent Spring* might be described more legitimately than *Man and Nature* as a book that changed the world. Yet neither of these books transformed the landscape of understanding, or caused a revolution *on its own*. Both contributed, unevenly, to developing forces of concern and conviction of which they were both reflections and parts. Still, the differences in their public impact and the sweep of their influence are worth contemplation by those who wish to intervene in current debates about the future of the earth, or more generally, to speak truth to power.

Ultimately, it seems to me that—context and all that that implies for the possibilities of dissemination aside—*Silent Spring* was a more powerful instrument of change than *Man and Nature* because of the nature of its story and the way it was told. Both Marsh and Carson were concerned about the ways in which the actions of members of their generation were despoiling the earth, but Carson seared her concern over the bioaccumulation of toxic chemicals into the public consciousness by lodging it in the very tissue of every human body, whereas Marsh emphasized the role of long-term physical processes reducing the fertility and utility of particular parts of the planet. Put simply, widespread threats to one's person (and one's children's persons) would seem more likely to move people to action than fears for the future of distant spaces or nearby places. Both *Man and Nature* and *Silent Spring* identified those responsible for the environmental challenges against which they railed, but they presented radically different possibilities for action against the ills they confronted because—in the broadest of terms—the villains revealed to readers of the former appeared as *us* and of the latter as *them*. Resolved to its essence, this contrast left *Man and Nature* hostage to what Garrett Hardin (1968) characterized as *The tragedy of the commons*—providing little incentive to individual action so long as there was no assurance that all others would act in accord—whereas *Silent Spring* offered up a clearly-identified and relatively small set of villains whose humankind-threatening deeds might be challenged and stopped. These are contrasts worth remembering in the twenty-first century as humankind struggles collectively, and in spite of an enormous accumulation of scientific knowledge, to address the "seething pottage" of climate change, even while noting the success of campaigns against the contamination of secondhand smoke or the use of plastic water bottles shown to contain endocrine disruptors.

All of this said, I have no doubt that both Marsh and Carson were great people—though I rest this judgment on their full lives led rather than their landmark books, and I cannot make of either of them Carlyle-style heroes or indispensable saviors of their epochs. Their lives were but pebbles in the sea of history—pebbles that fell with force and whose ripples fanned by acolyte winds continue to scud across the pond, but pebbles nonetheless. They made a difference, but within limits. And as for these writers of books preaching "to all men in all times and places," I confess to doubts on this score too. *Man and Nature* and *Silent Spring* have found readers around the globe and are indubitably mobile, but—like these and all other words—they have been and will be received and understood differently in the variety of contexts into which they are inserted. Nothing is immutable, not even the pedestals upon which preachers are elevated for reasons that often have as much to do with their acolytes as themselves.

References

Anon. (1875) Review of *The Earth as Modified by Human Action. A New Edition of Man and Nature*, By George P. Marsh. New York: Scribner, Armstrong & Co. *The International Review*, 2, 120–125.

Bairstow, David (Producer), & Gosnell, L. (Director). (1960). *Poisons, Pests and People*. National Film Board of Canada. Retrieved September 13, 2012, from http://beta.nfb.ca/film/Poisons_Pests_People

Bohman, J., & Rehg, W. (2009). Jürgen Habermas. In E. N. Zalta (Ed.), *The Stanford encyclopedia of philosophy*. Retrieved September 11, 2012, from http://plato.stanford.edu/archives/sum2009/entries/habermas/

Botkin, D. (1990). *Discordant harmonies: A new ecology for the twenty-first century*. New York: Oxford University Press.

Campaign Against Fraudulent Medical Research. (1996). The thalidomide tragedy: Another example of animal research misleading science. *CAFMR Newsletter*, Spring 1996. Retrieved September 13, 2012, from http://www.pnc.com.au/~cafmr/online/research/thalid2.html

Carlyle, T. (1840). *On heroes and hero worship and the heroic in history: Lecture I. The hero as divinity. Odin. Paganism: Scandinavian mythology*. Retrieved September 10, 2012, from http://www.gutenberg.org/files/1091/1091-h/1091-h.htm#2H_4_0002

Carlyle, T. (2007). *On heroes and hero worship and the heroic in history: Lecture V. The hero as man of letters: Johnson, Rousseau, Burns*. (Reprint of the "Sterling Edition" of Carlyle's Complete Works in 20 Volumes). Teddington, UK: The Echo Library. (Original work published 1841)

Carson, R. (1962). *Silent spring*. Boston: Houghton Mifflin.

Charbonneau, J. J., & Lyons, J. R. (1980). Hunting and fishing trends in the U.S. (General Technical Report NE-57). *Proceedings: 1980 National Outdoor Recreation Trends Symposium, Vol. 1*. U.S. Department of Agriculture, Forest Service.

Clark, A. H. (1954). Historical geography. In P. E. James & C. F. Jones (Eds.), *American geography: Inventory and prospect* (pp. 70–105). Syracuse, NY: Syracuse University Press.

Cronon, W. (2000). Foreword: Look back to look forward. In D. Lowenthal (Ed.), *George Perkins Marsh: Prophet of conservation* (pp. ix–xiv). Seattle: University of Washington Press.

Darnton, R. (1995). *The forbidden best-sellers of revolutionary France*. New York: W. W. Norton.

Darwin, C. (1859). *The origin of species by means of natural selection, or the preservation of favored races in the struggle for life*. London: John Murray.

Darwin, C. (1871). *The descent of man, and selection in relation to sex* (Vol. 2). London: John Murray.

Davis, K. S. (1971). The deadly dust: The unhappy history of DDT. *American Heritage, 22*, 44–47, 93.

De Buffon, M. (1782). *Histoire naturelle: Generale et particuliere* [Natural history: General and particular]. Paris: De l'Imprimerie royale.

Dunlap, T. (1974). *DDT: Scientists, citizens, and public policy*. Princeton, NJ: Princeton University Press.

Dunlap, T. (2008). *DDT, Silent Spring, and the rise of environmentalism*. Seattle, WA: University of Washington Press.

Egan, M. (2007). *Barry Commoner and the science of survival: The remaking of American environmentalism*. Cambridge, MA: MIT Press.

Ehrlich, P. R. (1968). *The population bomb*. New York: Ballantine Books.

Elton, C. (1942). *Voles, mice and lemmings: Problems in population dynamics*. Oxford, UK: The Clarendon Press.

Giles, H. (n.d). Quotation retrieved June 4, 2013, at: http://www.saidwhat.co.uk/quotes/favourite/henry_giles/the_silent_power_of_books_is_21525

Girard, M. F. (1990). Conservation and the Gospel of efficiency: Un modele de gestion de l'environnement venu d'Europe? [A model of environmental management in Europe]. *Histoire Sociale/Social History, 23*(45), 63–80.

Glacken, C. (1967). *Traces on the Rhodian shore: Nature and culture in Western thought from ancient times to the end of the eighteenth century*. Berkeley, CA: University of California Press.

Gopnik, A. (2009). *Angels and ages: A short book about Darwin, Lincoln, and modern life*. New York: Knopf.

Gore, A. (n.d.). *Introduction*. Retrieved September 12, 2012, from http://clinton2.nara.gov/WH/EOP/OVP/24hours/carson.html

Greeley, W. B. (1925). The relation of geography to timber supply. *Economic Geography, 1*, 1–11.

Grove, R. (1995). *Green imperialism: Colonial expansion, tropical island Edens and the origins of environmentalism*. Cambridge, UK: Cambridge University Press.

Habermas, J. (1984). *The theory of communicative action: Vol. 1. Reason and the rationalization of society* (T. McCarthy, Trans.). Boston: Beacon Press.

Habermas, J. (1987). *The theory of communicative action: Vol. 2. Lifeworld and system* (T. McCarthy, Trans.). Boston: Beacon Press.

Habermas, J. (1998). Some further clarifications of the concept of communicative rationality (Maeve Cooke, Trans.). In Maeve Cooke (Ed.), *On the pragmatics of communication* (pp. 307–342). Cambridge, MA: MIT Press.

Hall, M. (2005). *Earth repair: A transatlantic history of environmental restoration*. Charlottesville, VA: University of Virginia Press.

Hardin, G. (1968). The tragedy of the commons. *Science, 162*(3859), 1243–1248. doi:10.1126/science.162.3859.1243.

Hays, S. P. (1959). *Conservation and the gospel of efficiency: The progressive conservation movement, 1890–1920*. Cambridge, MA: Harvard University Press.

Hays, S. P. (1987). *Beauty, health and permanence: Environmental politics in the United States, 1955–1985*. Cambridge, UK: Cambridge University Press.

Hill, C. (1997). *Intellectual origins of the English revolution revisited* (Rev. ed.). Oxford, UK: Oxford University Press.

Hough, F. B. (1873, August). *On the duty of governments in the preservation of forests*. From the Proceedings of the American Association for the Advancement of Science. Portland Meeting. Retrieved September 11, 2012, from http://memory.loc.gov/cgi-bin/query/r?ammem/consrv:@field%28DOCID+@lit%28amrvgvg28div0%29%29

Humboldt, A. von (1858). *Cosmos: A sketch of a physical description of the universe* (Vol. 1, E C. Otte, Trans.). London: Harper & Brothers. (Original work published 1845) Quotation retrieved June 4, 2013, from http://www.gutenberg.org/files/14565/14565-8.txt

Hynes, H. P. (1989). *The recurring Silent Spring* (Athene series). New York: Pergamon Press.

Jarvis, A. A., Brown, J. R., & Tiefenbach, B. (1963, January 19). Strontium-89 and Strontium-90 levels in breast milk and in mineral-supplement preparations. *Canadian Medical Association Journal, 88,* 136–139.

Judd, R. W. (1997). *Common lands, common people: The origins of conservation in northern New England.* Cambridge, MA: Harvard University Press.

Kant, I. (1802). *Immanuel Kant's physische Geographie. Auf Verlangen des Verfassers aus seiner Handschrift herausgegeben und zum Theil bearbeitet von D. Friedrich Theodor Rink* [Immanuel Kant's physical geography: Compiled and partially edited by Friedrich Theodor Rink from Kant's lectures by order of Kant]. Königsberg: Göbbels und Unzer.

Kelly, W. (1972). *Pogo: We have met the enemy and he is us.* New York: Simon and Schuster.

Koelsch, W. A. (2012). The legendary "rediscovery" of George Perkins Marsh. *Geographical Review, 102*(4), 510–524.

Lear, L. (1992). Bombshell in Beltsville: The USDA and the challenge of "Silent Spring". *Agricultural History, 66,* 151–170.

Lowenthal, D. (1958). *George Perkins Marsh: Versatile Vermonter.* New York: Columbia University Press.

Lowenthal, D. (2000a). *George Perkins Marsh: Prophet of conservation.* Seattle, WA: University of Washington Press.

Lowenthal, D. (2000b). Nature and morality from George Perkins Marsh to the millennium. *Journal of Historical Geography, 26,* 3–23. doi:10.1006/jhge.1999.0188.

Lowenthal, D. (2013). Marsh and Sauer: Reexamining the rediscovery. *Geographical Review, 103*(3), 409–414.

Lutts, R. (1985). Chemical fallout: Rachel Carson's Silent Spring, radioactive fallout and the environmental movement. *Environmental Review, 9,* 210–225. doi:10.2307/3984231.

Lytle, M. H. (2007). *The gentle subversive: Rachel Carson, Silent Spring, and the rise of the environmental movement.* New York: Oxford University Press.

Marsh, G. P. (1864). *Man and nature; or, physical geography as modified by human action.* New York: Charles Scribner & Co. A digital copy (retrieved June 4, 2013) is available at: http://books.google.ca/books?id=Sw8EAAAAQAAJ&printsec=frontcover&dq=inauthor:%22George+Perkins+Marsh%22&hl=en&sa=X&ei=BDCuUez6GKqLiwLanoGgBA&ved=0CDYQ6AEwAQ#v=onepage&q&f=false

Marsh, G. P. (1965). *Man and nature; or, Physical geography as modified by human action.* (D. Lowenthal, Ed.). Cambridge, MA: The Belknap Press of Harvard University Press. (Original work published 1864)

McNeill, J. R. (2000). *Something new under the sun: An environmental history of the twentieth century.* London: W. W. Norton.

Miller, C. (2001). *Gifford Pinchot and the making of modern environmentalism.* Washington, D.C.: Island Press.

Mills, C. W. (1951). *White collar: The American middle classes.* New York: Oxford University Press.

Mumford, L. (1955). *Sticks and stones: A study of American architecture and civilization.* New York: Dover Publications (Original work published 1924).

Mumford, L. (1979). *My works and days: A personal chronicle.* New York: Harcourt Brace Jovanovich.

Murphy, P. C. (2005). *What a book can do: The publication and reception of Silent Spring.* Amherst, MA: University of Massachusetts Press.

Novak, F., Jr. (1995). *Lewis Mumford and Patrick Geddes: The correspondence.* London: Routledge.

Packard, V. (1959). *The status seekers: An exploration of class behavior in America and the hidden barriers that affect you, your community, your future.* New York: D. McKay.

Pinchot, G. (1947). *Breaking new ground.* New York: Harcourt Brace.

Rajan, S. R. (2006). *Modernizing nature: Forestry and imperial eco-development 1800–1950.* New York: Oxford University Press.

Reclus, E. (1868). *La terre* [The world]. Paris: L. Hachette.

Rich, A. J. (1891). Henry Giles. *The Unitarian Review, 36,* 276–285.

Riesman, D. (1950). *The lonely crowd: A study of the changing American character.* New Haven, CT: Yale University Press.

Rome, A. W. (2001). *The Bulldozer in the countryside: Suburban sprawl and the rise of American environmentalism.* Cambridge, UK: Cambridge University Press.

Roszak, T. (1969). *The making of a counter culture: Reflections on the technocratic society and its youthful opposition.* Garden City, NY: Doubleday.

Ruskin, J. (1851–1853). *The stones of Venice* (Vols 1–3). London: Smith, Elder, and Co.

Sachs, A. J. (2006). *The Humboldt current: Nineteenth-century exploration and the roots of American environmentalism.* New York: Viking.

Sauer, C. O. (1963). Theme of plant and animal destruction in economic history. In J. O. Leighly (Ed.), *Land and life: A selection from the writings of Carl Ortwin Sauer* (pp. 145–154). Berkeley, CA: University of California Press (Original work published 1938).

Smith, T. (1835). Conclusions on the results on the vegetation of Nova Scotia, and on vegetation in general, and on man in general, of certain natural and artificial causes deemed to actuate and affect them. *The Magazine of Natural History and Journal of Zoology, Botany, Mineralogy, Geology and Meteorology, 8,* 641–662.

Souder, W. (2012). *On a farther shore: The life and legacy of Rachel Carson.* New York: Crown.

Stott, R. (2012). *Darwin's ghosts: The secret history of evolution/in search of the first evolutionists.* New York: Spiegel and Grau.

Sutter, P. S. (2002). *Driven wild: How the fight against automobiles launched the modern wilderness movement.* Seattle: University of Washington Press.

Terres, J. K. (1946, March 25). Dynamite in DDT. *New Republic, 114,* 415–416.

The desolate year. (1962, October). *Monsanto Magazine,* 4–9. Retrieved June 4, 2013, at: http://iseethics.files.wordpress.com/2011/12/monsanto-magazine-1962-the-desolate-yeart.pdf

Thomas, W. L., Jr. (1956). *Man's role in changing the face of the earth.* Chicago: University of Chicago Press.

Udall, S. (1963). *The quiet crisis.* New York: Holt Rinehart and Winston.

Waddell, C. (2000). *And no birds sing: Rhetorical analyses of Rachel Carson's Silent Spring.* Carbondale: Southern Illinois University Press.

Williams, M. (1989). *Americans and their forests: A historical geography.* Cambridge, UK: Cambridge University Press.

White, G. (1789). *The natural history of Selborne.* London: Cassell & Company.

Whorton, J. C. (1975). *Before Silent Spring: Pesticides and public health in pre-DDT America.* Princeton, NJ: Princeton University Press.

Wynn, G. (1977). Conservation and society in late nineteenth century New Zealand. *New Zealand Journal of History, 11,* 124–136.

Wynn, G. (1979). Pioneers, politicians and the conservation of forests in early New Zealand. *Journal of Historical Geography, 5,* 171–188.

Wynn, G. (2002). Destruction under the guise of improvement? The forest 1840–1920. In E. Pawson & T. Brooking (Eds.), *Environmental histories of New Zealand* (pp. 100–116). Oxford, UK: Oxford University Press.

Wynn, G. (2004). On heroes, hero-worship, and the heroic in environmental history. *Environment and History, 10,* 133–151.

Wynn, G. (2008). Travels with George Perkins Marsh: Notes on a journey into environmental history. In A. MacEachern & W. J. Turkel (Eds.), *Method and meaning in Canadian environmental history* (pp. 2–23). Toronto, Canada: Thompson Nelson.

"Desk Killers": Walter Christaller, Central Place Theory, and the Nazis

Trevor J. Barnes

I live in the Managerial Age, in a world of "Admin." The greatest evil is not now done in those sordid "dens of crime" that Dickens loved to paint. It is not even done in concentration camps and labour camps. In those we see its final result. But it is conceived and ordered (moved, seconded, carried and minuted) in clean, carpeted, warmed and well-lighted offices, by quiet men with white collars and cut fingernails and smooth-shaven cheeks who do not need to raise their voices.

C. S. Lewis, *The Screwtape Letters*

Introduction

C. S. Lewis's (1942) Christian apologetic novel, *The Screwtape Letters*, consists of 31 epistles written by a head demon, Screwtape, to his junior demon nephew, Wormwood. They advise how best to secure the damnation of a British man, known in the book as only "the Patient." Screwtape counsels that to spread evil more effectively in the world, his nephew needs to get into management, to go into "Admin," to work behind a desk. C. S. Lewis wrote *The Screwtape Letters* in 1941. Already by that year, a number of German Nazi managers inhabiting the world of "Admin" had begun committing terrible evil acts, and the situation worsened in the following year when Hitler initiated the "Final Solution." By war's end, the Nazi "Admin" had dispatched millions of people to a frightful death.

Those managers were not usually raving monsters, psychopaths foaming at the mouth. Certainly, none had horns or a tail. Instead, as in Lewis's description, they

T.J. Barnes (✉)
Department of Geography, University of British Columbia,
1984 West Mall, Vancouver, BC V6T 1Z2, Canada
e-mail: tbarnes@geog.ubc.ca

© Springer Netherlands 2015
P. Meusburger et al. (eds.), *Geographies of Knowledge and Power*,
Knowledge and Space 7, DOI 10.1007/978-94-017-9960-7_9

were often "quiet men with white collars and cut fingernails and smooth-shaven cheeks who do not need to raise their voices." One example is Arendt's (1977) account of such a manager in her famous book *Eichmann in Jerusalem*. Adolf Eichmann joined the SS in 1932, and because of his administrative skills, particularly in logistics, he was given the task of deporting Austrian Jews after the 1938 *Anschluss* (annexation). His "success" resulted in an appointment at the Berlin branch of the Reich Main Security Office (RSHA) that dealt with Jewish affairs and evacuation. In 1942, Eichmann was promoted to Transportation Administrator for the Final Solution, responsible for coordinating the travel of millions of Jews across the Reich to the six death camps in Poland (Auschwitz alone had 44 separate lines of railway track leading into it, twice as many as New York's Penn Station; Clarke, Doel, & McDonough, 1996, p. 467). At the end of the war, Eichmann managed to evade detection by the Allies, secretly emigrating to Argentina in 1950. But no place was safe from the Mossad, the Israeli intelligence force. In 1960, they got their man, clandestinely capturing Eichmann in Buenos Aires and abducting him to Israel for a criminal trial. Found guilty of all 15 charges, including crimes against humanity, he was executed in May 1962.

Arendt's account of Eichmann is not of a wild-eyed, frenzied killer, "the Beast of Belsen." Rather, he comes across as an intensely ordinary person, "terribly and terrifyingly normal," as Arendt (1977, p. 276) describes it. Eichmann said at his defense, "I sat at my desk and did my work" (Papadatos, 1964, p. 29). Even one of the Israeli psychologists who examined Eichmann concluded, "This man is entirely normal … more normal at any rate than I am after examining him" (Arendt, 1977, p. 25). Consequently, there was an "incongruity," as Bruno Bettelheim reflected, "between all the horrors recounted, and this man in the dock, when essentially all he did was talk to people, write memoranda, receive and give orders from behind a desk" (quoted in Cole, 2000, p. 69). That same incongruity also struck Arendt, leading her to coin the now well-known phrase that forms the subtitle of her book, "the banality of evil." It conveys both the ordinariness and the awfulness of Eichmann's work.[1]

Certainly, one should never forget the awfulness. The memoranda that Eichmann wrote produced dreadful consequences. "Death by memoranda," as Cole (2000, p. 69) puts it. Gideon Hausner, Israel's attorney general and the chief prosecutor of Eichmann, said in his opening remarks in court:

> In this trial we shall … encounter a new kind of killer, the kind that exercises his bloody craft behind a desk … it was [Eichmann's] word that put gas chambers into action; he lifted

---

[1] Arendt's thesis is contested in Lozowick's (2002) book *Hitler's Bureaucrats*. Drawing on detailed archival sources, Lozowick examines the intentions of an elite group of Nazi SS administrators that included Eichmann. He finds that rather than passively sitting back, simply passing on orders from above as mere functionaries, Nazi managers actively participated in the design of the Final Solution, marshaling resources and ensuring its maximal efficiency. As Lozowick (p. 279) writes, Hitler's bureaucrats "worked hard, thought hard, took the lead over many years. They were the alpinists of Evil."

the telephone, and railroad cars left for the extermination centres; his signature it was that sealed the doom of thousands and tens of thousands.[2]

He was a "desk killer" (*Schreibtischtäter*) (Milchman & Rosenberg, 1992). The purpose of this chapter is to explore further the notion of a "desk killer," relating the idea to another Nazi paper-pusher working in "Admin" for the SS, albeit someone much lower in the bureaucratic hierarchy than Eichmann, the geographer Walter Christaller (1893–1969). I am especially interested in how Christaller, who was fearful of the Nazis before the war began, and who became a communist after the war came to an end, could be a Nazi during the war. Christaller allowed himself and his work to be used for the most regressive political ends. He was never a "desk killer" in the same sense as Eichmann, but he participated at least as a bureaucrat, and even in a minor way as an architect, in the Nazi's "*Generalplan Ost*" (General Plan for the East). That plan did terrible things: Expelling non-Aryans from their homes in German-conquered Eastern territories (*Entfernung*); replacing them with "Germanized" immigrants; and physically transforming the acquired lands according to the aesthetics, values, and rationality of National Socialism. Power and knowledge came together starkly, and in a brutal way. I make my argument by drawing on especially the works of Burleigh (1988) and Bauman (1989), both of whom are concerned with outlining the crucial role and techniques of modern bureaucracy ("Admin") within the larger Nazi project in which the Holocaust was central.

Space, Modernity, and Nazi Academic Bureaucrats

The Nazi project, while it clearly changed over time, was nonetheless in its various guises bound inextricably to problems and issues of space. My argument will be that those problems and issues were worked out using modern bureaucratic management and techniques. That is, the Nazis drew upon modernity in part to solve their geographical problems (as well as non-geographical ones too). But here lay the paradox. The Nazi objectives which propelled those spatial issues, and which modernity was supposed to solve, were informed by deep-seated reactionary beliefs, frequently turning on racial purity, and representing the rankest anti-modernity. Herf (1984) labels this paradox, which he believes was at the heart of the Nazi project, "reactionary modernism."

---

[2] The court transcripts for the entire Eichmann trial are available online at the Nizkor Project website: http://www.nizkor.org/hweb/people/e/eichmann-adolf/transcripts/. The quotation is from Attorney General Gideon Hausner's opening remarks, Session No. 6, April 17, 1961; retrieved December 14, 2012, from http://www.nizkor.org/hweb/people/e/eichmann-adolf/transcripts/Sessions/Session-006-007-008-01.html

Space

The Nazi quest for Aryan racial purity produced at least two geographies, which became inseparable from the larger regime (Charlesworth, 1992; Clarke et al., 1996; Doel & Clarke, 1998; Gregory, 2009). The first was about defining the boundaries of Aryan space. For the Nazis, this space was defined by *Lebensraum* (living space), the idea that German Aryan people naturally required a specific amount of land and resources for their habitation. The notion of *Lebensraum* first emerged in the nineteenth century, and was associated in particular with the German geographer Friedrich Ratzel. It was elaborated in the early twentieth century by another German geographer, Karl Haushofer. In turn, Haushofer introduced the concept to Hitler in the mid-1920s, providing him with geographical instruction while he was imprisoned (with Rudolf Hess) following the failed 1923 Munich ("Beer Hall") putsch. Moreover, it was while Hitler was in prison that he wrote *Mein Kampf*, in which the concept of *Lebensraum* plays a role: "Germany must find the courage to gather our people and their strength for an advance along the road that will lead this people from its present restricted living space [*Lebensraum*] to new land and soil.... It is not in colonial acquisitions that we must see the solution of this problem, but exclusively in the acquisition of a territory for settlement."[3] In particular, Hitler saw territories in eastern Europe as part of Germany's *Lebensraum* ("Drang nach Osten"—a yearning for the East). *Lebensraum* justified the various Nazi German territorial expansions that began in the 1930s and culminated in the invasion of Poland in September 1939, sparking the Second World War.

Nazism, then, was about reterritorialization (especially of the East), enlarging the Reich through military conquest to an appropriate size for the Aryan people, as justified by the concept of *Lebensraum*. But there was a complementary (and second) geographical issue, deterritorialization. Here the problem was expelling, removing, and separating "inappropriate" people (i.e., non-Aryans) from the land they occupied, taking them elsewhere. Deterritorialization was about *Entfernung* (expulsion, removal), which in the process created "empty space" for reoccupation by Germanized people (Hitler's phrase in a 1937 speech given in secret was "volksloser Raum"; Doel & Clarke, 1998, p. 53). *Entfernung* began with the intimidation of Jews, which followed the long-established (European) precedent of the pogrom (e.g., *Kristallnacht* in Berlin in 1938). By 1940, the plan was ratcheted up to forced marches and ghettoization (e.g., in Warsaw). It culminated in the Final Solution, the extermination of non-Aryans that occurred on a mass scale at six death camps in Poland. With "inappropriate" people removed, the empty lands were available for settlement by *Volksdeutsche* and Germans from the *Reich*. *Volksdeutsche* were defined as people whose language and culture had German origins but who did not hold German citizenship and lived outside the German Reich.

---

[3] Adolf Hitler, *Mein Kampf*, vol. 2, chap. 14, "Eastern Orientations or Eastern Policy" (1926). An English translation of the two volumes is available online at http://www.crusader.net/texts/mk/index.html, from which the quotation is taken.

The great majority of these people lived in the Baltic states, Russia, Poland, Czechoslovakia, Ukraine, Hungary, Romania, Yugoslavia, Italy, France, Belgium, and the Netherlands.

Modernity

Spatial issues, then, were integrated into the very nature of the Nazi project, inseparable from its realization. But to realize a project of this vast scale required enormous energy and resources, the coordinated efforts of myriad different people and material objects, and a decisive organization and directed instrumental rationality. In short, it required modernity. Herf's (1984) reactionary modernism thesis partly speaks to this argument, but even more direct and pointed is Bauman's (1989) writing on modernity and the Holocaust. Bauman argues that "the social norms and institutions of modernity ... made the Holocaust feasible. Without modern civilization and its most central essential achievements, there would be no Holocaust" (p. 87).

Bauman interprets the Holocaust expansively, allowing him to consider both how the Nazi regime could conceive such a terrible purpose and how techniques and technologies were forged within the regime to realize it. For Bauman (1989, p. 91), Nazism is modernist because it set down a benchmark, however perverted, of a "perfect society" that it then rationally sought to "social[ly] engineer." The Nazi "perfect society" was a "pure" Aryan society, a society without Jews but also without other groups such as Slavs, Romani people, homosexuals, and the physically and mentally challenged (Gregory, 2009). Non-Aryans were removed not because their eradication permitted the acquisition of new resources and territory. Military funds were actually diverted away from such acquisitions in order to increase the capacity for killing non-Aryans. The murder of non-Aryans was the prime goal, creating for the Nazis an "objectively better world" (Bauman, 1989, p. 92).

The tasks that needed to be carried out to construct that dreadful "objectively better world" were gargantuan, requiring large-scale investments in infrastructure, knowledge, and labor. The killing of Jews and people in other groups represented a magnitude of mass murder never before historically attempted. It could not be done sporadically, haphazardly, or casually. If it were, it would never be completed. Instead, it required concerted effort, systematicity, purposeful institutions, and comprehensive formal rules and procedures. Sabini and Silver (1980, p. 330; quoted in Bauman, 1989, p. 90) write that to complete "thorough, comprehensive, exhaustive murder required the replacement of the mob with bureaucracy, the replacement of shared rage with obedience to authority." A hierarchy of decision-making responsibilities needed to be drawn up to develop large-scale plans and to gather, organize, control, and direct the means for their implementation.

Similarly, the machinery of death required substantial management and expertise. Killing was undertaken on a mass, Fordist scale, in assembly-line factories of murder, requiring a meticulous, functional division of labor, scientific management, exact timing, and logistical efficiency.[4] Labor and management practices were necessarily integrated with advanced technology, with machines, and with qualified scientists who produced both machines and specialized knowledge. Black (2001), for example, has examined how IBM, through its German subsidiary Dehomag and the scientists who worked there, provided cutting-edge technology (the Hollerith system) for reading punch cards and enabling cross tabulation of information. That technology and the expertise associated with it combined to produce the machinery of death: To identify Jews in censuses and registrations, to trace ethnic ancestry, to run the trains, to organize concentration and slave labor camps.

The larger point is that although these scientists, experts, and high-level bureaucrats were heirs to the Enlightenment tradition, they generally failed to raise critical questions about the dark political ends to which their modernist practices were directed. At best, there was complicit silence. At worst, there was active collusion, the initiation of newly concocted horrors, taking Germany ever closer to a moral *Stunde Null*. Bauman (1989) writes:

> With relish, German scientists boarded the train drawn by the Nazi locomotive towards the brave, new, racially purified and German-dominated world. Research projects grew more ambitious by the day, and research institutes grew more populous and resourceful by the hour. Little else mattered. (p. 109)

Nazi Academic Bureaucrats

As Bauman's point implies, the more Nazi ends became regressive and irrational, the more its bureaucracy charged with implementation became larger, more determined, more motivated. The aim was for a "technocracy," the "management of society by technical experts" (Renneberg & Walker, 1994, p. 4). Hence the need for academic administrators and their concomitant research institutes. The National Socialist project relied crucially on academic labor. Admittedly, some of those projects, such as a few of those carried out at Heinrich Himmler's *Das Ahnenerbe* (ancestral heritage) institute, were madcap. For example, the institute propounded *Glazial-Kosmogonie* ("world ice cosmogony"), the idea that the universe begins and ends as frozen water (Szöllösi-Janze, 2001, pp. 1–2). Or again, the "H-Special Commission" ("H" is for *Hexen* [witches]) inside the Reich Main Security Office was charged with documenting everything there was to know about witchcraft, compiling a "witch card index" of 33,000 entries (Szöllösi-Janze, 2001, p. 3). But such work was the exception, and clearly incapable of realizing National Socialist military and ideological objectives. But the work of ordinary, everyday academics—scientists, social scientists, and assorted technocrats—who were "largely rational,

---

[4]While it may seem that the metaphor of Fordist production is over the top, death camps were run by the Economic Administrative Section of the *Reichssicherheitshauptamt* and expected to make a profit. Train transportation for death camp victims was booked using ordinary travel agents, with discounts given for mass bookings, and children under four traveling for free.

and result oriented ... [and] not ideologically dogmatic" (Szöllösi-Janze, 2001, p. 12) could realize these objectives.

The National Socialist reliance on academics coincided with the general impulse of National Socialism toward a modernism based on expertise and rationality. It also reflected a specific cultural belief in the general superiority of German scholarship and intellectuality. If any group could achieve Nazi goals, it would be German academics. As Aly and Heim (2002, p. 3) write, "the National Socialist leadership sought to maximize the inputs for scientific policy advisors and used their research findings as an important basis for their decisions—including the decision to murder millions of human beings."

Burleigh (1988) provides a brilliant case study, which is germane to my examination of Walter Christaller, on German wartime scholars carrying out research on the newly colonized Eastern territories (generally known as *Ostforschung*—Eastern research), particularly in Poland, Czechoslovakia, and later the Soviet Union. With respect to this case, Burleigh writes:

> Exponents of the view that academics are without influence have to explain why hard-headed SS managers thought and acted otherwise. Rightly or wrongly the latter recognised that the domination of conquered populations ... could be achieved through research institutes in Berlin or Breslau.... As scholarly experts in the East, the *Ostforscher* had a distinctive contribution to make to the accurate "data base"—the statistical and cartographic location of persons—upon which all aspects of Nazi policy in the East, as elsewhere, ultimately rested. Deportations, resettlements, repatriations and mass murder were not sudden visitations from on high, requiring the adoption of some commensurate inscrutable, quasi-religious, meta-language, but the result of the exact, modern, "scientific" encompassing of practices with card indexes, card sorting machines, charts, graphs, maps and diagrams.... This was why [*Ostforschung*] received generous funding. (p. 10)

Their bosses, however, wanted only very particular kinds of academic knowledge, which brings us back to Bauman's point about complicity. According to Burleigh (1988), academic bureaucrats

> did not challenge existing stereotypes and misconceptions; they worked within their boundaries and reified them through empirical "evidence" ... This is not a history of a radicalized and opportunistic "lunatic" fringe but of a section of the established, educated élite ... The *Ostforscher* voluntarily and enthusiastically put their knowledge at the disposal of the Nazi regime ... taking on board as many aspects of Nazi racial dogma as were consistent with their own (limited) notions of scholarly propriety. (p. 9)

Walter Christaller: Reactionary-Modernist, Nazi, *Ostforscher*

Walter Christaller was an *Ostforscher*. He "voluntarily and enthusiastically" put his knowledge, in his case, central place theory—a spatial theory of settlement he devised in the early 1930s—"at the disposal of the Nazi regime." In doing so, his work necessarily took on "many aspects of Nazi racial dogma." The reterritorialization of the newly acquired German East was to be in accordance with the principles of central place theory, and involve both the expulsion of non-Aryans from that space and their replacement by *Volksdeutsche*, whose resettlement Christaller personally helped to arrange. Christaller as an academic bureaucrat was up to his neck in the nasty racial politics of German National Socialism. But, in line with Burleigh's argument, Christaller was never part of a lunatic fringe. In the early 1930s he opposed Hitler, even seeking political refuge in France because of fears for his safety from the Brownshirts. But in the end, like Eichmann, he sat at his desk in his office in Berlin's Dahlem district, working for the SS, and did his job.

Christaller and the Development of Central Place Theory

Christaller's central place theory had a long gestation period. When he was 8, Christaller (1972, p. 601) received an atlas as a Christmas present from a geographically enlightened aunt, and was instantly "bewitched." As Christaller recalled, eerily anticipating what he was to do as a grown-up, "I drew in new railroad lines, put a new city somewhere or other, [and] changed the borders of the nations, straightening them out or delineating them along mountain ranges … I designed new administrative divisions and calculated their populations" (p. 602). He broke into tears only when his father refused to purchase a statistical handbook to add greater veracity to his map doodling (p. 602).

Christaller's subsequent university education was interrupted by the First World War, in which he fought and was wounded. It took him 17 years variously studying in Heidelberg, Munich, Berlin, and Erlangen before in 1930 he finally received his diploma in economics (Hottes, Hottes, & Schöller, 1977). Hottes et al. (1977) suggest that Christaller's intention at Erlangen was to carry on with a PhD in economics, but because he "found no response from the economists" (p. 11), he returned to his childhood interests and asked the biogeographer Robert Gradmann in the geography department to supervise his dissertation. Gradmann accepted, and Christaller (1972, p. 607) returned to his "games with maps" and drawing "straight lines," subsequently seeing "six-sided figures (hexagons)" emerge on the southern German topographic landscape that he studied. The thesis was completed in 1932 in just 9 months, and published the following year as *Die zentralen Orte in Süddeutschland* (Central Places in Southern Germany).

An enormous amount has been written about the substance of Christaller's central place theory, especially since the second half of the 1950s.[5] For the purposes of this short chapter, I shall make only three brief points. First, it was a *spatial* theory, in this case about the geographical distribution of different-sized cities (central places) that ranged from traditional individual farms surrounding a rural hamlet to the largest, most modern metropolis jam-packed with factories. Central to that theorization was the peculiar geometry of the hexagon that Christaller (1972) thought he could see surfacing from the very landscape itself if he stared at it (and

---

[5] There are many excellent reviews of central place theory. Berry's (1967) and Beavon's (1977) are two of my favorites in what forms a vast body of literature. More than thirty years ago, Beavon (1977, p. 3) estimated that already "the total literature encompassed some 2,000 papers."

"hiked" in it) long enough (p. 610). Second, Christaller at least believed that he was putting forward a *modern* scientific theory based on underlying spatial laws. "My goal was staked out for me: To find laws according to which number, size, and distribution of cities are determined" (p. 607). Consequently, this theory was no old-time regional geography, à la Alfred Hettner's chorology. It was something new. It was modern. It was the future. Finally, and possibly of greatest interest to Christaller, central place theory was a planning tool, a technology for practicing instrumental rationality. That intent was already demonstrated in his doctoral thesis, laid out as three planning principles (K=3 [marketing], K=4 [transportation], and K=7 [administrative]). Later these principles were further refined in his 1938 *Habilitation* (in effect, a second PhD in the German system, allowing him to become a professor— which he never did). From 1940 onward, after joining the Nazi party, Christaller was finally able to put into practice his planning principles while serving on Konrad Meyer's staff, which was charged with transforming the newly acquired German East.

Konrad Meyer and Generalplan Ost

Konrad Meyer was one of the key academic bureaucrats employed by the Nazis. A member of the SS from 1933, he was also professor of agronomy at the University of Berlin. He had his administrative finger in a larger number of pies, including from 1936 the Reich Association for Area Research (*Reichsarbeitsgemeinschaft für Raumforschung*), in which Christaller, along with many other German geographers, undertook work (in Christaller's case, it was research on the "German Atlas for Living Spaces" [*Atlas des deutschen Lebensraumes*]; Rössler, 1989, p. 422). More important for the purposes of this chapter, in 1938 Meyer was appointed chief of the Planning and Soil Department (*Hauptabteilung Planung und Boden*) under the Himmler-led Reich Commission for German Resettlement and Population Policy (*Reichskommissariat für die Festigung deutschen Volkstums*, RKFDV). In 1940, Christaller began working in Meyer's main office, which was concerned with planning Germany's newly acquired Eastern territories and which later was to fold into *Generalplan Ost*.

Generalplan Ost was top secret, developed and overseen within the SS (Aly & Heim, 2002; Burleigh, 1988; Rössler, 1989). Much of the plan's documentation was deliberately destroyed just before the end of the war for fear of its incriminating nature. One of the plan's principal architects was Konrad Meyer. In spring 1941, Himmler charged Meyer with planning Polish territories annexed by Germany (Madajczyk, 1962, pp. 3–4). The invasion of Poland by Germany on September 1, 1939, resulted in Poland being divided into three regions: Western Poland was incorporated into the Third Reich, becoming the provinces of Wartheland (later known as Warthegau) and Danzig West Prussia; Central Poland became a German military-occupied territory known as General Government (*Generalgouvernement*); and Eastern Poland (Galicia) was ceded to the Soviet Union as part of the secret Molotov-Ribbentrop Pact signed a week before Germany's assault on Poland.

Himmler was pleased by Meyer's planning efforts for Poland, so, taking an opportunity to impress again, Meyer submitted to Himmler just 3 weeks after the German invasion of the Soviet Union in June 1941 an even more expansive plan that applied not only to Poland, but to all subsequent German Eastern conquests (Madajczyk, 1962, p. 4).[6] Himmler approved, ordering Meyer in January 1942 to set out the full legal, political, and geographical foundations necessary for the reconstruction of the East, which Meyer did on May 28, 1942 (Burleigh, 2000, p. 547).

The Generalplan involved the two geographical pivots of the Nazi regime: *Lebensraum* and *Entfernung*. As Meyer said in a speech on January 28, 1942, "The *Ostaufgabe* [task in the East] is the unique opportunity to realize the National Socialist will, and unconditionally to let it become action" (quoted in Deichmann & Müller-Hill, 1994, p. 176–177). Action was to be effected by applying modernist planning principles along with the associated bureaucracy of experts and practitioners. Once land and resources were acquired, permitting Germany to fulfill the imperative of *Lebensraum*, those spaces would be Germanized by bringing in people of Aryan heritage. The plan estimated that resettlement would require more than four and a half million *Volksdeutsche* over a 30-year period (later revised upward to ten million). In contrast, *Entfernung* was the fate of most of the original inhabitants of the East, Slavs and Jews, who did not fit the Nazi Germanic ideal racial type. That could mean being dumped at a train station somewhere in *Generalgouvernement*; expulsion to the Warsaw Ghetto; incarceration in a slave labor or concentration camp; forced inclusion on a "death march"; or execution by firing squad, mobile gas van, or at one of the six Nazi death camps, all of which were located in the East, with two in annexed Poland and four in *Generalgouvernement* (Gregory, 2009). The number of planned expulsions varied from a low of 30 million to a high of 65 million (Burleigh, 2000, p. 547).

Christaller, Central Place Theory, and Generalplan Ost

Christaller's central place theory may have been given the cold shoulder by economists, and it certainly was no traditional Hettnerian regional chorology, but it was perfect theory for the Nazis. The theory was fundamentally about spatial relations, speaking to key aspects of the Nazi project. It was seemingly modernist (rational, law-seeking, scientific), but also made overtures to tradition and the past. Theoretically, its starting point was individual farmers surrounding the smallest urban unit, the village (*Dorf*), emphasizing rural community, people, and soil, or *Volksgemeinschaft*. But the culmination of the hierarchy was modernity, leading to industrial urban behemoths such as Dortmund, Essen, Bochum, and, the ultimate, Berlin. Finally, central place theory came as a ready-made planning tool. Christaller's detailed maps, figures, and plans needed only to be unfurled, the bulldozers brought in, and the East became "central places in southern Germany." As Rössler (1994)

---

[6] Various versions of Generalplan Ost existed from 1940 onward; but after some wayward arithmetic in earlier incarnations, "the more practiced Meyer" got the job (Burleigh, 2000, p. 547).

notes, the "aim was the transformation of the East into German land and as German landscape" (p. 134). That is exactly what Christaller's model did.

Preston (2009), who has examined Christaller's various wartime contributions existent in German archives, concludes that while working for Meyer, Christaller "contributed directly to plans facilitating German *Lebensraum* [search for living space] policy, on the one hand, and Himmler's RKFDV [Germanisation], on the other" (p. 6).

The first of these roles was associated with Christaller's application of central place theory initially used in annexed Poland, or, more specifically, Warthegau. Warthegau would be the "workshop" for the Reich, as Joseph Umlauf, a colleague of Christaller in Meyer's Planning and Soil Department, put it (quoted in Fehl, 1992, p. 96). Christaller shared this view. Writing in 1940, he said:

> Because of the destruction of the Polish state and the integration of its western parts into the German Empire, everything is again fluid.... Our task will be to create in a short time all the spatial units, large and small, that normally develop slowly by themselves … so that they will be functioning as vital parts of the German Empire as soon as possible. (translated and quoted in Preston, 2009, p. 23)[7]

A year later, Christaller was more strident and more specific.

> The aim of regional planning … is to introduce order into impractical, outdated and arbitrary urban forms or transport networks, and this order can only be achieved on the basis of an ideal plan—which means in spatial terms a geometrical schema … central places will be spaced an equal distance apart, so that they form equilateral triangles. These triangles will in turn form regular hexagons, with the central place in the middle of these hexagons assuming a greater importance … (quoted in Aly & Heim, 2002, p. 97)[8]

Consequently, parts of Warthegau were redesigned, "completely changing the face of the countryside," as Himmler had demanded in 1940 (quoted in Aly & Heim, 2002, p. 74). For example, the district of Kutno, in northeast Warthegau, was made over on paper at least according to Christaller's "geometrical schema."

But clearly there was work to do in making the world conform to the "ideal plan." Christaller wrote in the same 1941 planning document quoted above: "[where] it seemed absolutely essential … that a new town of at least 25,000 inhabitants" be built, then a new town would be "created from scratch" (quoted in Aly & Heim, 2002, p. 97). If Upper Silesia needed "a Duesseldorf or Cologne" of 450,000 people "to provide a cultural centre," then so be it (quoted in Aly & Heim, 2002, p. 97). If "Posen … has the power and potential to develop into a town of 450,000 [from 350,000]," it should (quoted in Aly & Heim, 2002, p. 97). More specifically, Christaller planned 36 new *Hauptdörfer* for Warthegau. Each one came, as Rössler

---

[7] The quotation is from an article that Christaller (1940) published in *Raumforschung und Raumordnung*, "Die Kultur- und Marktbereiche der zentralen Orte im Deutschen Ostraum und die Gliederung der Verwaltung" (Cultural and Market Segments of Central Places in the German East and the Structure of Administration).

[8] This translated quotation is originally from Christaller (1941), *Die Zentralen Orte in den Ostgebieten und ihre Kultur- und Marktbereiche* (Central Places in the Eastern Territories and Their Cultural and Market Segments).

(1994) notes, with a "National Socialist celebration hall, buildings for the Hitler Youth or a central parade square, in other words the visible buildings of the model for National Socialist society" (p. 134).

Before this could happen, however, many of the non-Aryan residents had to go—560,000 Jews and 3.4 million Slavs. Only 1.1 million of the existing population were thought to be Germanized enough to stay. Given the large expulsion, 3.4 million Germanized settlers needed to be brought in. This goal defined Christaller's second role, to assist in the migration of *Volksdeutsche* from various places in Europe so as to strengthen Germandom, which now included Poland. As Christaller put it, this goal provided another reason to construct a new central place system: "To give settlers roots so they can really feel at home" (quoted and translated by Preston, 2009, p. 21).[9]

Conclusion

Walter Christaller used to be a household name, at least for a period in the 1960s and 1970s in Anglo-American human geography. His central place theory was perhaps the only indigenously devised formal geographical theory in the discipline. It would have been scandalous to have called Christaller a "desk killer." There was rarely mention of his entanglements or the entanglements of his theory with the Nazis and the Second World War. Bunge (1977), who dedicated his book *Theoretical Geography* (1966) to Christaller, even maintained that Christaller "was not a fascist." Rather, Christaller was "a man of science" (1977, p. 84). His central place theory was neat and pure, the tidy arrangement of an unsullied logic. For this reason, Bunge was dumbfounded that Christaller was never offered a professorship in Germany.

Of course, logic is never unsullied, never separated from history and geography. There is no realm of knowledge that is hermetically sealed from the context of its production, and—most germane for the essays collected in this book—there is no realm of knowledge that is removed from the appropriation, distribution, and circulation of the concomitant imbricated social power. Michel Foucault, of course, famously joined knowledge and power in his hyphenated couplet, "power-knowledge." The hyphen is perhaps the most important element, connoting a single term. It is not knowledge on the one hand, social power on the other; or science on the one hand, the state on the other. It is mutual inherence. Power is exercised, asserted, denoted, and applied through knowledge, just as knowledge relies upon, demands, is manifest as, and takes up social power.

The Nazi regime was a regime of power-knowledge of an extreme kind. Its "Admin" departments shockingly exemplified the power-knowledge nexus. They provided data, records, typological criteria, anthropological assessments, planning

---

[9] This quotation is originally from Christaller's (1942) article "Land und Stadt in der Deutschen Volksordnung" (Country and City in the German National Order), published in the journal *Deutsche Agrarpolitik*.

precepts, and so much more. But this wasn't just information to be selectively picked over, haphazardly taken up, and discarded. It came with tremendous social force to direct action, to unfurl on the ground, and in the process to make multiple concrete conjunctions, sometimes of a very bad kind. The Gestapo arrive to search Anne Frank's hideaway attic in an Amsterdam apartment complex. Romanian *Volksdeutsche* take over now empty farmhouses in Kutno, Warthegau. The train pulls in at Auschwitz.

As Foucault makes clear, no one escapes such forces, certainly not Walter Christaller. There is no "outside." Christaller at first was against Hitler and National Socialism. Accused of sympathizing with the Communist Party, Christaller had been investigated in 1934 by the Gestapo. He bicycled to France to become a political refugee; friends helped him return (Wardenga, Henniges, Brogiato, & Schelhaas, 2011, p. 21). In the end, the disciplining force of power-knowledge was too strong; it was a temptation he could not resist: Christaller joined the National Socialist party in 1940 (Wardenga et al., 2011, p. 33). Christaller did not want to become part of the Nazi war machine, but he could not help himself. He needed a job; he sought academic credibility and relevance; he wanted to show that his ideas were not mere childhood squiggles on atlases but capable of remaking the world. Moreover, the SS gave him not a piece of paper on which to draw, but Warthegau, a whole conquered territory of 44,000 km². He couldn't resist the offer. Power-knowledge overwhelmed. This decision might explain why Christaller joined the Communist Party after the war, and from 1951 to 1952 represented the Communist Party as municipal councilor in Jugenheim (Kegler, 2008, p. 92), although he left the party in 1953 following accusations that he was an East German informant (the charges were never formally made, however).

The larger point, which is applicable to a number of Nazi bureaucrats (Lozowick, 2002): Although during the war Christaller may have just sat at his desk in "clean, carpeted, warmed and well-lighted offices," and he may never have "raised [his] voice," what he and they did was hellish.

References

Aly, G., & Heim, S. (2002). *Architects of annihilation: Auschwitz and the logic of destruction* (A. G. Blunden, Trans). London: Weidenfeld and Nicolson. (Original work published 1991)

Arendt, H. (1977). *Eichmann in Jerusalem: A report on the banality of evil.* Harmondsworth, UK: Penguin.

Bauman, Z. (1989). *Modernity and the Holocaust.* Ithaca, NY: Cornell University Press.

Beavon, K. S. O. (1977). *Central place theory: A reinterpretation.* New York: Longman.

Berry, B. J. L. (1967). *Geography of market centers and retail distribution.* Englewood Cliffs, NJ: Prentice-Hall.

Black, E. (2001). *IBM and the Holocaust: The strategic alliance between Nazi Germany and America's most powerful corporation.* New York: Crown Publishers.

Bunge, W. (1977). Walter Christaller was not a fascist. *Ontario Geography, 11,* 84–86.

Burleigh, M. (1988). *Germany turns eastwards: A study of Ostforschung in the Third Reich.* Cambridge, UK: Cambridge University Press.

Burleigh, M. (2000). *The Third Reich: A new history.* New York: Hill and Wang.

Charlesworth, A. (1992). Review article: Towards a geography of the Shoah. *Journal of Historical Geography, 18*, 464–469. doi:10.1016/0305-7488(92)90242-2.

Christaller, W. (1940). Die Kultur- und Marktbereiche der zentralen Orte im Deutschen Ostraum und die Gliederung der Verwaltung [Cultural and market segments in central places of the German East and the structure of administration]. *Raumforschung und Raumordnung, 4*, 498–503.

Christaller, W. (1941). *Die Zentralen Orte in den Ostgebieten und ihre Kultur- und Marktbereiche* [Central places in the Eastern territories and their cultural and market segments]. Leipzig, Germany: Verlag Koehler.

Christaller, W. (1942). Land und Stadt in der Deutschen Volksordnung [Country and city in the German national order]. *Deutsche Agrarpolitik, 1*, 53–56.

Christaller, W. (1972). How I discovered the theory of central places: A report about the origin of central places. In P. W. English & R. C. Mayfield (Eds.), *Man, space, and environment: Concepts in contemporary human geography* (pp. 601–610). New York: Oxford University Press. (Original work published 1968)

Clarke, D., Doel, M. A., & McDonough, F. X. (1996). Holocaust topologies: Singularity, politics, space. *Political Geography, 15*, 457–489. doi:10.1016/0962-6298(96)00027-3.

Cole, T. (2000). *Selling the Holocaust: From Auschwitz to Schindler: How history is bought and sold.* New York: Routledge.

Deichmann, U., & Müller-Hill, B. (1994). Biological research at universities and Kaiser Wilhelm Institutes in Nazi Germany. In M. Renneberg & M. Walker (Eds.), *Science, technology and National Socialism* (pp. 160–183). Cambridge, UK: Cambridge University Press.

Doel, M. A., & Clarke, D. (1998). Figuring the Holocaust: Singularity, and the particularity of space. In G. O'Thuathail & S. Dalby (Eds.), *Rethinking geopolitics* (pp. 39–61). London: Routledge.

Fehl, G. (1992). The Nazi garden city. In S. V. Ward (Ed.), *The garden city: Past, present, and future* (pp. 88–105). London: E. & F. N. Spon.

Gregory, D. (2009). Holocaust. In D. Gregory, R. Johnston, G. Pratt, M. Watts, & S. Whatmore (Eds.), *The dictionary of human geography* (pp. 337–340). Chichester, UK: Wiley-Blackwell.

Herf, J. (1984). *Reactionary modernism: Technology, culture and politics in Weimar and the Third Reich.* Cambridge, UK: Cambridge University Press.

Hottes, K., Hottes, R., & Schöller, P. (1977). Walter Christaller 1893–1969. In T. W. Freeman, M. Oughton, & P. Pinchemel (Eds.), *Geographers: Biobibliographical studies 7* (pp. 11–18). London: Mansell Information Publishing.

Kegler, K. R. (2008). Walter Christaller. In I. Haar & M. Fahlbusch (Eds.), *Handbuch der völkischen Wissenschaften: Personen, Institutionen, Forschungsprogramme, Stiftungen* (pp. 89–93). Munich, Germany: Saur.

Lewis, C. S. (1942). *The Screwtape letters.* London: Centenary Press.

Lozowick, Y. (2002). *Hitler's bureaucrats: The Nazi security police and the banality of evil.* London: Continuum.

Madajczyk, C. (1962). General Plan East: Hitler's master plan for expansion. *Polish Western Affairs, 3*(2), 391–442. Retrieved December 14, 2012, from World Future Fund website: http://www.worldfuturefund.org/wffmaster/Reading/GPO/gpoarticle.htm

Milchman, A., & Rosenberg, A. (1992). Hannah Arendt and the etiology of the desk killer: The Holocaust as portent. *History of European Ideas, 14*, 213–226. doi:10.1016/0191-6599(92)90249-C.

Papadatos, P. (1964). *The Eichmann trial.* New York: Frederick A. Praeger.

Preston, R. E. (2009). *Walter Christaller's research on regional and rural development planning during World War II* (METAR–Papers in Metropolitan Studies, Vol. 52). Berlin: Freie Universität Berlin, Institut für Geographische Wissenschaften.

Renneberg, M., & Walker, M. (1994). Scientists, engineers and National Socialism. In M. Renneberg & M. Walker (Eds.), *Science, technology and National Socialism* (pp. 1–35). Cambridge, UK: Cambridge University Press.

Rössler, M. (1989). Applied geography and area research in Nazi society: Central place theory and planning, 1933 to 1945. *Environment and Planning D: Society and Space, 7*, 419–431. doi:10.1068/d070419.

Rössler, M. (1994). "Area research" and "spatial planning" from the Weimar Republic to the German Federal Republic: Creating a society with a spatial order under National Socialism. In M. Renneberg & M. Walker (Eds.), *Science, technology and National Socialism* (pp. 126–138). Cambridge, UK: Cambridge University Press.

Sabini, J. P., & Silver, M. (1980). Destroying the innocent with a clear conscience: A sociopsychology of the Holocaust as survivors, victims and perpetrators. In J. E. Dimsdale (Ed.), *Survivors, victims, and perpetrators: Essays on the Nazi Holocaust* (pp. 329–358). Washington, DC: Hemisphere.

Szöllösi-Janze, M. (2001). National Socialism and the sciences: Reflections, conclusions, and historical perspectives. In M. Szöllösi-Janze (Ed.), *Science in the Third Reich* (pp. 1–35). Oxford, UK: Berg.

Wardenga, U., Henniges, N., Brogiato, H. P., & Schelhaas, B. (2011). *Der Verband deutscher Berufsgeographen 1950–1979: Eine sozialgeschichtliche Studie zur Frühphase des DVAG* [The Association of German Applied Geographers, 1950–1979: A social history of the early period of the DVAG]. Forum IfL 16. Leipzig, Germany: Leibniz-Institut für Länderkunde.

Knowledge and Power in Sovietized Hungarian Geography

Róbert Győri and Ferenc Gyuris

> *We should state clearly that no Marxist economic geographer wishes to "locate" the old, reactionary, capitalism-serving human geography, neither some nor any of its branches, in Marxist economic geography. It is no aim at all to rename the child. There are some unscrupulous people who, proceeding from an erroneous theoretical foundation, are afraid that we are throwing out the baby with the bathwater. In my opinion, we should just throw out the child.*
>
> Markos (1955, p. 365)

Introduction

The history and geography of science offer ample evidence of how those in power try to control knowledge and education, how certain regimes tried to manipulate scientific disciplines to benefit their own interests, how some disciplines adapted to radical changes in political systems and adjusted their theoretical concepts to new ideologies, and what efforts these disciplines made to appear "useful" to those in power. This chapter examines the means used by the Communist regime in Hungary after World War II to "conquer" science and colonize geography. Researchers have richly documented how Central and Eastern Europe became objects of "Soviet

R. Győri (✉)
Department of Social and Economic Geography, Eötvös Loránd University,
Pázmány Péter Street 1/c., 1117 Budapest, Hungary
e-mail: gyorirobert@caesar.elte.hu

F. Gyuris
Department of Regional Science, Eötvös Loránd University,
Pázmány Péter Street 1/c., 1117 Budapest, Hungary
e-mail: gyurisf@caesar.elte.hu

© Springer Netherlands 2015
P. Meusburger et al. (eds.), *Geographies of Knowledge and Power*,
Knowledge and Space 7, DOI 10.1007/978-94-017-9960-7_10

colonialism" (Chioni Moore, 2001); how these countries were turned into economic fiefdoms of the Soviet empire, with economic production undertaken on a command basis and trade permissible only through the Communist alliance; and what consequences this development had on various fields (cf. Chioni Moore, 2001, p. 114; Romsics, 1999; for international power relations within the Soviet bloc, see Bunce, 1985). But scholars working in the history and geography of science still pay little attention to the intellectual transformation that took place in the discipline of geography in these countries as of the late 1940s.

For this reason, we aim in this chapter to contribute to a better understanding of these issues by revealing how Hungarian geography was colonized during the 1950s. We show how the Communist system crushed "the old geography" in order to establish Hungarian Marxist-Leninist geography. We reveal how geographic knowledge, like knowledge in general, became "a form of power, and by implication violence" (McEwan, 2009, p. 26). We describe different epistemological cultures that influenced and determined the approaches, methods, social tasks, and educational role of Hungarian geography between the world wars and after World War II. We analyze the effect that Marxist-Leninist ideology had on Soviet geography in this period. Furthermore, we investigate how a colonizing ideology dominated Hungarian geography, how the institutional structure of geography was transformed, and how the career paths of the "old" geographers continued. We also outline who became the "new" geographers and how, and describe the new tasks set for Marxist-Leninist geography in Hungary.

Hungarian Geography Before World War II

The dramatic changes that occurred in Hungarian geography during the 1950s cannot be understood without knowledge of the discipline's role in Hungarian society and academia before then. The story begins at the end of World War I. As a consequence of the 1920 Treaty of Trianon (Paris Peace Conference), the country surrendered two thirds of its area, a large part of its industrial resources, and 60 % of its population (and one third of all native Hungarian speakers) to Romania, Czechoslovakia, the Kingdom of Serbs, Croats, and Slovenes, and Austria (Hajdú, 1998). Hungarians were shocked by these territorial, economic, and population losses. One of the main goals of postwar governments in Hungary was the revision of the peace treaty with respect to territorial losses. Support was given to disciplines that served revisionist aims and that promoted the strengthening of national identity. Geography—together with ethnography, history, and statistics—held a privileged position among such disciplines.

Although the peace talks failed to meet Hungarian expectations, geography gained a high reputation among the public and decision-makers involved in science and education policy. During the 1920s, the institutional development of the discipline saw new departments and research institutes being opened. The role of the geographer underwent remarkable changes, perhaps best illustrated by the scientific and political career of Pál Teleki (1879–1941), a prominent figure in Hungarian geography in those years (Fig. 10.1).

Teleki's career path reflects the interwoven nature of geography and national politics during the first decades of the twentieth century. He came from one of the

Fig. 10.1 Pál Teleki
(1879–1941), geographer,
prime minister of Hungary
(1920–1921, 1939–1941)
(Source: From the Archive
of Eötvös József Collegium.
Copyright by Eötvös József
Collegium)

most respected noble families of Hungary, and began his work on the history of
cartography. He became interested in French *géographie humaine* in the 1910s,
when he was a member of parliament. Teleki, after serving as prime minister in
1920–1921, was the head of the Department of Economic Geography at the Faculty
of Economics of the Hungarian Royal Pázmány Péter University in Budapest during
the 1920s and 1930s. He was elected the superintendent of Eötvös József Collegium,
a leading institution of national elite education established according to the
principles of the École normale supérieure in Paris. Moreover, he functioned as
chief scout of the Hungarian Scout Movement. Teleki was appointed minister (first
minister of religion and education, and then minister of foreign affairs), and became
prime minister for the second time in 1938, holding this position until 1941
(Ablonczy, 2007). In Teleki's career, the revisionary goals of Hungarian foreign
politics, national identity, and geographical research were strongly intertwined.

Given the privileged position of geography as a discipline, the vast majority of
the geographers allied themselves with the "official" conservative-national ideology
of the era and internalized the political goals of the regime. Hungarian geographers
dismissed or ignored left-wing movements criticizing the overall social and
institutional order of the country and the state. The interwar period witnessed the
"golden age" of regional geography in accord with national political goals. Almost
all monographs on the geography of Hungary focused on the geography of Greater
Hungary. Geographers sought to emphasize that the borders set by the Treaty of
Trianon were temporary ones. As Ferenc Fodor (1924), a disciple and colleague of
Teleki, wrote in his 1924 book on the economic geography of Hungary, "Describing

Fig. 10.2 Geography as
nationalist propaganda: One
of the best-known emblems
of interwar Hungarian
revisionist propaganda
(Source: From *Igazságot
Magyarországnak!* [Justice
for Hungary!], by O. Légrády
(Ed.), 1930, Budapest,
Hungary: Pesti Hírlap)

the economic geography of 'Truncated Hungary' is per se a contradiction." (p. 9).[1]
For Hungary's interwar geographers, the new borders of the country did not coincide
with any physical, social, or economic boundaries; they were considered the result
of an arbitrary decision forced on the country. Even physical form was made to
reflect this political moment. Gyula Prinz, a respected geologist and geographer,
published his *Tisia concept* on the tectonic development of the Carpathian Basin in
1926, and again, in a revised form, 10 years later. Prinz's purely tectonic model,
according to which tectonism had "folded up" the Carpathian Mountains, was used
to delineate the physical boundaries of a unitary country (Keményfi, 2006).

At the same time, everyday life was infiltrated by geographical discourse to a
much greater extent than ever before. The defense of national space was basically a
geographical issue. Geographical symbols appeared in schoolbooks, newspapers,
speeches, operettas, and songs of the period. It was popular, for example, to christen
new streets and squares after cities, mountains, and rivers of the lost territories. The
best-known emblem of the period might well be the map depicting the borders after
the Treaty of Trianon within those of Greater Hungary, with the text in the margin,
"Nem, nem, soha!" ("No, no, never!") or "Igazságot Magyarországnak!" ("Justice
for Hungary!"; see Fig. 10.2). Not only did geography infiltrate revisionist discourse;
revisionist rhetoric was also geographical.

---

[1] The translation of this quotation as well as all other Hungarian texts into English are by the
authors of this chapter, unless otherwise noted.

Hungarian revisionist foreign policy managed to achieve considerable success, though not until the late 1930s (Hajdú, 1998). Such territorial expansion was a national success, as was the success of Hungarian geography. However, Hungary had to pay a high price for these achievements. The country was becoming more and more obligated to the Axis powers, and the pressure on Hungary to enter the war was also growing. In June of 1941, Hungary declared war on the Soviet Union, thus entering World War II on the side of the Axis powers. The die was cast. To gain stronger control over the country, the German army occupied Hungary in the spring of 1944, and a fascist government serving the interests of Nazi Germany was formed that autumn. The war ultimately left Hungary in ruins.

The Soviet Colonization of Hungarian Geography After World War II

After 1945, Hungary became a part of the Soviet occupation zone. A brief provisional period with multiparty elections between 1945 and 1948 was followed by the violent establishment of the Communist regime. As Soviet pressure increased, the Soviets' reckoning with Hungarian fascism turned to a reckoning with the whole of conservative-bourgeois Hungary. It was Erzsébet Andics, a leading ideologist of the new system, who stated that Hungary had been a fascist state not only in the last year of the war but during the 1920s and 1930s as well (Andics, 1945). This view referred also to geography's place within the previous regime. Attempts at territorial revision were identified as the main reasons for entering the war. Against the scientific background of revision (and revisionist propaganda), the whole of geographical science was found guilty.

In Communist Hungary, geography, now stigmatized, fell from grace. The old research institutes were dissolved or ideologically "cleansed," and the geographers from the former staff were expelled. The heaviest casualty was the Hungarian Geographical Society, which was dissolved by decree of the Ministry of the Interior in 1949. The proscription was obviously motivated by the desire to quash "reaction-ary" geography: "Circumstances seemed not to guarantee the development of the society's work in a Marxist-Leninist spirit" (Koch, 1952, p. 884). The disbanding of the society also meant the end of its journal *Földrajzi Közlemények* (Geographical Review), published since 1872. Hungarian geography remained without a published forum for some years. The society's activities were stopped until the Hungarian Academy of Sciences, the organ for controlling science,[2] initiated the revocation of

---

[2] After World War II, the Hungarian Academy of Sciences, the leading non-university institution of Hungarian science since its formation in 1825, was transformed along Stalinist principles. With this change, the academy became the paramount institution in the hierarchy of Sovietized science: even the professional and administrative control of universities was placed in its hands. "Important" scientific research was removed from universities and concentrated in research institutes subordinate to the Hungarian Academy of Sciences. Universities were debarred from awarding doctor's degrees; candidate of sciences and doctor of sciences degrees were issued by the academy, and scientific societies were also subject to its supervision (Péteri, 1998).

the ban by the Ministry of the Interior in 1952. This development was possible because Hungarian geography was assessed as integrated into the Soviet-style scientific system. The justification provided enumerates nearly every step of scientific colonization: "Hungarian geographers have made big advances in the application of Marxist dialectic, and have familiarized themselves with the findings of Soviet geographical science, and Hungarian geography has gained new Marxist cadres" (Koch, 1952, p. 884).

The transformations affecting the whole discipline would not have been possible without changes in personnel. The staffing policy of the new system obviously followed Lenin's (1960) instructions on how to organize a revolutionary movement. As he put it in his pamphlet *What Is To Be Done?* "Such an organisation must consist chiefly of people professionally engaged in revolutionary activity." For him, this prerequisite was crucial to establish "a stable organisation of leaders," which "maintains continuity" and enables the structure to "endure" (p. 464). The realization of these principles in practice took various forms. As for the "old" geographers, some of them were pensioned off or exiled from academia. Others were driven to the periphery, where they could keep their job but not their former rank or position. Some researchers were forced to compromise with the system (at least formally). In the meantime, all new appointments of the transformed institutional structure were filled by politically reliable figures, some of whom possessed neither an education in geography nor a university degree. Their involvement was crucial in helping realize the "great ideological turn": Converting Marxist-Leninist principles into an unquestionable paradigm.

With the Communist party transforming the country ever more radically, "old" geographers' prospects became progressively worse. In 1949, after the "year of the turn," Communist science policy expelled all fellows of the academy who did not "fit" the new system. This "cleansing," one step in the transformation of the Hungarian Academy of Sciences, exerted a strong influence on social sciences overall. Fifty-four percent of all fellows were expelled from the academy. Almost two thirds of them were involved in the humanities or social sciences, and a bit more than one third in natural and applied sciences (Péteri, 1998). Geography suffered especially. All four geographers who were fellows of the academy were expelled. The scientific work of most "old regime" geographers was discussed and evaluated negatively from a Marxist-Leninist point of view (Abella, 1956, 1961; Koch, 1956; Markos, 1955; "Vitaülés," 1954). Members of the old regime staff were hindered from obtaining the newly introduced Soviet-style scientific titles and from having their articles and books published, and their disciples were expelled from universities.

The strategies of adaptation left few doors open for such "old" geographers, and for those who did have options remaining, the possibilities on offer for physical and human geographers were quite different. Although none of the "old" geographers became a supporter of the new system, learning and applying Marxist-Leninist ideology did present opportunities for physical and human geographers. Joining the Communist party might guarantee some measure of tranquility (although no real intellectual freedom) and the opportunity to reclaim former positions.

Because their field of research was politically more sensitive, the possibilities for human geographers were more limited. The economic geographer Ferenc Koch, a disciple of Teleki, compromised with the system—presumably to ensure his survival (Probáld, 2001). The urban geographer Tibor Mendöl could not, however, defend himself from attacks through his "passive resistance." Mendöl, having been the head of the Department of Human Geography at the University of Budapest since 1940, lost all of his disciples and his close colleagues as they were expelled from the university. He struggled to have his works published and to receive his doctor of sciences degree, the highest rank in science in the Soviet-style academic system (Győri, 2009). The fact that he also tried to reformulate some of his works along Marxist-Leninist principles (Mendöl, 1954) was not enough. As one of his critics, who understood the main point of his work, remarked, "Nothing in this work allows Mendöl to say anything new from the perspective of urban geography; [he] just repeats his old approach in a new form" (Abella, 1961, p. 124). If such "old regime" human geographers, even at the price of serious losses and unfair treatment, could retain some of their authority, the younger generation taught by them had virtually no such prospects.

After "solving the problem" of "old" geographers, "new" geography was built on the ground of well-tested Communists. The leading ideologist in Sovietized geography and a constant presence in scientific debates was György Markos (1902–1976), the initiator of the Marxist-Leninist approach in Hungarian geography (Fig. 10.3). Markos had neither a formal education in geography nor a university

Fig. 10.3 Strangers within: The "new" geographers – György Markos (1902–1976) (Source: From the Hungarian National Museum, Historical Photographic Collection, Ltsz. 78.942. Copyright by the Hungarian National Museum. Reprinted with permission)

degree. He had, however, formerly played a significant role in the labor movement. As a student, he participated in the Hungarian Soviet Republic of 1919, a short-lived Communist dictatorship established by the Party of Communists from Hungary led by the internationally known Bolshevik revolutionary Béla Kun.[3] Later on, he spent most of the interwar period as an émigré in the West, remaining a member of the movement but working as a publicist and caricaturist. Before World War II, he returned to Hungary, where several newspaper articles by him were published, together with two populist works on economic history. During the war he was imprisoned for antifascist activity. After 1945, he worked at several jobs that were important for the party (e.g., in the Central Planning Office). Later on, he was appointed head of Pál Teleki's former department (renamed the Department of Economic Geography) at the Marx Károly University of Economics, and he became the vice president of the re-established Hungarian Geographical Society in 1952 (Tatai, 2004). Markos, although he had no prior connection with Hungarian geography, used his authority rapidly. His articles applying Marxist-Leninist ideology to geography illustrated the new way not only for economic but also for physical geographers. In debates, he confronted practically all leading geographers of the former era of geographical science.

After a thorough change of staff, Markos shaped Teleki's former department to make it the leading workshop of Marxist-Leninist economic geography in Hungary. Three department heads of the socialist era began their scientific career under his aegis (Bernát, 2004). Markos's department soon became the most important "truth spot" in Hungarian economic geography. The dissemination of the new knowledge was the task of Markos's disciples, who, like U.S. "space cadets" (Barnes, 2004), began to work in the leading centers of scientific life, or gained high positions in state government after receiving their doctorates. Many of them joined the new hot-spots of science production, the socialist "centers of calculation" (Latour, 1987). The most important examples were the Geographical Research Institute of the academy, the Department of General Economic Geography at Eötvös Loránd University, the Scientific and Planning Institute of Urban Construction (VÁTI), the Central Planning Office, and the Party Academy of the Hungarian Socialist Workers' Party. The essence of Markos's life was succinctly summarized by one of his disciples in the special issue of *Földrajzi Értesítő* published on the occasion of Markos's retirement: "Markos was a revolutionary, a conscious Marxist with high standards in every situation" (Enyedi, 1968, p. 406). Markos was a revolutionary, indeed. His work had considerable influence on the function and objectives of science, on the theoretical framework for research, and on the lives of geographers, and

---

[3]The Hungarian Soviet Republic, emerging in the politically turbulent period after Austria-Hungary's defeat in World War I and the empire's dissolution, sought to achieve a thorough transformation of Hungarian society along Communist principles. For this reason, the new Communist leaders proclaimed the "dictatorship of the proletariat" and used open terror. However, due to the military intervention of neighboring countries with strong support from France and Britain, the Hungarian Soviet Republic, also challenged by widespread contempt among a broad spectrum of Hungarian society, collapsed after 19 weeks.

Fig. 10.4 Strangers within:
The "new" geographers –
Sándor Radó (1899–1981)
(Source: From Wikipedia
(http://upload.wikimedia.org/
wikipedia/hu/1/10/Rado_
shandor.gif))

led overall to thoroughgoing changes in Hungarian geography whose implications are still felt.

Perhaps an even more curious career was that of Sándor Radó (1899–1981; see Fig. 10.4), who succeeded Markos as head of the department after 1958. Markos had been transferred from the University of Economics after the Hungarian Revolution of 1956; nevertheless, he was able to pursue his scientific work at the academy's Geographical Research Institute. Like Markos, Sándor Radó had played an active role in the international labor movement and, as a law student, had been a political officer of the Hungarian Red Army during the Hungarian Soviet Republic of 1919. After the downfall of that republic, he emigrated to Vienna, and then to Germany. He studied geography and history at the universities of Jena and Leipzig, but official university documents prove that he did not complete his studies. After spending a semester in Jena in 1922/23 (Universitätsarchiv Jena (UAJ). Bestand Studienkartei (ca. 1915–1935)), Radó moved to Leipzig, where he began studies in the same disciplines, but he was expelled in 1925 on account of "not attending lectures" (Universitätsarchiv Leipzig, Sheet 486; see Fig. 10.5).

Finally, Radó went to the Soviet Union, where he gained a reputation as a cartographer (K L, 1960), and, according to a CIA report, was trained there for service with Soviet military intelligence (Thomas, 1968). Following some years in the USSR, he moved to Germany, then to Paris. From 1936, he lived in Geneva until 1944, where he was a secret agent of Soviet intelligence under the umbrella of the news agency Geopress. (Radó wrote an autobiographical fiction [Radó, 1971] about his service for the Soviet intelligence, which was brought to screen during his lifetime; see Fig. 10.6.) In 1945, he was evacuated to the Soviet Union, where he was accused of working for the British as a double agent and sentenced to 10 years of forced labor in 1946 (Trom, 2006). He was not released until November of 1954, although according to U.S. intelligence he spent only a short time in a Siberian coal

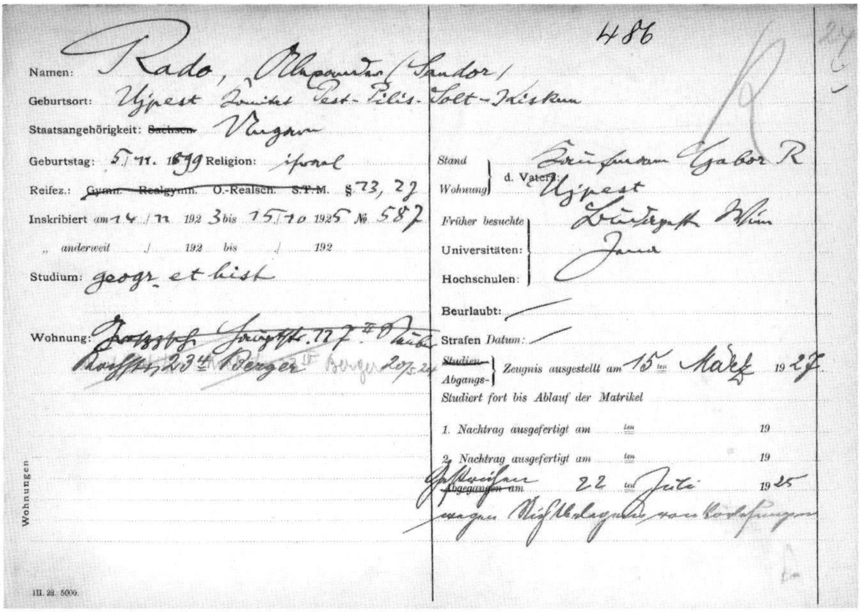
486

Namen: Rado, Alexander (Sandor)

Geburtsort: Ujpest

Staatsangehörigkeit: ~~Sachsen~~ Ungarn

Geburtstag: 5.XI.1899 Religion: israel

Reifez.: ~~Gymn. Realgymn.~~ O.-Realsch. ~~S.T.M.~~

Inskribiert am 14.XI.1923 bis 15.VII.1925 № 587

„ anderweit . . . 192 . . bis . . . 192

Studium: geogr. et hist.

Wohnung:

Stand } d. Vaters:
Wohnung } Ujpest

Früher besuchte Universitäten: Hochschulen:

Beurlaubt:

Strafen Datum:

~~Studien-~~ Abgangs- } Zeugnis ausgestellt am 15ten März 1927

Studiert fort bis Ablauf der Matrikel

1. Nachtrag ausgefertigt am . . . ten . . . 19

2. Nachtrag ausgefertigt am . . . ten . . . 19

~~Abgegangen am~~ 22ten Juli 1925

Wohnungen

III. 23. 5000.

Fig. 10.5 Radó's efforts to earn a university degree ended in failure. He was expelled from Leipzig, the last university he visited, on account of "not attending lectures." (Source: From Universitätsarchiv Leipzig, Quästur, Sheet 486, Alexander Rado. Copyright by Universitätsarchiv Leipzig, Quästur)

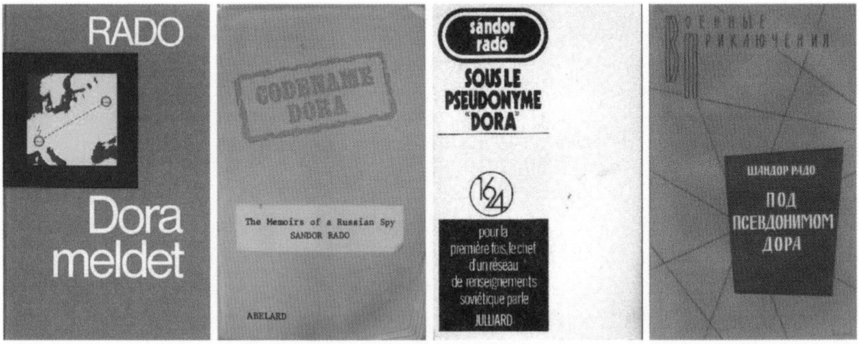

Fig. 10.6 Radó as a Communist hero: His autobiographical fiction was translated into over a dozen languages (Sources: From (**a**) *Dora meldet*, by S. Radó, 1974, Berlin: Militärverlag der DDR; (**b**) *Codename Dora: The memoirs of a Russian spy*, by S. Rado, 1977, London: Abelard; (**c**) *Sous le pseudonyme "Dora"*, by S. Radó, 1972, Paris: Julliard; (**d**) *Pod psevdonimom Dora*, by S. Rado, 1973, Moscow: Voenizdat. Copyright: No information available. Permission to reprint: "With friendly allowance of Militär Verlag, Berlin." The editors have made every effort to track down all owners of the image copyrights. However, some of the publishers no longer exist. Should such identity not have been ascertained, the customary fee will be paid by the editors if valid evidence of copyright ownership is submitted to the editors)

mine, where he managed teams of workers and thus was not subject to hard physical labor. Thereafter, he was transferred to a geographical observatory near Moscow as a "prisoner with privileges" (the CIA assumed that Radó's transfer and special treatment were the result of "string-pulling by friends") (Thomas, 1968).

Radó returned to Hungary in 1955. He was appointed head of the national cartographic office. Here, using his former international connections, he collected cartographic material with possible military-strategic relevance from around the world, a fact that concerned U.S. intelligence (Thomas, 1968). He was the head of the Department of Economic Geography from 1958 to 1966, and became the president of the Hungarian Geographical Society in 1973. In addition to receiving numerous prestigious Hungarian and Soviet awards, he was elected honorary member of several (e.g., Soviet, French, East German, and Bulgarian) geographical societies (Ormeling, 1982) and honorary doctor of the Lomonosov Moscow State University (Pécsi, 1982). Furthermore, he was elected honorary member of the International Cartographic Association (Ormeling, 1982) and became a commission member of the International Geographical Union (Papp-Váry, 1998). Radó, having been a Communist adventurer, continued the work in economic geography begun by Markos.

This radical transformation of geography and, actually, the whole of science was only possible due to the highly centralized power structure of the Communist dictatorship. Top party leadership, who were in fact puppies of the Soviet empire, could push through virtually all of their notions. The new leadership not only was able to suit science to its needs; it also had a decisive interest in doing so. Because politics (the power) and science are always dependent on each other, their reciprocity is hardly surprising. On the one hand, power requires perpetual legitimization, which is best served by science with its "neutral," "objective" standpoint. On the other hand, representatives of science require continual support in both a material and a moral sense, which they can best receive from a power that both appreciates and needs them (Meusburger, 2005, 2007; see also his chapter in this volume). This mutual dependence was especially strong in Soviet science. Communist power aimed at a radical transformation of society, and Marxist-Leninist scientists followed an ideology totally incompatible with that of their predecessors. Therefore, both groups needed strong support from each other, which led to them becoming almost perfectly intertwined.

The radical changes in science were realized rather quickly, so that the era of jockeying for position in Hungarian geography ended even before the mid-1960s. The remaining "old" geographers had by then retired or died, and few of their disciples or followers continued to pursue their research issues. A kind of personal (as well as thematic) continuity could be revealed in physical geography between the interwar and socialist epochs. However, human geography (or, in Marxist-Leninist terminology, economic geography) was distinguished by interruption and break (Fig. 10.7). From the 1960s on, all important positions in Hungarian economic geography were dominated by "newcomers" loyal to the system who regarded the "old" Hungarian human geography as a reprehensible, outdated, bourgeois-reactionary science.

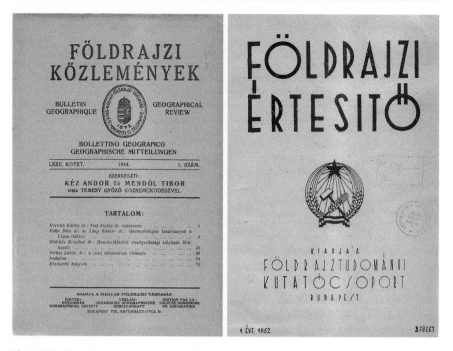

Fig. 10.7 Old and new geography: *Földrajzi Közlemények* (Geographical Review [1872–1948, 1953–]) and *Földrajzi Értesítő* (Geographical Bulletin [1952–])

New Geography – New Theory

The changes in the general context of science had major implications for geography. As the discipline's role and its basic approach were redefined, the inner structures and divisions of the discipline, and the relative weight and the content of party disciplines also changed. This process emulated the Soviet example. Hungarian geographers were expected to imitate the Marxist-Leninist approach to geography in the USSR, which had emerged in a very specific context that was thoroughly different from that of Hungary after World War II.

The first event that gave momentum to the creation of Marxist-Leninist geography was less theoretical than practical in nature. World War I and the Russian Civil War had brought the country to the brink of ruin. Lenin's aim was to revitalize the country's economy and to transform it from a small-peasant economy to a country with large-scale industrialization (Lenin, 1966). For him, the first prerequisite for this transformation was electrification, which led to the creation of GOELRO (Gosudarstvennaya Komissiya pa Elektrifikatsii Rossii, or the "State Commission for Electrification of Russia"), the first general economic/industrialization plan for the Soviet economy (Horváth, 2008). Moreover, the realization of this project necessitated the establishment of *economic rayons*, the spatial units of socialist economic planning (Radó, 1957a). Because the work on GOELRO and the creation of *economic rayons* involved numerous Russian geographers, these projects played a key role in the formation of Marxist-Leninist geography (Radó, 1957b).

In the meantime, because Soviet geography was basically propelled by the interests of bureaucratic planning, theoretical conceptualization lagged far behind practical work. The first definition of the "fundamental object" of Marxist-Leninist economic geography was given no earlier than 1926 by Nikolay Baranskiy, in his book *Economic Geography of the U.S.S.R.* Baranskiy, an active member of the revolutionary movement since his student days and a key contributor to "the foundation of new Soviet economic geography" (Saushkin, 1962), saw the task of the discipline in "study[ing] the distribution and spatial combination of the productive forces, i.e., of the main factors that are required for production—the means of production, as well as the men themselves with their production experience and skill" (Baranskiy, 1956, p. 7).

However, the point was not only to describe the world, but also to change it. As Baranskiy (1956) put it, economic geography not only was aimed at the "fullest and strictest consideration of different natural conditions" and "the utilization of natural resources." It also was to carry out "a radical transformation of nature" (p. 8) in order to contribute to the construction of socialism. For Baranskiy, "economic geography of the U.S.S.R. [was] an 'active geography,' involved in the transformation of nature carried out … *under the leadership of the Communist Party of the Soviet Union*" (Baranskiy, 1956; italics added). This stance was based on a peculiar Marxist-Leninist interpretation of the human–nature relationship. Soviet geographers internalized Marx's opinion that "the most basic connection between society and nature … is production." For them, production was a process through which the human affects and changes nature, while also changing its own nature through this process (Markos, 1952b, p. 271; Marx, 1949, pp. 191–192). But, as Stalin, refining Marx, emphasized, "The change and development of society is incomparably faster than the change and development of nature" (Stalin, 1950, pp. 648–649). The Marxist-Leninist point of view was, therefore, not that of geographical determinism but of *economic determinism.* In this approach, it was the mode of production that determined the human–nature relationship.

The belief in "a radical transformation of nature" also necessitated a firm belief in the omnipotence of science, which was a characteristic feature of Marxist-Leninist ideology. As Sergey Vavilov, the president of the USSR Academy of Sciences between 1945 and 1951 put it, "The starting point for the philosophical materialism of Marxism is that the world and its laws can be understood … that there are no things in the world not to be revealed" (Vavilov, 1950, pp. 20–21). This attitude was common among Soviet leaders, and through various forms of mass media it also found its way into virtually every stratum of society.

Marxist-Leninists, however, were also convinced that the enormous potential of science should be exploited only if science was made to serve practical needs. Thus, scientists were expected to focus on practical issues. As Lenin stated, "Our science shall not remain a dead letter or a fashionable phrase … science shall really become flesh and blood" ("Lenin and science," 1970, p. 130). The same was propagated by his successor, Joseph Stalin, for whom "the guiding star of the proletariat's party" was "the link between science and practical activity, the link between theory and practice, and their unity" ("A Szovjetunió Kommunista," 1949, p. 123). This concept led to an extremely practical orientation in all of Soviet science, and wiped out all initiatives concentrating on "purely theoretical" issues (cf. Ivanov, 2002).

With all of these characteristics, Soviet geography was not a direct successor of any earlier geographical traditions in Russia. The so-called branch-statistical school of V. Den in St. Petersburg had been based on German bourgeois political economy and had produced what Saushkin (1954, p. 96) called "barren, formal, metaphysical economic statistics." Thus, it could not be tolerated by Marxist-Leninists. Other prominent figures in Russian human geography during the czarist period were usually regarded in the Soviet era as followers of Friedrich Ratzel, Alfred Hettner, or Paul Vidal de la Blache (Radó, 1957a; Saushkin, 1954, 1962). Soviet geographers often lumped these scientists together as representatives of "bourgeois" geography, who were blamed for the "scientific substantiation" of the expansive politics of "imperialistic states," and thus for serving "imperialistic" elites (Dobrov, 1952).

Given these considerations, it is easier to understand the structure and terminology of Marxist-Leninist (and Stalinist) geography. The economic determinist view of the discipline and Stalin's concept of the different speeds of natural and social "development" suggested that natural and social processes were not to follow the same regularities. Marxist-Leninist geographers rejected the "bourgeois" concept of geographical monism which, for them, "tries to expand the effect and validity of natural rules to human society" (Radó, 1962, p. 227). In their opinion, this argumentation only aimed to provide scientific substantiation for the expansion and conquering wars of the "imperialistic" states (Dobrov, 1952). Instead, Soviet geographers distinguished "two geographies", *physical geography* and *economic geography*. The former, considered a natural science, was to investigate the regularities propelling the development of the geographical environment. The latter was regarded a social science, focusing on the rules that determine the spatial allocation of population and production (Gerasimov, 1959). Although it was emphasized that this dual structure entailed a dialectical—not a discrete—relation of the two geographies, this point was often ignored by Marxist-Leninist geographers who wanted to avoid being perceived as determinist, "bourgeois," or "reactionary." As a consequence, cooperation between physical and economic geography became extremely weak by the mid-1950s.

The Marxist-Leninist Turn in Hungarian Geography

The transformation of Hungarian geography was carried through in accord with the prevailing theories of Soviet geography in the postwar period. It affected every subdiscipline, although the turn had especially far-reaching implications for economic geography.

Before 1939, economic geography in Hungary was considered a branch of human geography, and its basic principles were in line with the conceptual framework of the French *géographie humaine* (Győri, 2001). Pál Teleki (1922) identified the goal of economic geography as presenting human economic life as part of all life on Earth and in analyzing the relationship between economic life and life as a whole. After the Communist turn, human and physical geography were separated in accord with the Soviet practice, and human geography was renamed. From then on, the term "economic geography" embraced all aspects of the discipline which concerned society. This shift in perspective was grasped by one of the "new" geographers, Csaba Kovács (1954), who formulated it much like Baranskiy had: "The essence and main task of economic geography is the analysis of the geographical division of labor" (p. 417).

The introduction of the new term "economic geography" supported Marxist-Leninist doctrine by putting production to the fore, which was an issue of fierce debate. At a 1954 session of the academy's scientific committee, Tibor Mendöl argued that neither population nor urban geography could be wholly regarded as a part of economic geography. He instead proposed the use of "social geography" as a general term for issues in the discipline that did not belong to physical geography (Bulla, 1955a). Mendöl's endeavor was also supported by Béla Bulla, who became a physical geographer in the interwar period and was a personal friend of Mendöl. The idea, however, was firmly opposed by György Markos, who argued that Mendöl and Bulla were trying to bring back the old Hungarian human geography under the cover of "social geography" (Markos, 1955).

The autocracy of Marxist-Leninist economic geography led to the dismantling of several disciplines that had played a key role in the interwar period. In the case of political, ethnic, and historical geography, the direct or indirect link with such revisionist endeavors was obvious; thus, the dismantling of these branches (and their exile from canonized scientific vocabulary) did not require detailed explanation. Theoretical issues were marginalized in the new economic geography because their research results had no "practical utilization"; they did not serve production or the more efficient organization of the geographical division of labor in a direct way. As for population and urban geography, their survival was ensured to the extent that their reformulated, practice-oriented scientific goals were integrated into the tight framework of the all-embracing economic geography. These new tasks were precisely formulated by the urban geographer Március Matejka, who had returned from the Soviet Union. Population and settlement geography were considered the branches "which have as their subject the spatial allocation of the most important force of production—that of humans" (Abella, 1961, p. 123).

Such approaches were alien to the tradition of Hungarian urban geography. Humans had never before been reduced to a "force of production," and the practical (planning) orientation of the new approach was also unprecedented. The old Hungarian urban geography had had three special interests during the 1930s. First, researchers had investigated towns and villages as the smallest kinds of landscape (along methodological principles of the French *géographie humaine*). Second, they had analyzed the regularities and the development of the urban network. Third, they had dealt with urban morphology (Győri, 2009). The latter two topics were based on German settlement geography. None of these fields of research was incorporated into the new urban geography in the 1950s. Just as in the Soviet Union, regional geographical research along the lines of Vidal de la Blache's work was considered erroneous in Communist Hungary. From a Marxist-Leninist perspective, such research was rooted in harmful theory because it related social phenomena to physical factors. It was said to make no more than "minor corrections" on "pure geographical determinism" (Dobrov, 1952, p. 7). The quantitative and, in general, positivistic research on urban networks was thought to be deductive speculation. For Marxist-Leninist geographers, it displayed "abstract forms, geometric shapes, schemes," which hid the real reasons behind social disparities ("Vitaülés," 1954, pp. 780–781).

The apolitical urban morphology paradigm came under the most severe attack. The main accusation leveled at it was that morphology is an empty, "formalist," art-for-art's-sake investigation with no connection to practical issues such as urban network planning. According to Antal Vörösmarti, it was an especially serious misapprehension that Hungarian urban geography (i.e., Tibor Mendöl) linked social and economic characteristics of urban population to morphological types of urban layout. In the eyes of the "new" geographers, this method gave the false impression that morphological and functional research can be joined up (Abella, 1961, pp. 124–125). Another important reason for its rejection was that Mendöl—erroneously—tried to make statements about "essential" structures on the basis of investigation of the surface only. This approach was diametrically opposed to Marxist-Leninist logic. Markos had noted some years earlier:

> The unitary bourgeois geography is formalist and objectivist in all of its details, as it can serve capitalism best in this way.... [It] makes do with never ending investigations of details, with analysis of small formal questions, does not see and does not desire to see the content and the process behind form; thus, it necessarily becomes formalist. (Markos, 1955, p. 362)

As a consequence, such morphological analysis was a "bourgeois trick"; its conscious aim was to divert the attention of the scientific community or broader society away from essential questions about the severe contradictions and crisis of capitalism.

Morphological studies became problematic not only in urban geography but in physical geography as well. After 1945, Hungarian physical geography had to distance itself from the morphology of Davis and Penck, as these theories traced surface development back to cyclical processes (Davis) and to quantitative change (Penck). These approaches contradicted Marxist-Leninist teachings, which regarded the concept of linear development as dogma. Béla Bulla made an attempt to reformulate the principles of geomorphology, fitting them to the needs of the times and to Marxist-Leninist dialectic. In his view, surface development is a "necessarily rhythmic process" revealed through "the realization of the dialectically controversial development of the surface and the interpretation of the essence of development" (Bulla, 1955a, p. 104). Bulla's endeavor to find a place for geomorphology among the sciences that was based on Marxism-Leninism was not successful. György Markos rejected a major section of Bulla's article, stigmatizing it as formalist and emphasizing that physical geography should also have a practical orientation. For Markos (1955), much like for Baranskiy,

The point is not only to interpret forms on the surface of the Earth, but to utilize and, if necessary, change them for the sake of society. The excess of morphology in physical geography is a bourgeois heritage here, and the task is not to perceive but to eliminate it. (p. 362)

Socialism in the Making: Practical Goals of Marxist-Leninist Geography in Hungary

The Hungarian Communist leadership was keen to emulate the Stalinist model in order to make Hungary the "best disciple" of the Soviet Union within the Communist bloc. While pursuing this aim, the chief party leader Mátyás Rákosi and his right-hand man Ernő Gerő, the minister of state, strongly argued against any divergence from the Soviet model. As they put it, "The basic features of socialist construction in the Soviet Union are universally valid," so "there are no specific national roads to socialism" (Spriano, 1985, p. 304). Thus, Hungarian science was expected to strive for the same goals as its Soviet counterpart. However, the leadership believed this objective would be possible only after a thorough transformation of science in Hungary. In Gerő's words, the "old" Hungarian science often "diverged from real life," and "closed itself within its narrow ivory tower" (Gerő, 1950b, p. 345). For him, the People's Republic of Hungary needed a science that regarded "efficient participation in the realization of our five-year plan and ten-year electrification and irrigation plans and in the ascension of our country as its decisive tasks" (p. 348).

In other words, Hungarian science—like Soviet science—had to contribute to the realization of big Communist goals. Geography was no exception. Physical geography, after identifying and understanding the rules behind processes in the geographical environment, had to change these processes in order to transform nature in relation to the needs of society. Its aim was the improvement of society's productive forces. Economic geography was responsible for the rational allocation of the population and production in space—thus, for scientifically substantiated spatial economic planning (Abella, 1956). This task was made explicit by János Kolta, who became an economic geographer after the Communist turn, having been a rural lawyer. For him, it was the role of economic geography "to ensure the scientific substantiation and scientific character of national economic planning ... through its deep-drilling analyses, through considering the principles of the maximal development of the productive forces and of the 'priority of production'" (Kolta, 1954, p. 200). In the case of Hungarian (economic) geography, the main aims were threefold. First, the development of Hungarian urban and rural systems, with a special emphasis on the issue of "scattered farms." Second, scientific determination of the economic regions of the country (so-called *rayonization*). Third, the transformation of nature in order to improve agricultural production. The political leadership had high expectations, as Gerő (1950a) stressed, "We [aimed to] change the socioeconomic map of our country" (p. 576).

Objective 1: Socialist Transformation of the Urban Network

In accord with "new" geography's main objectives, Communist urban and rural development policy in Hungary was responsible for creating a "more rational" spatial framework for production. But it was also considered a tool for radical and voluntaristic transformation of society. The main aims were the creation of "socialist towns" (new industrial or newly industrialized centers dominated by the working class), the gradual disappearance of the urban–rural divide, and the "socialist transformation" of villages—following that of cities (Hajdú, 1992). Emphasis was placed on the development of new industrial towns, which was seen as a precondition for accomplishing the first Five-Year Plan (1950–1955). (Especially big efforts were made in Sztálinváros [Stalintown], renamed Dunaújváros in 1961, as the planned new center of Hungarian iron and steel production and the symbol of the Stalinist approach to development.) This plan pushed forward the rapid industrialization of the country. As Mátyás Rákosi, radiating trust in the omnipotence of Marxist-Leninist science, pronounced in 1949, "This plan aims to develop Hungarian industry in a 5-year period as much as it grew in the 50-year period preceding it" (Rákosi, 1951, p. 14).

The most pressing issue for urban geography to solve, however, was the problem of scattered farms (*tanyas*) on the Great Hungarian Plain. The emergence of these small settlements can be traced back to the decades after Hungary's liberation from the Ottoman occupation in the late seventeenth century. The century and a half of occupation had left vast areas of the Great Plain deserted. Several villages disappeared because their inhabitants fled to the few towns. After the end of Ottoman rule, a gradual resettlement of the deserted areas began. Peasants, although remaining inhabitants of the rural towns, established small farmsteads on the property they owned (Beluszky, 2001). According to the statistics of the Communist planning institutions, almost 900,000 people were living in scattered farms at the end of the 1940s (Hajdú, 1992).

The issue of scattered farms was a serious challenge for the Communist system. Although instructive scientific debates on the issue took place during the interwar years, no real steps were made. After the 1945 land reform, the number of inhabitants living in scattered farms dramatically increased. Furthermore, it was a main aim of the postwar political regimes (even those before the "Communist turn" in 1948) to carry out a thorough reform of the administrative system. Thus, solutions to the question could not be delayed for long.

Some initial steps from 1945 to 1948 were the creation of new local administrative units from groups of scattered farms formerly belonging to nearby agricultural towns. Then the Ministry of the Interior established the Preparatory Scattered Farm Committee in 1948. The committee and its successor from 1949 on, the Scattered Farm Council, were responsible for the solution of the "scattered farm" problem. Ferenc Erdei, who made the transition from moderate left-wing politician to Communist, and who was appointed the minister for agriculture in 1949, was personally asked by Mátyás Rákosi, the leader of the Communist Hungarian state, to lead the council (Hajdú, 1990–1991). Erdei accepted the offer, but nevertheless the council (in accordance with Soviet notions of urban development) had to follow a strict policy of demolishing scattered farms and organizing them into villages. Such a solution was diametrically opposed to Erdei's analysis and proposals between the wars. In fact, it was more similar to a suggestion from Tibor Mendöl, who, incidentally, was suppressed by the Communist system and whose disparagement was partly due to Erdei (Győri, 2009).

Yet the council's initial plans for the infrastructural development of the new villages mostly remained unfulfilled. After several years of gradual decline due to a lack of proper coordination and waning interest among political leaders, the council was officially disbanded in 1954. Nevertheless, it played a decisive role in opening a new, explicitly "anti-scattered farm" (and anti-rural) chapter in the history of Hungarian urban development. This development strongly influenced related scientific concepts in the decades that followed (Hajdú, 1990–1991). For instance, an official planning document from 1951 on the Hungarian urban network argued for a total ban on any kind of investment in almost half of the Hungarian settlements, thus implicitly aiming at their gradual physical decay and destruction over the long term (Hajdú, 1992). Because Ferenc Erdei, the well-known and respected sociologist, took up leadership of the council, the "socialist solution to the scattered farm issue" (i.e., their destruction) could be legitimized as "the scientific solution" to the question (Hajdú, 1990–1991, pp. 120–121).

Objective 2: Establishing a Spatial Framework for Socialist Planning

Besides the "socialist planning" of the urban network, another practical issue of Hungarian economic geography was to identify the economic regions (*rayons* in Marxist-Leninist terminology) of the country. These regions were intended to become the effectively functioning spatial units of production. As mentioned earlier, *rayonization* had a strong tradition in the USSR. It helped bring Marxist-Leninist economic geography into being in the Soviet Union during the early 1920s, and it gained in importance there even before 1939. On the one hand, this practice was rooted in indisputably rational economic interests. The identification of economic regions, together with the review of their environmental conditions and economic potential, was a crucial prerequisite for the long-term development of the USSR. On the other hand, *rayonization* also served propagandistic goals by emphasizing the "conscious" and "methodical" character, and thus the superiority, of the Communist regime. *Rayons* were regarded as the means of improving efficient cooperation among units of production, which were obviously characterized by different conditions. To improve the spatial division of labor, *rayons* were expected to exhibit two features simultaneously: A kind of specialization of production, but also complexity. Specialization meant that each *rayon* had to contribute to the output of that or those branches which they had optimal natural conditions for to a larger extent than other regions with less favorable conditions. Complexity reflected another goal, namely, that complementary activities, supplying the basic needs of the local population, should not be totally disregarded, but the "proportionate development of branches of production" had to be ensured (Krajkó, 1982). Yet it remained unclear to what extent *rayons* should be "complex." Some authors argued for nothing less than a strict autarky, including self-sufficiency in food production and basic consumer goods (cf. Beluszky, 1982; Enyedi, 1961). In practice, the realization of these seemingly contradictory objectives was possible only at different geographical levels. The big *rayons*, whose number never exceeded 32 for the entire Soviet Union, had to become complex units with a broad variety of economic activities. Meanwhile, the improvement of smaller subregions involved only a few, or even just a single branch or a single mammoth company (Horváth, 2008).

About three years after the "Communist turn" in 1948, the basic principles of *rayonization* were also introduced into the Hungarian geographical discourse by György Markos. He laid down the theoretical principles of the issue in 1951, followed by his hypothetical *rayon* system for Hungary one year later (Kolta, 1954; Markos, 1952a). Markos followed the relevant Soviet ideas in all respects. In his interpretation, *rayons* were intended as "adequate spatial units of production for spatial planning" (Kolta, 1954, p. 201). Other supporters of *rayonization* went even further. János Kolta argued that after a while *rayons* should also become administrative units "unconditionally" (p. 203). The issue of economic regionalization was introduced into Hungarian economic geography very quickly. Thanks to this rapid change, to the country's real administrative challenges, and to the political pressures prevailing in scientific life, the next 10–15 years can be characterized as "the decade of *rayonization*" in Hungary (Beluszky, 1982, p. 4). In these years, any economic geographer who wanted to matter in the discipline developed his or her own concept or at least tried to contribute to the discourse (Beluszky, 1982).

Rayonization was, however, never a successful feature of Hungarian geography. The first serious problems emerged in the early years. Some geographers disputed whether it was possible in such a relatively small country to identify "specialized" *and* "complex" economic regions similar to those in the USSR. In their opinion, the whole of Hungary can be regarded as one (complex) *rayon*. The main proponent of this argument was Béla Bulla, who temperately but unambiguously criticized Markos for the precipitous introduction of the issue. As he stressed, "In the absence of the necessary theoretical and practical foundation, it has been impossible to succeed in the creation of a plan that is acceptable for national economic planning" (Bulla, 1955a, p. 110). In fact, this criticism was common in several East European Communist states. For instance, the East German economic and political geographer Heinz Sanke, later a member of the academy of the German Democratic Republic, was of the same view. And so was Anastas Beshkov, the Bulgarian economic geographer and fellow of the Bulgarian academy (Bulla, 1955b). Nonetheless, others were convinced of the opposite. The most sophisticated counterargument in Hungary was made by Gyula Krajkó, a key supporter of the *rayonist* concept. Krajkó underlined that what was important was neither territorial extension nor the number of branches of production determining complexity, but rather the relations of production and the development of productive forces. For him, even a small country could be divided into complex economic subunits, at least if it was a social-ist one. In capitalist countries, however, according to Krajkó, complex economic regions could not emerge because of the existence of private property and the lack of "planned, proportionate development of the national economy" (Krajkó, 1961, pp. 224–225).

In general, several theoretical questions remained open, and results were contradictory. The numerous studies attempting to identify *rayons* in Hungary were full of remarkable peculiarities and did not share similar findings. The number of *rayons*, for instance, varied on a broad scale from 6 to 13 (Beluszky, 1982). Due to these issues and perhaps to the fact that the question of further significant transformation in the spatial framework of public administration was dropped, most research on the matter was no longer pursued (Beluszky, 1982). Although a university research group, led by Krajkó, continued with *rayonization* at the University of Szeged, and a few other experiments were conducted, the issue surfaced only once more—as part of a special issue in 1982 (Beluszky, 1982). Even then, authors did not reach a consensus on the issue, but researchers pointed to a fact that had not been taken into consideration previously. It was that "specialized" and "complex" *rayons* can be identified only in countries with specific circumstances: Either in a country in a relatively earlier phase of spatial division of labor (the phase of emerging large-scale heavy industry) or in special cases where new economic centers are created in formerly untouched regions, as happened in several Siberian districts (Beluszky, 1982; Enyedi, 1982). Overall, *rayonization* in Hungary and in other East European countries was a highly doubtful scientific project which completely ignored the economic conditions of the Communist "satellite states."

Objective 3: The Transformation of Nature

The third big practical task that Hungarian Marxist-Leninist geography was set to tackle was the transformation of the country's natural environment in order to improve agricultural production. This endeavor focused on three goals: Grandiose irrigation projects, the creation of forest belts protecting the soil from wind erosion, and the naturalization of new species of plants. The initiative was influenced by the Stalin Plan for the Transformation of Nature, which was introduced in the Soviet Union in 1948 (Brain, 2010; Hajdú, 2006). In a theoretical sense, all three goals were based on a kind of economic determinism which dominated Soviet geographical thought, and on a firm belief in science. This theoretical position was totally accepted and internalized by the Hungarian Communist leadership. For Mátyás Rákosi,

> The country of socialism is the country of unlimited possibilities.… Where is the upper limit in its construction? I gave the answer: The sky is the upper limit! … The planned construction of socialism does not have the limits of capitalism. (as cited by Hajdú, 2006, p. 250)

Marxist-Leninist geographers were keen to give scientific substantiation to leading politicians' ideas. György Markos again played a crucial role in this respect. In 1952, he gave a detailed interpretation, from the perspective of Hungarian science, of Stalin's theories on human–nature relations and on the transformation of nature (Markos, 1952b).

The National Planning Office was assigned to prepare a ten-year irrigation plan of Hungary as early as 1948. The plan mainly focused on the Great Plain, which is the most fertile agricultural region in the country, though it frequently experiences droughts during the summer. Particular emphasis was placed on the transformation of physical conditions in the Hortobágy region, the driest in the Great Plain. In order to solve the problems of this region, a Planning Committee for the Transformation of Nature in the Tiszántúl region (Tiszántúli Természetátalakító Tervbizottság) was established in 1952. One of the committee's members was Ferenc Erdei, who at the time also served as president of the Scattered Farm Council. Such irrigation plans were not new, earlier plans had aimed at the construction of three dams, and irrigation projects in the Hortobágy region had been underway since 1937/38. The main difference between the old and the new projects was their scale and the plan for their realization. Because the process of construction could not meet the irrational expectations of the Communist political leaders, the irrigation project could not be fully completed (Hajdú, 2006). Similar but bigger challenges were faced by the large-scale afforestation proposals: Directives were unrealistic, and the project lacked adequate theoretical preparation (Hajdú, 2006).

Still, although these projects proved impossible to carry out on account of their unrealistic scope, their overall objective was technically realistic and failed only because of a lack of money, labor, and equipment. Some other initiatives of the Stalinist regime were, however, incompatible with natural conditions that human agency cannot alter radically. The most significant example was without a doubt the naturalization of new plants. Although experimentation with the introduction of new plants has a long tradition in the history of agriculture, and attempts in Hungary had already been made before 1939, the initial phase of small-scale experimentation had always been slow and cautious. But where economic profit had motivated these smaller schemes, the Communist regime considered the naturalization of new plants a crucial political issue, and devoted considerable financial and institutional resources to its success.

Such massive effort can be seen clearly in the case of cotton, an emblematic plant in the first decade of Communism in Hungary. For economic reasons, small-scale experiments with the production of cotton had been conducted during the interwar period, but were soon ended. The issue of naturalizing cotton emerged again in the late 1940s, and became a main goal of the new regime. After the decree of the Council of Ministers in 1948, in the following year the so-called Council for Cotton Production was established. Experimentation began on some 850 acres, with a planned increase of the sown area to more than 140,000 acres in 1950 (Hajdú, 2006). The project was unique in Stalinist economic policy given that—unlike many other initiatives which were too grandiose but technically realistic (such as the creation of shelterbelts and irrigation infrastructure)—the naturalization of cotton and several other plants (e.g., citrus fruits, peanuts, and tea; see Gyenes, 1952, 1954) was profoundly incompatible with natural conditions in Hungary. Yet enormous resources were invested in these projects, and in order to inform the people about the goals and "achievements" of socialist agrobiology, massive propaganda campaigns were launched (Hajdú, 2006). Science was also mobilized to assist in realizing these ends. At the Academy of Sciences, new committees such as the Agrobiological Committee, the Crop Production Committee, and the Lemon Committee were established for the purpose of scientifically substantiating the grandiose political aims (Hajdú, 2006).

Although it was mostly agronomists and biologists who contributed to this work, physical geographers were also involved. Their task was to identify those regions of the country with feasible terrain and climatic conditions. The first issue of the newly established journal of the GRI HAS, the *Földrajzi Értesítő*, devoted more than 30 pages to the question of the economic promise of new plants. The author, Lajos Gyenes, was a geographer of the "new generation." At the same time, the subject of geography in primary and secondary education became an important tool for popularizing the new "socialist methods" in agriculture; national competitions for pupils contained several exercises on the issue (Simon, 1955).

Given such "scientific preparation," the production of new crops gained strong impetus in 1950. Although agrobiological experiments failed, the hot weather of that year resulted in a relatively good crop yield, convincing the party leadership of the correctness of their goals. Their new initiative urged doubling the production area given over to cotton. After further progress in the likewise remarkably hot summer of 1951 (Figs. 10.8 and 10.9), Hungarian cotton production soon ended in failure. In 1953, as a result of the economic failure and of the changed political landscape given the death of Stalin, the political leadership began to give up its grand schemes on the "transformation of nature" (Hajdú, 2006), which by then was ignoring issues of physical geography and of profitability. The exception was rice, as experiments to increase its production met with significant success. The Communist regime overplayed its role in this success story, however, as naturalization and production of this crop had already begun in the interwar period (Hajdú, 2006).

As most Hungarian scientists had never become convinced supporters of the initiative, there was a greater willingness to express negative opinions following 1953. In 1956, József Bognár, the chief secretary of the Hungarian Academy of Sciences, strongly criticized the project, and Ferenc Erdei, while evaluating the scientific work of the academy's Agricultural Sciences Section in 1957, said nothing about the issue of new plants (Hajdú, 2006). In geography, arguments for and against the large-scale production of new crops afford insight into the inner politics and structure of Hungarian science and geography. In 1954, the Economic Geographical Session of the Hungarian Geographical Society hosted a lecture by Lajos Gyenes on this issue. The lecture, together with a draft review of the comments from the audience, was published in an issue of *Földrajzi Értesítő* (Gyenes, 1954).

At the lecture, Gyenes, as the strongest advocate among Hungarian geographers of schemes for the naturalization of new plants, argued strongly for experimentation with new crops. He referred to Marxism-Leninism in making his point, arguing that science's task lay in contributing to the construction of socialism and, thus, to the improvement of the people's living conditions. In his words, "These new plants, serving the national economy, national healthcare, and the workers … can significantly contribute to the improvement of our agriculture, the standard of living of our working people, and the healthcare maintenance of our workers" (Gyenes,

Fig. 10.8 Drying of cotton close to the town of Békéscsaba (Source: "Forgatással szárítják az asszonyok a betakarított gyapotot" (Women drying harvested cotton by turning it), by Pál Jónás. Magyar Fotó. Békéscsaba, October 19, 1951. Copyright by MTI Hungarian News Agency Corp., Media Service Support and Asset Management Fund. Reprinted with permission)

1954, pp. 102–103). In his eyes, experimentation with new plants and the participation of science in such projects was a necessity: "Soberly, but courageously, experiments shall be made! It is the very thing science wishes and waits for from us. It is the very thing being wished and awaited from us by our working people." (p. 102). Gyenes continued, giving a long and detailed description of the physical geographical requisites of numerous "new plants," while emphasizing their economic benefits.

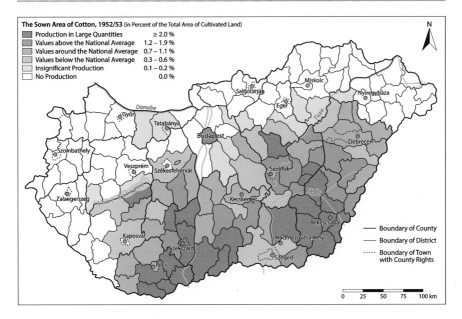

Fig. 10.9 "We aim to change the socioeconomic map of our country." Main regions of cotton production in Hungary (Source: From *Magyarország mezőgazdasági földrajza* [An agricultural geography of Hungary] (Appendix 31), by L. Görög, 1954, Budapest, Hungary: Tervgazdasági Könyvkiadó. Map design: László Görög. Cartographer: Daniel Söder (redrawn 2013))

Other main participants at the lecture (some as representatives of other disciplines) were not at all convinced. Members of the audience who were involved in interwar academic life criticized Gyenes's Stalinist approach because of his relative ignorance of physical geographical factors and of considerations of profitability.

An Implicit Objective: Manufacturing Political Propaganda

Several Marxist-Leninist geographers actively participated in politically motivated, grandiose planning projects, but the discipline's practitioners did even more than contribute to such practical endeavors. Geography also played a significant role in the propaganda of "constructing socialism." This role was especially evident in geographical education in primary and secondary schools. Pupils were expected to use theoretical knowledge in the solution of practical issues. Geographically relevant questions of economic planning (the naturalization of new plants and the optimal spatial allocation of the forces of production in Hungary) enjoyed a dominant place in the curriculum (Korzsov, 1955; Simon, 1955). At the same time, Soviet researchers' popular science articles in Hungarian translation were published in richly illustrated books such as *A szovjet nép átalakítja a természetet* (1951; "The Soviet people transform nature") or *A sztálini korszak nagy építkezései* (1951; "Large constructions of the Stalinist era"). These works were sent to libraries throughout the country in

Fig. 10.10 Geography as socialist propaganda: The atlas of the Three-Year Plan, written by György Markos (Source: From *Magyarország gazdasága és a hároméves terv* [The economy of Hungary and the Three-Year Plan], by G. Markos (Ed.), 1948, Budapest, Hungary: Szikra)

order to inform the masses about the "achievements" of the socialist state and to indoctrinate them in Communist ideology. The same was true for publications reporting on the goals of the economic plans, maps were used for propagandistic reasons. As the foreword of one such book emphasized, "There is nothing more convenient for letting the broad masses know and evaluate the Plan than geographical representation, which projects our economic resources and the prospect of their methodical development onto the map" (Berei, 1948, p. 2; see Fig. 10.10).

Marxist-Leninist geography thus not only contributed to practical projects, but also was a tool for propagating official ideology. In other words, although official propaganda defined the goal of science as producing factual knowledge, in actuality all disciplines were also expected to produce and disseminate orientation knowledge for propagandistic goals (cf. Meusburger, 2005). Geography was no exception, its role was not only to contribute to the realization of big projects, but to mediate Marxist-Leninist ideology and, thus, to legitimize the ruling order.

Conclusion

Hungarian geography both before and after World War II is a characteristic example of how politics (the power) and science are intertwined in specific contexts. Because Hungary suffered major territorial losses after World War I, geography became a highly respected discipline in the eyes of political leaders because it was regarded as an important tool for the scientific substantiation of arguments for territorial revision. For the same reason, the discipline was seen as "guilty" by the newly emerging postwar Communist regime. Because it was regarded as having served "fascist" interests in the interwar period, it was an explicit goal of the new system to demolish the "old," "reactionary," and "bourgeois" geography and to build up a new, Marxist-Leninist one following the example of Soviet geography. This aim led to changes in the discipline's institutional setting. Some of the leading personalities of interwar geography were suppressed, others given only limited authority. In their place, a number of "new" geographers—the most loyal supporters of the Communist regime, many of whom lacked a formal university education in geography—were given prominent positions in the discipline.

The rapid Sovietization of Hungarian geography occasioned dramatic changes in the latter's theoretical approach and in the lives of those who worked in the field. The discipline was transformed in line with Marxist-Leninist expectations, in line with the Soviet example. The terms "human geography," "social geography," and "cultural geography" were erased from the new discourse, and their successor christened "economic geography." This massive theoretical transformation left few or no places for formerly flourishing fields of interest. Geographical research on politics, religion, ethnicity, or social disparities, for instance, was banned. Several topics were criticized for concentrating on the form instead of its essence, for engaging in a "bourgeois trick" "serving capitalist interests." Thus, urban morphology was affected, but so too was geomorphology.

After exiling "bourgeois" elements, geography absorbed Marxist-Leninist ideology and rigid scientism. At the same time, Marxist-Leninist geography was to actively contribute to the "construction of socialism." For this reason, Hungarian geography became involved in the problem of urban network planning, large-scale economic planning (through the creation of *rayons*, or economic regions), and the transformation of nature in order to improve agricultural production. It was deemed important for geography to participate in the propaganda of these practical goals, through mass education as well as in the literature of popular science.

As for science, geography became a mouthpiece of the Communist leadership, and its "new" representatives received full support from the political leadership. In this way, geography as a science and politics (the power) became even more closely intertwined than before. This mutual support was crucial for both sides. Because Marxist-Leninists had no influence on Hungarian geography before the Communist turn, they were extremely reliant on political power as the only possible source of their legitimacy. The political leadership, however, also required strong support from science to justify its much propagated goal of transforming society. It was this constellation of interests that opened the way for radical change in both spheres of life. Although Hungarian Marxist-Leninist geography of the Stalinist period did not succeed in realizing all of its objectives, its implications were far-reaching and proved to be long-lasting in the discipline.

Acknowledgements We would like to thank our friends Márton Czirfusz, Zoltán Gyimesi, and Énok Oláh for the many inspiring talks that we had while writing this chapter. We are also grateful to Trevor Barnes, Peter Meusburger, and especially Charles Withers for their comments, thoughtful criticism, and help on the early versions of the manuscript. Róbert Győri appreciates the scholarship of the cofunded International Exchange Programme of the Royal Society of Edinburgh and the Hungarian Academy of Sciences, which was taken at the University of Edinburgh. Ferenc Gyuris would like to acknowledge the University of Heidelberg Graduate Academy for the Landesgraduiertenförderung scholarship. This chapter was also supported by the János Bolyai Research Scholarship of the Hungarian Academy of Sciences.

References

A szovjet nép átalakítja a természetet [The Soviet people transform nature]. (1951). Budapest, Hungary: Szikra.

A Szovjetunió Kommunista (bolsevik) Pártjának története [History of the Communist (Bolshevik) Party of the Soviet Union]. (1949). Budapest, Hungary: Szikra.

A sztálini korszak nagy építkezései [Large constructions of the Stalinist era]. (1951). Budapest, Hungary: Szikra.

Abella, M. (1956). Vita a földrajzi tudományok filozófiai problémáiról [Debate on philosophical challenges of geography]. *Földrajzi Értesítő, 5*, 462–466.

Abella, M. (1961). Ankét a településföldrajz helyzetéről és feladatairól [Workshop on the state and tasks of settlement geography]. *Földrajzi Értesítő, 10*, 121–127.

Ablonczy, B. (2007). *Pál Teleki: The life of a controversial Hungarian politician.* Wayne, NJ: Hungarian Studies Publications.

Andics, E. (1945). *Fasizmus és reakció Magyarországon* [Fascism and reaction in Hungary]. Budapest, Hungary: Szikra.

Baranskiy, N. N. (1956). *Economic geography of the U.S.S.R.* (S. Belsky, Trans.). Moscow: Foreign Languages Publication House. (Original work published in 1926)

Barnes, T. (2004). Placing ideas: Genius loci, heterotopia and geography's quantitative revolution. *Progress in Human Geography, 28*, 565–595. doi:10.1191/0309132504ph506oa.

Beluszky, P. (1982). A szerkesztő vitaindítója – A komplex gazdasági körzet – hit, remény vagy valóság? [The author's keynote: Complex economic district—belief, hope, or reality?]. In P. Beluszky & T. T. Sikos (Eds.), *Területi Kutatások* (Vol. 5, pp. 3–24). Budapest, Hungary: MTA FKI.

Beluszky, P. (2001). *A Nagyalföld történeti földrajza* [Historical geography of the Great Plain]. Budapest, Hungary: Dialóg Campus.

Berei, A. (1948). Előszó [Preface]. In G. Markos (Ed.), *Magyarország gazdasága és a hároméves terv* [The economy of Hungary and the Three-Year Plan] (p. 2). Budapest, Hungary: Szikra.

Bernát, T. (2004). Emlékezés Markos Györgyre, az iskolateremtő geográfusra halálának 100. évfordulóján [Remembrance of György Markos, the school's founding geographer, upon the 100th anniversary of his death] (sic). *Földrajzi Közlemények, 128*, 180–182.

Brain, S. (2010). The great Stalin plan for the transformation of nature. *Environmental History, 15*, 670–700. doi:10.1093/envhis/emq091.

Bulla, B. (1955a). A magyar földrajztudomány útja a felszabadulás után [The path of Hungarian geography since the liberation]. *Földrajzi Közlemények, 79*, 93–114.

Bulla, B. (1955b). Válasz Markos Györgynek [Reply to György Markos]. *Földrajzi Közlemények, 3*, 367–371.

Bunce, V. (1985). The empire strikes back: The evolution of the Eastern bloc from a Soviet asset to a Soviet liability. *International Organization, 39*, 1–46. doi:10.1017/S0020818300004859.

Chioni Moore, D. (2001). Is the post in postcolonial the post in post-Soviet? Notes toward a global postcolonial critique. *PMLA, 116*, 111–128.

Dobrov, A. S. (1952). Az imperializmus védelmének alapvető módszerei a burzsoá gazdasági földrajzban [Basic methods of defending imperialism in bourgeois economic geography]. *Földrajzi Értesítő, 1*, 1–32.

Enyedi, G. (1961). Vita a gazdasági rayonkutatás elméleti és módszertani kérdéseiről [Debate on the theoretical and methodical issues of economic *rayon* research]. *Földrajzi Értesítő, 10*, 117–121.

Enyedi, G. (1968). Markos György. *Földrajzi Értesítő, 17*, 405–406.

Enyedi, G. (1982). Temetés vagy a tetszhalott ébresztése [Burial or vivifying the undead]. In P. Beluszky & T. T. Sikos (Eds.), *Területi Kutatások* (Vol. 5, pp. 52–54). Budapest, Hungary: MTA FKI.

Fodor, F. (1924). *Magyarország gazdasági földrajza* [Economic geography of Hungary]. Budapest, Hungary: Franklin Társulat.

Gerasimov, I. P. (1959). A földrajz a Szovjetunióban [Geography in the Soviet Union]. *Földrajzi Értesítő, 8*, 406–415.

Gerő, E. (1950a). Az ötéves tervvel a szocializmus felé. (Beszéd az Országgyűlésben, 1949. december 9-én) [With the Five-Year Plan toward socialism. (Address in the Parliament on December 9, 1949)]. In E. Gerő (Ed.), *Harcban a szocialista népgazdaságért, Válogatott beszédek és cikkek 1944–1950* (pp. 554–592). Budapest, Hungary: Szikra.

Gerő, E. (1950b). A Magyar Tudományos Tanács feladatairól [On the tasks of the Hungarian Scientific Council] (Gerő's speech on the statutory meeting of the Hungarian Scientific Council on February 25, 1949). In E. Gerő (Ed.), *Harcban a szocialista népgazdaságért, Válogatott beszédek és cikkek 1944–1950* (pp. 341–354). Budapest, Hungary: Szikra.

Gyenes, L. (1952). A citrusfélék hazai termelésének gazdaságföldrajzi vonatkozásai. (Beszámoló az MTA Állandó Földrajzi Bizottsága által 1952. március 7-én rendezett szakülésről) [Economic geographical features of producing citrus fruits in Hungary. (Report on the workshop organized by the Permanent Geographical Committee of the Hungarian Academy of Sciences on March 7, 1952)]. *Földrajzi Értesítő, 1*, 85–118.

Gyenes, L. (1954). Új gazdasági növényeink [Our new economic plants]. *Földrajzi Értesítő, 3*, 101–138.

Győri, R. (2001). A magyar gazdaságföldrajz a két világháború között [Hungarian economic geography between the world wars]. In J. Nemes Nagy (Ed.), *Geográfia az ezredfordulón* (pp. 61–83). Budapest, Hungary: ELTE TTK RFT.

Győri, R. (2009). Tibor Mendöl. In H. Lorimer & C. W. J. Withers (Eds.), *Geographers: Biobibliographical studies* (Vol. 28, pp. 39–54). London: Continuum.

Hajdú, Z. (1990–1991). A Tanyai Tanács története (A "szocialista tanyapolitika" alapvetése és a tanyakérdés megoldásának radikális, voluntarisztikus kísérlete, 1949–1954) [The history of the Scattered Farm Council (The fundaments of "socialist scattered farm policy" and the radical, voluntaristic experiment of solving the scattered farm issue, 1949–1954)]. *Alföldi Tanulmányok, 14*, 105–124.

Hajdú, Z. (1992). Település- és településhálózat-fejlesztési politika Magyarországon az államszocializmus időszakában [Urban and urban network development policy in Hungary during the era of state socialism]. *Földrajzi Közlemények, 116*, 29–37.

Hajdú, Z. (1998). *Changes in the politico-geographical position of Hungary in the 20th century* (Discussion Papers 22). Pécs, Hungary: CRS HAS.

Hajdú, Z. (2006). A szocialista természetátalakítás kérdései Magyarországon, 1948–1956 [Issues of the socialist transformation of nature in Hungary, 1948–1956]. In A. Kiss, G. Mezősi, & Z. Sümeghy (Eds.), *Táj, környezet és társadalom: ünnepi tanulmányok Keveiné Bárány Ilona professzor asszony tiszteletére* (pp. 245–257). Szeged, Hungary: University of Szeged, Department of Climatology and Landscape Ecology; Department of Geoinformatics.

Horváth, G. (2008). Regionális átalakulás Oroszországban [Regional transformation in Russia]. In G. Horváth (Ed.), *Regionális fejlődés és politika az átalakuló Oroszországban* (pp. 11–62). Pécs, Hungary: MTA RKK.

Ivanov, K. (2002). Science after Stalin: Forging a new image of science. *Science in Context, 15*, 317–338. doi:10.1017/S0269889702000467.

K L. (1960). Radó Sándor, a Magyar Földrajzi Társaság Társelnöke 60 éves [Sándor Radó, copresident of the Hungarian Geographical Society, is sixty years old]. *Földrajzi Közlemények, 84*, 220–223.

Keményfi, R. (2006). Egységes magyar államtér alatt egységes kőzetalap: A Tisia-masszívum mítosza [Unitary lithological base beneath unitary Hungarian national space: The myth of Tisia Massive]. In R. Győri & Z. Hajdú (Eds.), *Kárpát-medence: települések, tájak, régiók, térstruktúrák* (pp. 418–438). Budapest, Hungary: Dialóg Campus.

Koch, F. (1952). A Magyar Földrajzi Társaság újjáalakulása [Reestablishment of the Hungarian Geographical Society]. *Földrajzi Értesítő, 1*, 884–887.

Koch, F. (1956). Teleki Pál gazdaságföldrajzi munkásságának bírálata [Critique of Pál Teleki's contribution to economic geography]. *MTA Társadalmi – Történeti Tudományos Osztályának Közleményei, 8*, 89–122.

Kolta, J. (1954). A gazdaságföldrajzi rayonírozás néhány elméleti kérdése és adatok Baranya megye rayonbeosztásához [Some theoretical issues of *rayonization* in economic geography and some data for the *rayonization* of Baranya county]. *Földrajzi Közlemények, 2*, 199–219.

Korzsov, N. I. (1955). Gazdasági földrajzi kirándulások a középiskolában [Economic geographical excursions in secondary school]. *Földrajzi Közlemények, 3*, 51–61.

Kovács, C. (1954). A gazdasági földrajz néhány elméleti kérdéséről [On some theoretical issues of economic geography]. *Földrajzi Értesítő, 3*, 414–432.

Krajkó, G. (1961). A gazdasági körzetesítés néhány elvi problémája [Some conceptual problems of economic regionalization]. *Földrajzi Közlemények, 3*, 223–246.

Krajkó, G. (1982). A gazdasági körzetek néhány fontosabb vonása [Some main characteristics of economic districts]. In P. Beluszky & T. T. Sikos (Eds.), *Területi Kutatások* (Vol. 5, pp. 25–41). Budapest, Hungary: MTA FKI.

Latour, B. (1987). *Science in action: How to follow scientists and engineers through society.* Cambridge, MA: Harvard University Press.

Lenin, V. I. (1960). *Collected works* (Vol. 5). Moscow: Progress Publishers. (Original works published in May 1901–February 1902)

Lenin, V. I. (1966). *Collected works* (Vol. 31). Moscow: Progress Publishers. (Original work published in April–December 1920)

Lenin and science. (1970). *Journal of Mining Science, 6*, 129–131.

Markos, G. (1952a). Magyarország gazdasági körzetbeosztása: A Magyar Tudományos Akadémia Állandó Földrajzi Bizottságának 1952. június 13-án rendezett vitaindító előadása [Division of economic districts in Hungary: Keynote speech at the Permanent Geographical Committee of the Hungarian Academy of Sciences on June 13, 1952]. *Földrajzi Értesítő, 1*, 582–634.

Markos, G. (1952b). A természeti földrajzi környezet hatása különböző társadalmi formák között [The implication of physical geographical environment for various social forms]. *Földrajzi Értesítő, 1*, 271–287.

Markos, G. (1955). Reflexiók egy beszámolóhoz. (A földrajzi tudományok rendszertani alapjairól) [Reflections on a report: On the taxonomical basis of geographical sciences]. *Földrajzi Közlemények, 79*, 359–365.

Marx, K. (1949). *A tőke* [Capital] (Vol. 1). Budapest, Hungary: Szikra. (Original work published in 1867)

McEwan, C. (2009). *Postcolonialism and development.* London: Routledge.

Mendöl, T. (1954). A szocialista településföldrajz problémái [Challenges of socialist urban geography]. *MTA Társadalmi – Történeti Tudományos Osztályának Közleményei, 5*, 599–627.

Meusburger, P. (2005). Sachwissen und Orientierungswissen als Machtinstrument und Konfliktfeld: Zur Bedeutung von Worten, Bildern und Orten bei der Manipulation des Wissens [Factual knowledge and orientation knowledge as instruments of power and fields of conflict: On the significance of words, images, and places in the manipulation of knowledge]. *Geographische Zeitschrift, 93*, 148–164.

Meusburger, P. (2007). Macht, Wissen und die Persistenz von räumlichen Disparitäten [Power, knowledge, and the persistence of spatial disparities]. In I. Kretschmer (Ed.), *Das Jubiläum der*

Österreichischen Geographischen Gesellschaft: 150 Jahre (1856–2006) (pp. 99–124). Vienna: Österreichische Geographische Gesellschaft.

Ormeling, F. J. (1982). Az ICA megemlékezik Radó Sándor professzorról [The ICA commemorates Professor Sándor Radó]. *Geodézia és Kartográfia, 34*, 298–299.

Papp-Váry, Á. (1998). Radó Sándor (1899–1981). *Földrajzi Közlemények, 122*, 99–101.

Pécsi, M. (1982). Búcsúbeszéd Radó Sándor (1899–1981) ravatalánál [Farewell speech at the catafalque of Sándor Radó (1899–1981)]. *Földrajzi Közlemények, 106*, 290–292.

Péteri, G. (1998). *Academia and state socialism*. New York: Columbia University Press.

Probáld, F. (2001). Száz esztendeje született a Regionális Földrajzi Tanszék alapító professzora: Koch Ferenc [Ferenc Koch, the founding professor of the Department of Regional Geography, was born 100 years ago]. In J. Nemes Nagy (Ed.), *Geográfia az ezredfordulón* (pp. 85–88). Budapest, Hungary: ELTE TTK RFT.

Radó, S. (1957a). Adatok a szovjet gazdaságföldrajz történetéhez [Some data on the history of Soviet economic geography]. *Földrajzi Értesítő, 6*, 490–492.

Radó, S. (1957b). A szovjet földrajztudomány 40 éve [Forty years of Soviet geography]. *Földrajzi Közlemények, 81*, 305–318.

Radó, S. (1962). A kommunizmus építése és a földrajzi tudományok [The construction of Communism and geographical sciences]. *Földrajzi Közlemények, 86*, 225–232.

Radó, S. (1971). *Dóra jelenti*. Budapest, Hungary: Kossuth.

Rákosi, M. (1951). Május 15-én egy emberként a népfront mellé! (A Népfront választási gyűlésén, Celldömölkön, 1949. május 8-án mondott beszéd) [Together on the side of the People's Front on May 15 (Address made at the electing convention of the People's Front in Celldömölk village, May 8, 1949)]. In M. Rákosi (Ed.), *A békéért és a szocializmus építéséért* (pp. 5–28). Budapest, Hungary: Szikra.

Romsics, I. (1999). *Hungary in the twentieth century*. Budapest, Hungary: Corvina.

Saushkin, J. G. (1954). A Szovjetunió gazdasági földrajzának alapvető kérdései [Basic issues of economic geography of the Soviet Union]. *Földrajzi Értesítő, 3*, 86–100.

Saushkin, J. G. (1962). Economic geography in the U.S.S.R. *Economic Geography, 38*, 28–37. doi:10.2307/142323.

Simon, L. (1955). Földrajzoktatásunk néhány alapkérdése a "Rákosi Mátyás" tanulmányi verseny eredményeinek tükrében [Some general issues of our geographical education in the light of results at the "Mátyás Rákosi" competition for pupils]. *Földrajzi Közlemények, 3*, 157–166.

Spriano, P. (1985). *Stalin and the European Communists* (J. Rothschild, Trans.). London: Verso.

Stalin, J. V. (1950). *A leninizmus kérdései* [Issues of Leninism]. Budapest, Hungary: Szikra.

Tatai, Z. (2004). Markos György. *Comitatus, 14*, 71–80.

Teleki, P. (1922). Térkép és statisztika a gazdasági földrajzban [Map and statistics in economic geography]. *Földrajzi Közlemények, 50*, 74–91.

Thomas, L. (1968). Alexander Rado. *Studies in Intelligence, 12*, 41–61.

Trom, A. (2006). A *teljes* Dóra jelenti margójára [To the margin of the *entire* Codename Dora]. In S. Radó (Ed.), *Dóra jelenti* (Ötödik, bővített, javított és átszerkesztett kiadás) (pp. 5–15). Budapest, Hungary: Kossuth Kiadó.

Universitätsarchiv Jena (UAJ). Bestand Studienkartei (ca. 1915–1935), Alexander Rado.

Universitätsarchiv Leipzig (UAL), Quästur. Sheet 486, Alexander Rado.

Vavilov, S. I. (1950). *A sztálini korszak tudománya* [Science in the Stalinist era]. Budapest, Hungary: Szikra.

Vitaülés objektív gazdasági törvényszerűségek feltárásáról [Public debate on revealing objective economic regularities]. (1954). *Földrajzi Értesítő, 3*, 779–783.

The Geopolitics of Knowledge About World Politics: A Case Study in U.S. Hegemony

11

John Agnew

Much of our knowledge about world politics involves the universalizing of what can be called "doubtful particularisms." These interpretive projections are from the knowledge experiences of specific places and times onto all places and times. By knowledge I mean explanatory schemes, frames of reference, crucial sets of assumptions, narrative traditions, and theories. A great deal of interpretive projection is the result of the imposition of intellectual and political hegemonies from some places onto others. Thus, much of what today goes for "international relations theory" is the projection onto the world–at large of U.S.-originated academic ideas about the nature of statehood and the world economy derived from a mixture of largely mid-twentieth-century European premises about states and American ones about economies. The theory reflects the application of ideas about how best to model a presumably hostile world, which are drawn from selected aspects of U.S. experience and a U.S.-based reading of world history. In this chapter I propose a way of thinking about this geopolitics of knowledge by using the example of theories of world politics.

My point is not so much that knowledge of world politics is simply a coercive imposition of the view from some places onto others as that the dominant ways in which intellectuals and political elites around the world have come to think about world politics are not the result of either an open "search" for the best perspective or theory or a reflection of an essentially "local" perspective. The most prestigious repertoires of thinking about world politics represent the historical emergence of theoretical genres intimately associated with specific times and places which circulate and adapt in association with the spheres of influence of schools and authors with the best reputations and which in turn reflect the current geopolitical order.

J. Agnew (✉)
Department of Geography, University of California,
1171 Bunche Hall, Box 951524, Los Angeles, CA 90095, USA
e-mail: jagnew@geog.ucla.edu

© Springer Netherlands 2015
P. Meusburger et al. (eds.), *Geographies of Knowledge and Power*,
Knowledge and Space 7, DOI 10.1007/978-94-017-9960-7_11

The idea that there is some sort of "geography" of knowledge is increasingly seen as helpful in understanding the production and circulation of knowledge of all kinds. Shapin (1998, p. 9) has best expressed perhaps the basic intuition upon which a much larger theoretical edifice can be constructed: "We are importantly attracted to, or repelled by, ideas as they are embodied in familiar others—kin, teachers, colleagues, neighbors." From this primitive sociological premise about the geographical bias built into knowledge creation and dissemination, it is possible to hypothesize about which ideas crop up where, how ideas adapt as they circulate, and why some ideas never quite make it into wider circulation, to name just a few of the ways in which geography in an expansive sense shapes knowledge.

After providing a brief summary of various ways of conceiving the geography of knowledge, I present four premises for what I am calling the geopolitics of knowledge. I then consider the specific case of how a particular theoretical perspective of peculiarly American provenance came to dominate much academic thinking about world politics outside the United States. A short conclusion summarizes the main points of the chapter.

Geographies of Knowledge

I have previously surveyed some of the ways in which "the geography of knowledge" can be brought into the study of world politics (Agnew, 2007). The purpose was to review this developing field and what it can offer to students of world politics. I suggested that there are five ways in which the geography of knowledge can be conceived and related to world politics.

The first way of conceiving of the geography of knowledge is the *ethnographic*, by which I mean approaches that conceive of knowledge as inherently plural and focus on the venues and sites in which knowledge is produced and consumed. The focus lies in either rehabilitating what are sometimes called "indigenous knowledges" or pointing out how "science" is culturally inflected. A good example of this approach is Nader's (1996) collection of studies of how scientific experiments on the same topic are conducted in different ways in different countries. A related but distinctive position tends to privilege the role of *coloniality* or the effects of colonialism on knowledge hierarchies. This approach is, of course, closely associated with the name of Said (1978), but others, such as Mignolo (2000), have developed it much further. A third derives more immediately from the philosophies of *phenomenology* which emphasize the intimate relations between particular geographical contexts of "being," on the one hand, and knowledge acquisition, on the other. In historical geography, a classic work of this genre would be that of Lowenthal (1961) on "geographical epistemology."

While also seeing knowledge as produced locally, a fourth approach emphasizes *how the local becomes the global* given the rise and fall of ideas as their political or intellectual sponsors undergo a similar process. A good example of such a process is the spread of neoliberal modes of economics under U.S. influence and as a result of the hegemony exercised by U.S.-based economics since the 1970s (Biersteker,

1995). Finally, emphasis has shifted somewhat in some recent accounts from simply knowledge production to knowledge circulation and consumption in the form of highlighting what is called by Livingstone (2005), one of its main proponents, the *geography of reading*. This approach assumes that similar ideas circulate widely but generate distinctive readings in different places, thus potentially creating different perspectives that then inform different practices.

In this chapter I focus my attention primarily on the fourth of these approaches, how the local becomes the global, with special attention given to theoretical thinking about world politics. My reason for doing so is that world politics is itself fundamentally an outcome of a basic hierarchy among states and between world regions (Agnew, 2005). It is not that the other conceptions of the geography of knowledge are irrelevant—the fifth is also in play to a certain extent, as the examples will show—but that in this context they are secondary to the primary one. The presumption of my approach is that global structures of political inequality underwrite whose imagination gets to dominate globally in theorizing about world politics. This conception in turn has obvious implications for any liberatory politics. In other words, thinking about world politics reflects the relative hierarchy of power within world politics. Yet much of the dominant thinking about world politics usually makes claims that either obscure or limit the degree to which world politics is hierarchical. I first provide some premises upon which the argument is based, and then use U.S., English, Russian, and Chinese examples of thinking about world politics to illustrate the argument.

The Geopolitics of Knowledge

The first premise is that the marketplace of ideas is never a level playing field. There is a geopolitics to knowledge production and circulation. Which knowledge becomes "normalized" or dominant and which knowledge is marginalized has something to do with who is doing the proposing and where they are located (Agnew, 2005). In the context of world politics, all knowledge, including that claiming the mantle of science, is socially conditioned by the rituals, routines, and recruitment practices of powerful educational and research institutions. On a global scale perhaps the outstanding feature of past centuries has been the way most places have been incorporated into flows of knowledge dominated by Europeans and extensions of Europe overseas, such as the United States. This phenomenon is the story, in Wolf's (1982) evocative phrase, of "Europe and the people without history."

The second premise is that, as Geertz (1996, p. 262) said, "No one lives in the world in general." Actual places, both as experienced and as imagined, serve to anchor conceptions of how the world is structured politically, who is in charge, where, and with what effects, as well as what matters to us in any given place in question. Thus, for example, Americans and U.S. policy-makers bring to their actions in the world a whole set of presuppositions about the world that emanate from their experiences as "Americans," particularly narratives about U.S. history and the U.S. "mission" in the world, which are often occluded by academic debates about "theories" that fail to take into account such crucial background geographical conditioning. As Anderson (2003, p. 90) has noted, much of the "liberal tradition" that has shaped social science in the United States has had "a geographical, territorial association." She quotes Prewitt (2002) in support of this idea:

> The project of American social science has been America. This project, to be sure, has been in some tension with a different project—to build a *science* of politics or economics or psychology. But I believe that a close reading of disciplinary history would demonstrate that the "American project" has time and again taken precedence over the "science project" and that our claims to universal truths are, empirically, very much about the experience of this society in this historical period. (p. 2)

Of course, the very idea of requiring a "scientific" theory of politics may itself be seen as arising out of a specifically American desire to account for the United States and its place in the world in such terms.

Third, universalizing creeds must recruit adherents beyond their places of origin in order to become hegemonic. Gramsci's (1992) concept of "hegemony" is helpful in trying to understand how elites (and populations) accept and even laud ideas and practices about world politics and their place in it that they import from more powerful countries and organizations. If part of American hegemony in the contemporary world, for example, is about "enrolling" others into American practices of consumption and a market mentality (and, crucially, supplying intellectual justifications for them, such as those provided by various management gurus and journalists), it also adapts as it enrolls by adjusting to local norms and practices (Agnew, 2005). This facility is part of its "genius." During the Cold War, the Soviet alternative always risked political fission among adherents because it involved adopting a checklist of political-economic measures rather than a marketing package that could be customized to local circumstances as long as it met certain minimal criteria of conformity to governing norms. Today, the conflict between militant Islam and the United States government is largely about resisting the siren call of an American hegemony associated with globalization that is increasingly detached from direct U.S. sponsorship and that has many advocates and passive supporters within the Muslim world itself.

Fourth, knowledge about world politics (or anything else) from one place is not necessarily incommensurable or unintelligible relative to knowledge produced elsewhere. Cross-cultural communication goes on all the time without everything being lost in translation. Cultures in the modern world never exist in isolation and are themselves assemblages of people with often cross-cutting identities and commitments (Lukes, 2000). From this viewpoint, culture is "an idiom or vehicle of inter-subjective life, but not its foundation or final cause" (Jackson, 2002, p. 125). Be that as it may, knowledge creation and dissemination are never innocent of at least weak ontological commitments, be they related to nation, class, gender, or something else. But the history of knowledge circulation suggests that rarely are ideas simply restricted within rigid cultural boundaries. Rather, with powerful sponsors, international and transnational networks arise to carry and embed ideas from place to place (e.g., Sapiro, 2009).

Taken together, these premises make the case for referring to the geopolitics of knowledge: The question of *where* brings together under the rubric of spatial difference a wide range of potential ontological effects. At the same time, however, massive sociopolitical changes in the world are shaping how we (whomever and wherever *we* are) engage in how knowledge is ordered and circulated. Cross-global linkages are arguably more important today than at any time in human history, not so much in terms of the conventional story of producing places that are ever more alike, but more especially in terms of creating opportunities for interaction between local and long-distance effects on the constitution of knowledge. As a result, anomalies in established dominant theories can be exposed as the world unleashes surprises. The subsequent limits to the conventional theoretical terms in which social science theories have been organized—states versus markets, West versus rest, religion versus secularism, past versus present, the telos of history versus perpetual flux—pose serious challenges to the disciplinary codes that have long dominated thinking about world politics.

Perhaps the most serious issue concerns the continuing relevance of the idiographic–nomothetic (particulars–universals) opposition that has afflicted Western social science since the *Methodenstreit* of the late nineteenth century. Knowledge is always made somewhere by particular persons reflecting their place's historical experience. "Universals" often arise by projecting these experiences onto the world at large (Seth, 2000). What is needed are ways of understanding how this process occurs and drawing attention to the need to negotiate across perspectives so that world politics in itself can be less the outcome of hegemonic impositions and more the result of the recognition and understanding of differences, both cultural and intellectual (Agnew, 2009).

Geopolitics of Theories of World Politics

Much of what goes for international relations theory today is the projection onto the world at large of U.S.-originated academic ideas about the nature of statehood and the world economy derived from a mixture of mid-twentieth-century European premises about states and American ones about economies even when these ideas can often depart quite remarkably from the apparent contemporary sources of U.S. foreign-policy conduct. The theory reflects the application of ideas about how best to model a presumably hostile world, which are drawn more from selected aspects of U.S. experience and a U.S. reading of world history than from fidelity to how actual U.S. policies are constituted from a mix of domestic interests and foreign-policy inclinations. Contrast the predictions of a defensive U.S. neorealism, for example, which might counsel prudence in invading other countries without a set of clear objectives and an "exit strategy," with recent U.S. foreign policy in the Middle East driven by what Connolly (2005) calls a domestic alliance in the United States between "cowboy capitalism" and evangelical Protestantism.

The intellectually dominant realist tradition of U.S. international relations theory (even its opponents, including liberals and idealists, share many of its assumptions) is based on a central assumption of "anarchy" beyond state borders (Agnew, 1994; Powell, 1994). This conception is not a straightforward objective fact about the world but a claim socially constructed by theorists and actors operating in conditioning sites and venues (premier universities, think tanks, government offices, etc.) who unthinkingly reproduce the assumption, drawing on particular interpretations of unimpeachable intellectual precursors (such as the early modern European thinkers Machiavelli and Hobbes) irrespective of its empirical "truth" status. Other related ideas, such as those of a world irretrievably divided into territorial "nation-states" organized along a global continuum of development, and even ideas often presumed to challenge the mainstream view such as "rational choice" and "hegemonic succession," can be thought of similarly as reflecting social and political experiences of particular theorists in specific places more than as objective truth about the world per se. If believed, of course, and if in the hands of those powerful enough, they can become guides to action that make their own reality (Agnew, 2003).

The constitutive ideas of so-called realism as developed by Machiavelli, Hobbes, and others have taken on a very different form in the hands of the German refugee scholars in the United States, such as Hans Morgenthau, most responsible in the early Cold War years for creating the realist perspective, and then in the hands of more Americanized theorists, such as Robert Gilpin, than the originals might initially suggest could ever be the case (Inayatullah & Rupert, 1994). Most notably, what became in the 1970s and 1980s the main consensus position, so-called neorealism, combines elements of classical political realism and liberal economics that have traveled some intellectual distance from their geographical roots in, respectively, Renaissance Italy (with Machiavelli) and late eighteenth-century Scotland (with Adam Smith) (Donnelly, 1995). This American synthesis and related emphases have ruled the academic roost in international relations much as the neoclassical synthesis has in U.S. academic economics.

Realist theory was both a reaction against the behavioral trend in U.S. political science in the 1940s and 1950s, which presumed a science of politics could be founded entirely on the basis of rational principles of individual behavior, and the result of the desire to maintain close connections between the practitioners and the academic study of world politics in a furthering of *Staatslehre*, or the proffering of advice to political leaders on the basis of profound and presumably unchangeable truths about human nature and the state system (Guilhot, 2008). It was to be a "special field" separate from the other social sciences. With support from the Rockefeller Foundation and the powerful example of Hans J. Morgenthau with his influential textbook, *Politics Among Nations* (the systematic confusion between nations and states is suggestive of the overall orientation), this vision became ensconced widely in U.S. academia particularly through the influence of academics at Columbia University and the University of Chicago. Relative unease over whether or not "international relations" constituted or could constitute a separate "discipline" (Kaplan, 1961) was never paralleled until recently by fears that it might well be a "science" based largely on projecting American views onto the world at large (Gareau, 1981; Grunberg, 1990; Kahler, 1993; Kripendorff, 1989; Smith, 1987). Eventually even the behavioralists at Princeton University melded into the pot by bringing their ideas of modernization (basically, following in American footsteps) into the mix of what rapidly evolved into neorealism. The myriad students from all over the world who go off to do a preprofessional master's in International Relations at Tufts, Harvard, Chicago, and elsewhere (prior to working in the practice of foreign policy) find that most of their teachers subscribe to this theory of world politics, even if they also sometimes review other theoretical options, such as liberalism and constructivism. The most systematic study of research and teaching trends I am aware of (Long, Maliniak, Peterson, & Tierney, 2005) uses the terms "realism," "liberalism," "constructivism," "Marxism," and so on (terms with special definitions in the field of international relations that all seem to share many of the assumptions referred to in the next paragraph) to show, by means of a coding of research articles and a survey of teachers, that realism has declined relative to liberalism in research but remains dominant, if less so more recently, in syllabi and classroom teaching.

In this understanding, states stand as naturalized abstract individuals, the equivalent of individual persons in the realm of "international relations"; the distribution of technological and other economic advantages drives communication, competition, and cooperation; central or hegemonic states rise and fall as they succeed or fail in capturing the economic benefits of hierarchy; and the overall dynamic as far as each state is concerned is of gaining improvement in "advantage," either absolute (typically realist) or relative (typically liberal), within the overall system (Agnew, 2003). The heart of the perspective is a conception of a state of nature in the world in which the pursuit of wealth and power is projected onto states as the only way of escaping from the grasp of anarchy. A Freudian egotism is translated from the realm of the individual to that of the state (e.g., Schuett, 2007). Thus, a particular cultural conception of life is projected onto the world at large (Inayatullah & Rupert, 1994, pp. 81–82). More specifically, the belief in spontaneous order long regarded in the American ethos as the persisting motif of Americanism, as individuals pursue their own goals unhindered by government and thereby reach a higher synthesis out of disparate intentions, is thus brought to bear in the broader global arena with states now substituting for persons, albeit now tinged with a Germanic-Lutheran pessimism that necessitates interventions by the United States as the most benign and public-minded of "powers" when the "best" order fails to arise spontaneously (Agnew, 2005, p. 97; Grunberg, 1990; Inayatullah, 1997; Nossal, 2001).

The connection with actual U.S. foreign-policy making is crucial. Though international relations has claimed both a basis in the eternal facts of human nature and/or the state-systemic constraints on political action and an advisory role to the U.S. government in pursuit of its particular interests, it has been the latter that has tended to dominate. As a putative policy field, international relations has long attracted adherents more through its putative practical appeal than through its intellectual rigor (Kahler, 1997). Kripendorff (1989, pp. 31–32) refers to this attraction as the "Kissinger syndrome" or the "ambition to be accepted by or adopted into the real world of policy making, to gain access to the inner halls of power." He sees this ambition as something specifically American in its desire to provide a fixed intellectual foundation for why international relations must remain the domain of a specialized elite rather than be subject to democratic discussion and critique. In his view, since the inception of the field following the Second Word War, the goal of international relations was the training of specialists and practitioners, not the creation of a "critical scholarly enterprise" (Kripendorff, 1989, p. 36).

In fact, considerable energy in academic international relations today in the United States and elsewhere focuses on the weaknesses of the neorealist synthesis even as the master's programs continue to churn out would-be practitioners often oblivious to the political and theoretical bases of the arcane debates among some of their teachers (Long et al., 2005). The continuing, even revived, appeal of the neorealist synthesis seems to lie in its ritual appeal to U.S. centrality to world politics (the "necessary nation," "the lender of last resort," etc.) and in the enhanced sense since the end of the Cold War and after 9/11 of a dangerous and threatening world that must be approached with trepidation and preparation for potential violent reaction and intervention as mandated by realist thinking. Yet in practice there is a massive gap between the predictions of such theorizing and what actually goes into the making of U.S. (or any other) foreign policy, much of which has to do with persisting geopolitical orderings of the world and domestic interests and their relative lobbying capacities (Hellmann, 2009; Oren, 2009).

International relations as a field around the world has followed largely in American footsteps. I can attest that my own introduction to it in the late 1960s in Britain involved reading textbooks that came overwhelmingly from the United States. Debate about the relative degrees of theoretical "pluralism" in the United States and Britain suggests that at least the modes of categorizing theories are somewhat less hidebound in the latter than in the former and that in recent years at least there has been some- thing of a parting of the ways across the Atlantic, with nonrealist views becoming much more widespread in British universities than in their American counterparts (Schmidt, 2008; Smith, 2008). More recently and elsewhere around the world, U.S. theories, particularly neorealism, have proved rather more pervasive and persistent. In Russia, for example, which one might not expect to be particularly congenial to U.S. ideas, the main academic journal about world politics, *Mehdunarodnyye protsessy* (International Trends), seems to adhere to ideas about international anarchy, nation-state developmentalism, and systemic constraints on state action that are remarkably similar to those represented by U.S. neorealism. Even the more liberal currents, reflecting on globalization and a less state-oriented world, mainly cite U.S. sources (Tsygankov & Tsygankov, 2007). Perhaps this tendency reflects the lack of local alternatives following the demise of official Marxist conceptions, dependence on funding from Western foundations, and a general disorientation following the collapse of the Soviet Union. It does not, however, entail much by way of support for U.S. foreign policy, only a similar theoretical logic in arriving in this case at Russian- centered positions (Müller, 2008). The recent revival of Eurasian geopolitical thinking perhaps is a harbinger of a more Russian-centric mode of thinking as an alternative to imported brands (Tsygankov & Tsygankov, 2004).

Given the cumulative crisis of the United States in world politics over the past two decades, one might expect to see some emerging alternative theoretical visions emanating from beyond U.S. shores. The so-called English School of international relations, associated in particular with the idea of "international society" but effectively realist in many respects, has recently undergone a concerted revival as an alternative to U.S. theories. It has certainly traveled well beyond Britain, even if with questionable success (e.g., Waever, 1992; Wendt, 1999). Zhang (2003) has examined how well it has traveled to China since Adam Roberts, one of its main advocates, visited Beijing in 1991. Lacking in equivalently talented entrepreneurs or salesmen and the institutionalized connections between U.S. and Chinese universities, the English School has had limited influence, according to Zhang, in comparison to the continuing dominance of U.S. scholars. But most of the main works are also not available in Chinese, and the major research institutes in China are run by people trained in the United States. To a large extent, therefore, academic Chinese knowledge of the "international" largely remains refracted through intellectual lenses made in the United States.

Within China, however, change is in the offing. Some Chinese academics write explicitly about what they term "international relations theory with Chinese characteristics" (Xinning, 2001). In other words, China has become involved in developing something akin to what happened in the United States in the 1940s and 1950s. What is this Chinese synthesis? According to Xinning (2001), there are two variants, with the second smaller but growing more quickly. The first borrows the phrase "Chinese characteristics" from Deng Xiaoping to indicate an international relations theory that centers on China's need to protect its sovereignty, engage in peaceful coexistence with other states, and use Chinese language, thought, and expression. The second asserts a more radically Chinese vision of the world with China's status at the center of a surrounding system, Confucian "benevolent governance," the winning of conflicts without resorting to war, and interests, not morality, as the basis of interstate behavior. In Xinning's words:

> After the Tiananmen Square incident of 1989, most social science disciplines (especially political science, sociology, and journalism) suffered a setback because of the government's campaign against the ideological liberalism of Chinese scholars and the so-called peaceful evolution initiated by the West. However, International Relations received a different treatment. Theoretical studies on IR continued to develop. The teaching of Western IR theories continued at key universities, and academic exchanges with the West in IR studies became more active. This was mainly because Chinese leaders worried more about China's isolation from the outside world than a "peaceful evolution." (Xinning, 2001, p. 62)

More recently, as Xinning makes clear, a new Chinese international relations is evolving which combines a range of elements (also see Yang & Li, 2009). As in the U.S. case, however, it is its connection to state policy that gives it special status. As in so many other features of the relationship between the United States and China, there is an almost mirror image in assumptions between the theory imported from the United States and what increasingly goes for "Chinese" international relations theory. *Plus ça change, plus c'est la même chose.*

Conclusion

In brief compass, I have tried to illustrate one facet of the geography of knowledge, what I have called the geopolitics of knowledge, in relation to one body of thinking, so-called international relations theory. I have emphasized its founding in the early postwar United States, its travels around the world as a function of American hegemony, and the story of two alternatives, the English School, to illustrate the limits of pluralism, and the rise of an IR theory with "Chinese characteristics," to show how an alternative with hegemonic potential can begin to emerge. Who knows, particularly if this latter, as Callahan (2001) has said in direct response to Xinning's (2001) essay on Chinese thinking about world politics, adjusts to the more globalized and transnational world that has seemed, at least until recently, to be in the offing, then we may actually end up with a theory of world politics that avoids the inside–outside views of sovereignty and the need for a single hegemonic power that so much of recent IR theory has been devoted to normalizing (Agnew, 2009). Don't bet your house on it. As long as we have global political hierarchy, we are likely to have parallel "theories" of world politics which naturalize that state of affairs.

References

Agnew, J. (1994). The territorial trap: The geographical assumptions of international relations theory. *Review of International Political Economy, 1*, 53–80. doi:10.1080/09692299408434268.

Agnew, J. (2003). *Geopolitics: Re-visioning world politics* (2nd ed.). London: Routledge.

Agnew, J. (2005). *Hegemony: The new shape of global power*. Philadelphia: Temple University Press.

Agnew, J. (2007). Know-where: Geographies of knowledge of world politics. *International Political Sociology, 1*, 138–148. doi:10.1111/j.1749-5687.2007.00009.x.

Agnew, J. (2009). *Globalization and sovereignty*. Lanham, MD: Rowman and Littlefield.

Anderson, L. (2003). *Pursuing truth, exercising power: Social science and public policy in the twenty-first century*. New York: Columbia University Press.

Biersteker, T. J. (1995). The "triumph" of liberal economic ideas in the developing world. In B. Stallings (Ed.), *Global change, regional responses: The new international context of development* (pp. 174–196). Cambridge, UK: Cambridge University Press. doi:10.1017/CBO9781139174336.006.

Callahan, W. A. (2001). China and the globalization of IR theory: Discussion of "building international relations theory with Chinese characteristics". *Journal of Contemporary China, 10*(26), 75–88.

Connolly, W. E. (2005). The evangelical-capitalist resonance machine. *Political Theory, 33*, 869–886. doi:10.1177/0090591705280376.

Donnelly, J. (1995). Realism and the academic study of international relations. In J. Farr, J. S. Dryzek, & S. T. Leonard (Eds.), *Political science in history: Research programs and political traditions* (pp. 175–197). Cambridge, New York: Cambridge University Press.

Gareau, F. H. (1981). The discipline of international relations: A multi-national perspective. *Journal of Politics, 43*, 779–802. doi:10.2307/2130637.

Geertz, C. (1996). Afterword. In S. Feld & K. H. Basso (Eds.), *Sense of place* (pp. 259–262). Santa Fe, NM: School of American Research.

Gramsci, A. (1992). *Prison notebooks* (J. A. Buttigieg, Ed. & Trans.). New York: Columbia University Press. (Original work *Quaderni del carcere* written between 1929 and 1935, but not published in Italian until after the Second World War). http://postcolonialstudies.emory.edu/hegemony-in-gramsci/#ixzz2P8T7ixfT

Grunberg, I. (1990). Exploring the "myth" of hegemonic stability. *International Organization, 44*, 431–477. doi:10.1017/S0020818300035372.

Guilhot, N. (2008). The realist gambit: Postwar American political science and the birth of IR theory. *International Political Sociology, 2*, 281–304. doi:10.1111/j.1749-5687.2008.00052.x.

Hellmann, G. (2009). Fatal attraction? German foreign policy and IR/foreign policy theory. *Journal of International Relations and Development, 12*, 257–292. doi:10.1057/jird.2009.11.

Inayatullah, N. (1997). Theories of spontaneous disorder. *Review of International Political Economy, 4*, 319–348. doi:10.1080/096922997347805.

Inayatullah, N., & Rupert, M. (1994). Hobbes, Smith, and the problem of mixed ontologies in neorealist IPE. In S. J. Rosow, N. Inayatullah, & M. Rupert (Eds.), *The global economy as political space* (pp. 61–85). Boulder, CO: Lynne Rienner.

Jackson, M. (2002). *The politics of storytelling: Violence, transgression and intersubjectivity.* Copenhagen, Denmark: Museum Tusculanum Press.

Kahler, M. (1993). International relations: Still an American social science? In L. B. Miller & M. J. Smith (Eds.), *Ideas and ideals: Essays on politics in honor of Stanley Hoffmann* (pp. 395–414). Boulder, CO: Westview Press.

Kahler, M. (1997). Inventing international relations: International relations theory after 1945. In M. W. Doyle & G. J. Ikenberry (Eds.), *New thinking in international relations theory* (pp. 20–53). Boulder, CO: Westview Press.

Kaplan, M. A. (1961). Is international relations a discipline? *Journal of Politics, 23*, 462–476. doi:10.2307/2127101.

Kripendorff, E. (1989). The dominance of American approaches to international relations. In H. C. Dyer & L. Mangasarian (Eds.), *The study of international relations* (pp. 28–39). London: Macmillan.

Livingstone, D. N. (2005). Science, text and space: Thoughts on the geography of reading. *Transactions of the Institute of British Geographers, 30*, 391–401. doi:10.1111/j.1475-5661.2005.00179.x.

Long, J. D., Maliniak, D., Peterson, S., & Tierney, M. J. (2005, March 1–5). *Teaching and research in international politics: Surveying trends in faculty opinion and publishing.* Paper presented at the 2005 Convention of the International Studies Association, Honolulu, Hawaii.

Lowenthal, D. (1961). Geography, experience, and imagination: Towards a geographical epistemology. *Annals of the Association of American Geographers, 51*, 241–260. doi:10.1111/j.1467-8306.1961.tb00377.x.

Lukes, S. (2000). Different cultures, different rationalities? *History of the Human Sciences, 13*(1), 5–18. doi:10.1177/09526950022120566.

Mignolo, W. D. (2000). *Local histories/global designs: Coloniality, subaltern knowledges, and border thinking.* Princeton, NJ: Princeton University Press.

Müller, M. (2008). Situating identities: Enacting and studying Europe at a Russian elite university. *Millennium, 37*, 3–25. doi:10.1177/0305829808093728.

Nader, L. (Ed.). (1996). *Naked science: Anthropological inquiry into boundaries, power, and knowledge.* London: Routledge.

Nossal, K. R. (2001). Tales that textbooks tell: Ethnocentricity and diversity in American introductions to international relations. In R. M. A. Crawford & D. S. L. Jarvis (Eds.), *International relations—still an American social science? Toward diversity in international thought* (pp. 167–186). Albany, NY: State University of New York Press.

Oren, I. (2009). The unrealism of contemporary realism: The tension between realist theory and realists' practice. *Perspectives on Politics, 7*, 283–301. doi:10.1017/S1537592709090823.

Powell, R. (1994). Anarchy in international relations theory: The neorealist-neoliberal debate. *International Organization, 48*, 313–344. doi:10.1017/S0020818300028204.

Prewitt, K. (2002). The social science project: Then, now and next. *Items and Issues, Social Science Research Council, 3*(3–4), 1–2.

Said, E. W. (1978). *Orientalism.* New York: Harper.

Sapiro, G. (Ed.) (2009). *L'espace intellectuel en Europe: De la formation des États-nations à la mondialisation XIX-XXI siècle* [Intellectual space in Europe: From the formation of the nation-state to globalization, 19th–21st centuries]. Paris: La Découverte.

Schmidt, B. (2008). International relations theory: Hegemony or pluralism? *Millennium, 36*, 295–310. doi:10.1177/03058298080360020601.

Schuett, R. (2007). Freudian roots of political realism: The importance of Sigmund Freud to Hans J. Morgenthau's theory of international power politics. *History of the Human Sciences, 20*(4), 53–78. doi:10.1177/0952695107082491.

Seth, S. (2000). A "postcolonial world"? In G. Fry & J. O'Hagan (Eds.), *Contending images of world politics* (pp. 214–226). Basingstoke, UK: Macmillan.

Shapin, S. (1998). Placing the view from nowhere: Historical and sociological problems in the location of science. *Transactions of the Institute of British Geographers, 23*, 5–12. doi:10.1111/j.0020-2754.1998.00005.x.

Smith, S. (1987). Paradigm dominance in international relations: The development of international relations as a social science. *Millennium, 16*, 189–206. doi:10.1177/03058298870160022501.

Smith, S. (2008). Debating Schmidt: Theoretical pluralism in IR. *Millennium, 36*, 305–310. doi:10.1177/03058298080360020701.

Tsygankov, A. P., & Tsygankov, P. A. (2004). New directions in Russian international studies: Pluralization, Westernization, and isolationism. *Communist and Post-Communist Studies, 37*, 1–17. doi:10.1016/j.postcomstud.2003.12.005.

Tsygankov, A. P., & Tsygankov, P. A. (2007). A sociology of dependence in international relations theory: The case of Russian liberal IR. *International Political Sociology, 1*, 307–324. doi:10.1111/j.1749-5687.2007.00023.x.

Waever, O. (1992). International society: Theoretical promises unfulfilled. *Cooperation and Conflict, 27*, 97–128. doi:10.1177/0010836792027001004.

Wendt, A. (1999). *Social theory of international politics*. Cambridge, UK: Cambridge University Press.

Wolf, E. (1982). *Europe and the people without history*. Berkeley, CA: University of California Press.

Xinning, S. (2001). Building international relations theory with Chinese characteristics. *Journal of Contemporary China, 10*(26), 61–74. doi:10.1080/10670560125339.

Yang, G., & Li, M. (2009). Western political science theories and the development of political theories in China. *Journal of Chinese Political Science, 14*, 275–297. doi:10.1007/s11366-009-9061-y.

Zhang, Y. (2003). The "English School" in China: A travelogue of ideas and their diffusion. *European Journal of International Relations, 9*, 87–114. doi:10.1177/1354066103009001003.

"Hot Spots, Dark-Side Dots, Tin Pots": The Uneven Internationalism of the Global Academic Market

12

Anssi Paasi

The international community of social scientists has become increasingly sensitive not only to the fact that language and context are crucially related in the construction of scientific accounts, but also to the forms of power (or geopolitics) involved in such relations (Canagarajah, 2002; Tietze & Dick, 2009). The role of language and the links between knowledge and power have also become important issues in human geography, where a number of scholars have challenged what has been labeled Anglo-American or Anglophonic hegemony (e.g., Kitchin, 2005). Particularly researchers working outside the English-speaking world or in the formerly colonized "peripheries" of this world have been worried about what can be described as "the uneven geographies of international publishing spaces" (Paasi, 2005). Key themes in recent debates have been: what is international geography, how should this idea be understood, and who are the actors with the power to define it (see e.g., Garcia-Ramon, 2003; Gregson, Simonsen, & Vaiou, 2003).

Scholars operating in small linguistic areas have always been dependent on the predominant academic languages. In many European states, for instance, the German language was crucial in academic interaction and publications until after World War II. A major question, therefore, is how a particular language, English, later gained its hegemonic position as a lingua franca, as a global synonym for "international". Many scholars have challenged an idea that is nowadays increasingly being taken for granted in the academic world; namely, that it is English-language publications, preferably produced in the United States and United Kingdom that are international, whereas publications in other languages are considered national or parochial. Paradoxically, this idea is currently strongly supported by many national ministries of science outside of the English-speaking world, for example, in Scandinavia and on the continent of Europe.

A. Paasi (✉)
Department of Geography, University of Oulu, Linnanmaa, Oulu 90014, Finland
e-mail: anssi.paasi@oulu.fi

© Springer Netherlands 2015
P. Meusburger et al. (eds.), *Geographies of Knowledge and Power*,
Knowledge and Space 7, DOI 10.1007/978-94-017-9960-7_12

Such debates have not emerged *in vacuo*, but reflect wider social, political, and economic tendencies associated with the globalization of science and the neo-liberalization of university life, as witnessed around the world since the 1990s (e.g., Albert, 2003; King, 2004). One contemporary feature that raises the question of academic *centers* and *peripheries* is the fact that the traditionally rather modest marketing of universities to attract students has been dramatically transformed into a fierce struggle over prestige, research money, and students. This has created an evaluation industry in many countries and a symbolic struggle that draws on assessments and rankings as well as on material and symbolic distinctions. The ranking of universities and the often one-sidedly mechanical measurement of research output and citations are major features of current academia (cf. Agnew, 2009). Citation counts are seen "as votes cast in an ongoing election over whose work matters" (Fuller, 2002, p. 207).

This paper will scrutinize how a context-bound social science—human geography—becomes interpreted as international, how internationality is understood in this new constellation, and how power relations and hegemony are structuring (and are structured in) practices and discourses related to internationality. This theme has become important not only in geography but also in science studies and social science fields, such as international relations, management studies, and postcolonial studies (see Canagarajah, 2002; Tietze & Dick, 2009).

I will look first at how current politico-economic and neoliberal pressures modify higher education and how the processes of globalization not only take place in an economic or cultural sense, but also manifest themselves in the management of the production of knowledge. National states have long regarded knowledge as a key factor in value production and an element of social reproduction for individual states. It now appears, however, that this function is being rescaled, leading to the increased homogenization of national science policies across borders as competition and corporatization become the dominant features of academic life around the world. This has dramatically impacted how relevant knowledge is defined, produced, and controlled, and, correspondingly, it has also changed the forms of publishing. To take but one example from human geography, some debate has recently emerged on the publishing significance of monographs versus articles in a journal, with this issue seeming to be topical both on the continent of Europe and in the English-speaking world (Harvey, 2006; Ward et al., 2009).

Secondly, I will scrutinize both the roots and increasing importance of the English language in the international academic market and note how this has become a particular challenge for social scientists operating outside the English-speaking world. The authorities responsible for academic governance and current neoliberal competition in many countries have raised claims that research should be published in English and in the "best" journals, which more often than not means journals that have been classified by one North American company, Thomson Reuters, which has a monopoly in compiling the Web of Science (WoS) citation data and in the choice of the journals that are represented in these data. Competing databases, such as Scopus, published by the Elsevier publishing company, have not achieved similar status in academic evaluations. Furthermore, Scopus provides different citation results because it is based in part on different journals.

The fact that WoS data are now acknowledged around the world as a synonym for quality and excellence further increases the impact of these listings and can also modify what is considered relevant science. Neoliberal pressure related to impact factors (a measurement showing the average number of times a journal's articles have been cited) or research money can thus easily lead to power struggles over prestige between academic fields. One might expect that the representatives of those fields that gain some advantage (prestige, research money) through the use of such instruments of classification would support WoS apparatus. My experience at my own university suggests that researchers operating in universal fields, such as medicine, physics, mathematics, or biosciences, benefit more from the WoS apparatus than researchers in humanities or social science. Yet this discrepancy may not be as prevalent at the core universities of the Anglophone world because of the role of the English language in social sciences and humanities.

As I have shown, this is a complex situation for geographers, since the overwhelming majority of classified geography journals are published and edited in English-speaking countries (Paasi, 2005). Many European geographers, often coming from strong linguistic communities such as Spain or Germany, have commented that this concentration reduces their opportunities to perform research and present their results, stating that in order to have papers published, scholars are also compelled to adapt to agendas reflecting certain Anglophone research themes and theoretical orientations (see Garcia-Ramon, 2003; Gregson et al., 2003; Tietze & Dick, 2009). This situation has changed slightly in geography during the last few years in a formal sense, because some new journals from non-English-speaking countries have been accepted onto the WoS lists (Paasi 2013). This, however, raises new questions regarding the topic of internationalization. Do national journals published in English or other languages become international quality journals just through the expansion of the database? Or might this expansion of the database simply be a maneuver to keep scholars from the periphery—as well as the managers of universities—happy in the new, competitive market?

Thirdly, I will make some observations on the complex scalar geographies of current academic spaces of competition. I will develop metaphors, such as "hot spots," "dark-side dots," and "tin pots," to describe the current spaces of academic competition. The choice of these terms has to do with the ongoing international tendency to classify universities not only as competitive academic spaces but also as spaces competing with each other. Perhaps one of the best examples of this trend is Shanghai Jiao Tong University's Shanghai List, which ranks 500 of the world's universities. The ranking is compiled annually on the basis of indicators that generally favor the hard, or natural, sciences and the publishing of articles in journals classified by Thomson Reuters. The leaders of universities seem to pay increasing attention to such listings.

Trends in the Internationalization and Neoliberal Globalization of Academia

The practice of academic research involves hard work as well as a very complex constellation of power relations, practices, and discourses related to grant systems, publications, evaluation, and rankings, so it is no exaggeration to suggest that language plays a crucial role in this enterprise. Without giving any particular autonomy to texts, it is clear that scientific observations and facts, arguments, and theories are constructed, communicated, and evaluated mostly in the form of written statements, which is to say publications. This simply means that research work is, as to its end product, a largely literary and interpretative activity (Bourdieu, 2004). Science studies suggests that this activity is scaled in complex ways, both horizontally and vertically. Academic territories and tribes, to use the expression of Becher (1989), differ from each other with respect to their moral orders and epistemic cultures, and follow divergent strategies in warranting knowledge.

The state also makes a difference: national forms of academic socialization, specific rules, and pecking orders have by tradition produced different publication strategies, for example, and disparate understandings of how quality should be recognized (Becher, 1989). This has been particularly obvious in the social sciences and humanities, which have been by tradition important vehicles of nationalist and colonialist practices and discourses. Correspondingly, publishing in the national language has been standard practice in the social sciences and humanities in most countries. This pattern has been challenged fairly dramatically since the 1990s, when "international" and "internationalism" became keywords in higher education, in association with both the internationalization of science and the emergence of national science policies showing a high degree of congruence across national boundaries (King, 2004; Paasi, 2005). States have adopted increasingly similar views of science policy and of its instruments and forms of management, with the standardization of scientific practice and certification of quality appearing to be the methods preferred worldwide for administering globalizing science.

The current pressures regarding internationalization are one element of neoliberal globalization. These processes of globalization are being shaped in complex ways by transformations in the world economy and geopolitics. They involve many core forces of social life, such as changes in capitalist production, technological innovation in communications, rationalism's spread as a dominant knowledge framework, and various new forms of governance, which have enabled the establishment of new regulatory frameworks (Scholte, 2000). Often, the outcome has been the introduction of a competitive market orientation into higher education and the creation of a whole new vocabulary to depict the links between science, universities, and society, with terms such as McUniversity, academic capitalism, or triple helix gaining prominence (Castree & Sparke, 2000; Paasi, 2005). The dynamics of science and research is becoming an increasingly significant element in this process. This is because knowledge is so crucial to capitalism's forces of production and to internationalization, competition and governance, and the regulation of globalization. At the same time, the research culture itself is changing. A form of scientific nationalism is emerging within this new landscape of economic competition, with investment in science and its impact being compared strictly within a national framework. This often follows models created by the Organization for Economic Co-operation and Development (OECD), which originally brought science indicators into the international debate (the OECD was established in the early 1960s). Secondly, universities are simultaneously placed in a position of implied competition across national borders (e.g., in the Shanghai List). Thirdly, individual academic fields are being compared and ranked within a national framework in many countries (as in the Research Assessment Exercise, RAE, now the Research Excellence Framework, REF, in Britain).

Internationalization in academia is thus not leading to a borderless world in which academic ideas flow and interact, but to the rise of new, uneven spatial patterns and a fusing of the national-territorial and international-relational dimensions (Paasi, forthcoming). The OECD makes international comparisons of publication activities and citations between the nations, using the WoS data as a background, hence maintaining a competitive environment that national higher education systems and individual universities support and coordinate. National ministries and governments also employ the OECD reports to steer their science policies, and in many cases even commission such OECD reports to support their decisions (see Kallo, 2009). International comparison is rendered possible through the citation data produced by Thomson Reuters. The collection of this data was initiated by the Institute for Scientific Information in Philadelphia in the early 1960s, to produce information on new publications and citations. The social science publications tracked by Thomson Reuters's web-based database— the Web of Science—have, until recently, been overwhelmingly from the English-speaking world. This situation has placed scholars from non-English speaking countries under increasing pressure to publish in English in journals produced primarily in the United Kingdom or United States in order to fulfill the requirement of internationalization.

Language and Context: From an International to an English-Dominated Geography

It is thus non-English-speaking scholars who have placed the question of language on the agenda, although it has been Anglo-American human geography that has accentuated the dependence of contexts and texts in the production, introduction, and reproduction of the world since the 1980s. This focus was part of a trend that recognized the significance of language as a powerful medium in the production and reproduction of social systems (e.g., culture, economics, and politics). That development also had a more existential background, whereby language issues came to be viewed as crucial aspects deeply "caught up in one's sense of self and how one makes sense of the world" (Schoenberger, 2001, p. 366). There are many kinds of language users in all possible contexts, or to put it another way, linguistic communities are often contested. Scientists make up one very specific group in the broader linguistic markets, and even they are a diverse group. Language has an especially important role in science, but it is above all part of the wider intellectual context. The major guru of functionalist sociology, Talcott Parsons (1947/1966), wrote in his introduction to the translation of Max Weber's *The Theory of Social and Economic Organization* as follows:

> It is perhaps one of the most important canons of critical work that the critic should attempt so far as possible to see the work of an author in the perspective of the intellectual situation and tradition out of which it has developed. This is one of the best protections against the common fallacy of allowing superficial interpretation of verbal formulae to mislead one into unfair interpretations of ideas and inadequate formulations of problems (p. 8).

David Harvey (1984) reminded geographers 40 years ago of the role of context: "the history of our discipline cannot be understood independently of the history of the society in which the practices of geography are embedded" (p. 1). It may be asked, of course, what is "our discipline" if it is so dependent on society. Do different societies produce different disciplines, even if the name of the field is the same? And do they produce different context-bound histories? (Paasi, 2011). There seem to be multiple lines of inquiry, rather than just one: How has the diffusion of the idea of international occurred; how have certain ideas or contexts gained a hegemonic position in this diffusion; and what are the possible structural factors behind this? In other words, how have internationalism and the English language become largely synonymous?

The current dominance of the English language has long roots. The ties between English and the spheres of trade, business, and commerce have been close for centuries. Some authors have associated the spread of English with the spread of capitalism, and the process was certainly boosted by the rise of colonialism (Tietze & Dick, 2009). As far as Europe was concerned, the relationship with the United States that developed after World War II was significant. The emerging global divide and the rivalry between the capitalist and socialist camps created new forms of international interaction in the western world. The Marshall Plan was created in the United States after the war to help Western European states recover and to promote the ideals of freedom. The Organization for European Economic Cooperation (OEEC), predecessor of the OECD, was to some extent established to implement the Marshall Plan. The OECD, for its part, was the key economic adviser to the capitalist West in the Cold War struggle against the Soviet Union (Kallo, 2009). One significant institution that brought a number of U.S. scholars and Europeans into closer contact was a network of grant systems, which also included the American Fulbright Program of scholarships established in 1946. This system has been described as "a cultural variant of the Marshall Plan" (Frijhoff & Spies, 2004, p. 69). There were also a number of other U.S.-based science institutions that aimed at producing a "consensual hegemony" of scientific dominance by the United States (Krieger, 2006; Parmar, 2002). Such hegemony not only served the interests of post-war reconstruction in Europe, but also helped to maintain American leadership and to Americanize scientific practices in many fields, especially the hard sciences (Paasi, 2015). International science indeed became a vehicle to promote American values and interests in the post-war world, with U.S. authorities also engaging the concept to reconfigure the European scientific landscape (Krieger, 2006).

Human geography and other fields in the social sciences, such as political science, sociology, or social policy; and in the humanities (folklore and the writing of national histories) were originally very much bound up with the national contexts in which they were created. In the case of geography, the key mediator was geographical education in schools, where geography's nationalist and colonial associations often blossomed (e.g., Buttimer, Brunn, & Wardenga, 1999). A particularly interesting issue is how, in the post-war period, Anglo-American geography so rapidly gained its hegemonic position in defining what is considered relevant human geography, especially in view of how influential German and French geographers had been in laying the early foundations of geography in the United States and the United Kingdom before the war. Indeed, Ackerman (1945) was very worried about the linguistic skills of American geographers after World War II, and recognized "the unfamiliarity of most young American geographers with foreign geographic literature; their almost universal ignorance of foreign languages; their bibliographic ineptness." Barnes (2004) has shown how wartime service brought numerous geographers into contact with other fields of science, and how, together with the emerging positivist philosophies and quantitative methods, this gave rise to a new theoretical approach. Even at this stage, German influences were crucial, with the translations of the works of Walter Christaller and August Lösch published during the 1940s and 1950s being important sources of inspiration for establishing new perspectives (Paasi, 2011).

It can be argued without exaggeration that human geography—especially as far as the exchange and directions of flows of geographical ideas across national borders are concerned—was more international prior to the 1960s, with researchers apparently mastering more languages, or at least willing to use foreign languages in their research work to closely examine what was taking place in academic geography in other linguistic contexts. Evidence of this can be found in the lists of references in many highly important but historically contingent Anglo-American geography books, such as Hartshorne (1939) or Wright (1966). These scholars appear not only to have learned languages such as German and French, but also to have actively drawn on geographical literature produced in these contexts when developing their arguments. Hartshorne admitted his debt to continental European geography in his famous *The Nature of Geography*.

The situation is quite different today, with genuine dialogue having given way to influences in human geography that are more unidirectional. Many Europeans look to trends in Anglo-American human geography, and scholars in the United States and especially in the United Kingdom are certainly interested in the science being produced in continental Europe, for instance, but it is often sociological and philosophical works rather than geographical ones that serve as sources of inspiration. These are frequently read in English translation, with references to translated works by authorities such as Foucault, Derrida, Deleuze, Virilio, or Agamben now common in geographical literature. Indeed, the circulation of such "other ideas" seems to be accelerating. A recent book, *Geographical Thought: An Introduction to*

Ideas in Human Geography (Nayak & Jeffrey, 2011), is fitting evidence of the tendency to ignore, for example, the continental European developments in geography. In spite of its all-embracing title, the book largely neglects the long tradition of geographical ideas outside of the English-speaking world and basically serves as a useful introduction to Anglophonic geographic thought. At the same time, however, the book extensively cites many of the philosophical authorities mentioned above.

It is an interesting paradox that English-speaking geographers (and philosophers) are now circulating the ideas of continental European philosophers, particularly in the face of the growing influence of American pragmatism in German and French philosophy since the 1970s. This appears to be part of a dialectics whereby scholars who want to radically challenge what exists in their context simply draw on influences taken from outside (Shusterman, 2000). Yet there is also a more structural background: this circulation of ideas may be seen as part of an increasingly competitive academic environment, in which scholars must strive to create a distinct profile for themselves in order to achieve recognition.

This development, of course, raises the important question of novelty, which is an ongoing issue for both Anglophone and non-Anglophone geographers. Fuller (2002, p. 234) broached this topic, asking "how does one judge the relative merit of importing ideas and findings from another discipline into one's own vis-à-vis exporting ideas and findings from one's own discipline into another." For sure, geographers have been, with a few exceptions, much better in the import than the export sector. This issue is not new: Agnew and Duncan (1981) wrote 30 years ago of the need for geographers to display more critical acuity in borrowing ideas from outside the field. Although such borrowing provides an opportunity for the increasingly rapid—mostly unidirectional—adoption of "new" ideas (that have often been published in original languages much earlier), it is far removed from an exchange of conceptual ideas in a truly international science of geography (Paasi, 2011, 2015).

Participating in Uneven Publishing Spaces: Creating Cores and Peripheries

In response to growing pressure to carry out research considered international, more and more authors are operating in international publishing markets, where they will certainly have to confront the question of readership, a problem Fuller (2002) described, writing "the main reason most . . . academics cannot muster the attention of their colleagues to read their works has more to do with the fact that they write too much that interests too few" (p. 177). This problem is accentuated by the lack of homogeneity in the international publishing space, a situation that reflects another important aspect of science studies, that the world of science is spatially polarized, a fact that holds true particularly in the case of social science (Becher, 1989). This observation is also valid in regard to publishing opportunities, and it certainly affects how published research will find its audience and potential readers. Where one works and publishes makes a difference.

Indeed, the current interest in publication cultures and language has been part of a broader debate on the almost self-evident understanding that the geography practiced in the United States and United Kingdom is a product of the global core and that the same discipline practiced elsewhere is a product of the periphery. Many "authors from the periphery" have documented their experience of this tendency, including a sense of marginalization regarding ideas and observations coming from peripheries, the inability of those from the core or center to understand the Anglo-American hegemony, or even feelings of being marginalized or not heard (e.g., Minca, 2000). It is not only language competence, but also the knowledge of academic culture and ways of thinking that is crucial in such international markets (Tietze & Dick, 2009). The empirical evidence provided by an examination of the national backgrounds of authors and the members of international editorial boards further suggests that many geographical journals are narrowly Anglo-American rather than international. Yet some scholars have refused to view the world on the basis of such binary distinctions and have suggested that national scales, traces, and traditions are hybrids inextricably tangled up with other contexts (e.g., Samers & Sidaway, 2000). Similarly, some scholars—operating in English-speaking academic markets—have suggested that rather than being a hegemonic element, English is a vehicle for creating diversity in human geography (Rodriquez-Pose, 2004).

While the members of editorial boards and even the editors of many Anglo-American geographical journals now come from outside of the Anglophone world, acute problems related to publishing practice still remain. Several issues complicate international publishing for human geographers simultaneously and also widen the gap between human and physical geography. It is well known that publication practices in the social sciences differ from those in the natural sciences. Fuller (2002, pp. 204–205) suggests that the perceived "hardness" of a science directly affects the rate of acceptance of articles for publication in journals and the "harder" the science, the easier it is to get into print. This might sound surprising, but becomes understandable when Fuller illustrates the peer review practices that apply in the natural and social sciences. Firstly, the rejection rate for articles in the natural sciences is relatively low, when compared with the rate for articles in the social sciences. One reason for this may be that there is often a good deal of page space available in natural science journals, the most highly respected of which appear every week. In many equipment-intensive fields articles are written in groups, with novices and experienced scholars working together, which certainly helps newcomers to take their first steps in the publishing market. Secondly, Fuller suggests that the peerage criteria are clear in the natural sciences and unclear in the social sciences. Also, writing in the natural sciences is topic-neutral, and in the social sciences topic-sensitive. Furthermore, Fuller writes that the cause of rejection in the natural sciences is often incompetence, while in the social sciences it is often politics. Finally, the verdict in the case of the natural sciences is based on "professionalism" and that in the social sciences on "amateurism." These dimensions identified by Fuller are definitely somewhat stereotypic, but they do illustrate the informal way in which social science research is done in comparison with the more formal orientation in the natural sciences. On the other hand, what scholars in various fields regard as original research—a key criterion for the acceptance of a paper for publication in many journals—also seems to vary between academic fields (Guetzkow, Lamonot, & Mallard, 2004).

If we now combine the perspectives raised in Fuller's discussion with the fact that most highly respected journals in human geography (i.e., journals that are ranked highly in the WoS database) are published in English-speaking countries, it is not difficult to understand why the political dimension easily comes into play. Observations or arguments from place x, often written in English that is perhaps not perfect, may sound irrelevant, uninteresting, or parochial for a scholar coming from the powerhouses of Anglo-American academia. According to the observations of many participants in the debate on Anglo-American hegemony, it often seems that for research to be considered original it must be related in some way to the research streams or debates going on in the Anglophone world.

This tendency extends to other areas, as Fuller points out (2002, p. 234) in examining whether domestic or foreign peers decide on the international status of a piece of research. Science studies have shown that the patterns of internationalization and the motives for research activity considered international are broadly affected by the center-periphery dichotomy. This can be seen most clearly in the fact that scholars operating in peripheral regions are typically more dependent on cores than scholars in core regions are on peripheries. These observations on core/periphery relations and hegemony are not unique to geography (Kyvik & Larsen, 1997). This uneven constellation—founded in the fact that where one works makes a difference—has many consequences. What is understood as international science may often just be standard national science as practiced in the core region, but it nonetheless tends to shape the criteria for excellence in the peripheries as well. In the current globalizing world this influence manifests itself in the fact that national science institutions world-wide are developing strategies to have their national research recognized as part of the core. In some cases, research is thus deemed international simply as the result of a particular nation's science policy. In the social sciences, this development often means that researchers are forced to adapt to the intellectual, theoretical, and methodological demands of the centers.

Such requirements may also come explicitly from the centers. This is illustrated by the Finnish sociologist Alasuutari (2004), who reflects on his own role as an author of textbooks on qualitative methods in the social sciences. Textbook publishing is part of the international publishing business, which is dominated by the English-language markets, particularly the British and American ones. Alasuutari describes how a British publisher expressed concern about the fact that the manuscript of his book included numerous references to empirical studies published in Finland and related to Finnish social and cultural contexts and concluded that such a proportion of Finnish work cited in the text would not be "helpful" to British, American, or other readers. The publisher then asked him to replace these references with examples that are fairly well-known in the literature published in English. This is, of course, a very dramatic request when we think of the contextuality of the social sciences.

Hot Spots, Dark-Side Dots, and Tin Pots: The Uneven Geography of Academic Geography and Beyond

Cores are not solely located in the Western world and peripheries elsewhere in the world, but instead both are scaled vertically in complex ways within nations, universities, and even single departments. This means that symbolic capital and prestige are unevenly distributed between academic scholars, academic disciplines, universities, and states. Vertical scaling is especially obvious in the case of publication forums. It is well known that academic journals are ranked in many ways and that specific journals carry given levels of prestige (Paasi, forthcoming). This classification may be based on peer vision, but nowadays it is increasingly based on straightforward bibliometrics, that is on journal impact factors given by the WoS. The publishers of journals today are also aware of the power of impact factors and use them unashamedly to advertise certain journals and highlight their standing in such rankings. The fact that impact factors are assigned only to journals selected for the WoS databases, thus to publications that until now have predominately been published in English-speaking countries, cannot be disregarded (Paasi, 2005). This has been a particularly complicated issue in the social sciences, where about 85 % of all ISI-ranked journals are from English-speaking countries. Non-English speaking researchers therefore find themselves in a dilemma: their universities and national science policies increasingly require that they should publish in top international journals, but publishing in such journals often forces them to adapt to research agendas created in the Anglophone world.

With competition increasingly seeming to be the order of the day in many countries, university ranking lists, such as those published by the Shanghai Jiao Tong University have also become an indicator of quality widely used around the world (see http://en.wikipedia.org/wiki/College_and_university_rankings). In Finland, for example, university rectors, government ministries, and the media are guided by the images of excellence produced by such lists. Rankings like these compare only about 10,000 universities, less than half of the total in the world. A brief analysis of the Shanghai List also shows that the overwhelming majority of the 500 highest ranked universities are located in the English-speaking world and in the countries of continental Europe. The 100 top-ranked universities in 2007, for example, included 54 from the United States, 11 from the United Kingdom, 4 from Canada, and 2 from Australia, or more than 70 universities from English-speaking countries.

The context, however, makes a difference, which I will illustrate using several metaphors. Looking at the world's universities, we may distinguish hot spots, dark-side dots, and tin pots. Hot spots are not just the site of successful universities but may also be places of general interest regarding, for instance, nature, cultural aspects, history, urbanization, rural development, tourism, economics, and/or commerce. These locations may thus have important social, cultural, and physical contexts. Hot spots may also be important because they are the locations of universities or research centers that are considered to be well known, established, or of high quality and are capable of generating considerable symbolic capital for scholars operating in these contexts. In the most fortunate cases all or many of the above-mentioned symbolic dimensions coincide. Cambridge, Oxford, and Harvard, for example, are names that speak for themselves. The locality gives prestige to its university and the university prestige to its locality. If we put aside the fact that the Shanghai List favors the hard sciences and their modes of doing and publishing research, the best universities have one common feature: a long history of operating in their context. Few have existed for less than a hundred years. These universities have all thus had a relatively long period of time to develop and accumulate prestige and symbolic capital. Yet the fact remains that most of the world's universities are not located in such hot spots, most localities are not representative examples of *global futures*, and a number of universities are located in linguistic environments that do not belong to the world of hegemonic languages. In the increasingly competitive global scheme, less known universities and departments may become dark-side dots that are not recognized on the international scene, although their research may be nationally well respected. In the worst case they can even become tin pots, which may be disparaged for their poor quality performance in a national context, too. Individual departments within universities may also suffer or benefit from the overall ranking of their university.

These three metaphors point to the fact that most universities (and also geography departments) do not have a world-wide reputation and history, being located in seemingly uninteresting, less "sexy" places and in marginal linguistic contexts. Furthermore, most human geographers at most universities are not interested in universal, often theoretical or methodological questions and may also be obliged to be interested in applied themes related to local or regional development. And even if some scholars happen to be interested in such more general, theoretical themes, the surrounding academic community may be ready to discourage them for doing the wrong thing, for instance theoretical work when the community expects empirical research. It is also a harsh economic fact, that scholars in many developing countries have little time for reading up on and learning about the most recent theoretical, cutting-edge developments in research. They may also simply not be able to afford the costs of the major journals in their fields (see Canagarajah, 2002). Such factors can often mean that these scientists are doomed to work outside the core. Some scholars may indeed find international debates more inspiring than the ideas dominating certain national research communities, and therefore be willing to move to a core and to learn its rules and pecking orders.

Conclusions

Is there anything wrong then, with authors from various countries deciding to submit their work to leading international journals, even if these are mainly Anglophone? On an individual level, I think not. Why then, should the concentration of publishing activities and the increasing importance of English be considered a problem? I think this process is something that the scientific community should definitely be worried about, because the globalization of science appears to be leading to an unparalleled standardization and homogenization of scientific practice. My concern has to do with the fact that the English language is not only the key medium in this standardization but increasingly also a major source of standards—as a result of the WoS ranking of journals, an influential process that could lead to a dominance of Anglophone research agendas elsewhere as well. This outcome would without a doubt impoverish our understanding of languages and conceptual differences, and also call into question the diffusion of ideas and influences. Linguistic communities are not equally positioned in the global academic market. As I have shown, many institutions—national governments and ministries included—are struggling to transform the operation of the global scientific community according to one format, which increases the importance of the English language as a self-evident lingua franca still more.

The English-speaking community is in an advantageous position in publishing markets, because its representatives are "freewheelers" in the hegemonic linguistic market. The self-evident status of English as the dominant language of academic writing can manifest itself in ironical ways. A fitting example is Livingstone's (2003) *Putting Science in its Place: Geographies of Scientific Knowledge*. This excellent book provides a number of examples of spaces, places, and contexts where science has been practiced and created (a laboratory, museum, or botanical garden, and even a coffee house, an asylum, and the human body), and how these specific contexts have affected the knowledge generated. But Livingstone quite surprisingly does not problematize the role of language in the constitution of these contexts of science and knowledge, in spite of the fact that he suggests (Livingstone, 2003, p. 87) that "global forces are homogenizing our world" and does indeed discuss "the regional geographies of science." The author simply does not choose to recognize the importance of English as a medium for this homogenization. The issue of language is implicit in questions such as "does the space where scientific inquiry is engaged . . . have any bearing on whether a claim is accepted or rejected" (p. 3), comments such as "every social space has a range of possible, permissible, and intelligible utterances and actions: things that can be said, done and understood" (p. 7), or in relativistic statements such as "it will be wiser, therefore, to work with the assumption that in different spaces different kinds of science are practiced" (p. 15). The latter example is probably a painful reminder for those scholars who are operating outside the core of their science and are nonetheless required to publish in the universal markets defined by the entities governing globalizing science (e.g., WoS instruments). A further example of linguistic myopia is the fact that only *one* of the innumerable sources included in the book's 36-page "bibliographic essay" is in a language other than English, namely French. Many foreign references are translations into English. Imperialism related to the history of geography is discussed in the book on a number of occasions, but the linguistic imperialism rife in contemporary science remains unrecognized.

The idea that a close connection exists between language and context often implies another presupposition: that contexts differ radically from one another because of the languages used in them. This conclusion may be true to some extent. Language is always part of culture; indeed it is a medium for producing, reproducing, and transforming culture. But it is also one of the basic facts of the cultural sciences that most cultures are based on cultural loans—even scientific cultures. It is the task of researchers to problematize and expose this contextuality. If they fail to do this, much of social science and humanities will be in serious trouble, because international communication between scholars will then become impossible. Generalization and theoretization—key ideas in scientific research aimed at scrutinizing phenomena instead of making purely empirical observations—are crucial for international communication and interaction in our increasingly networking and globalizing world. A balanced and active exchange of ideas provides more benefit for science than unidirectional flows. One more example of the homogenization of science is greater importance currently attributed to articles in journals than to books or monographs, an issue that has also been raised in the field of geography (Harvey, 2006; Ward et al., 2009). This brings to mind the influential idea put forth by Thomas Kuhn (1962/1970, pp. 18–21) in *The Structure of Scientific Revolutions*. He suggested that fields that have passed the "pre-paradigmatic" stage normally publish their results in journals, and no longer in reports or monographs, which usually include long narratives concerning the context and background of the research. Accordingly, a scientist who writes books "is more likely to find his professional reputation impaired than enhanced" (p. 20). This implies that the social sciences must be less developed than the natural sciences, because representatives of the former still publish in monographs. It is an ironic paradox that Kuhn himself came originally from physics and that the monograph that actually made him famous was itself an example of such a pre-paradigmatic publishing practice. Views such as Kuhn's are an important component in understanding the relationship of between power and knowledge. Such exclusivity or hierarchical thinking must be questioned in the name of both pluralism and academic freedom (Paasi, forthcoming). Papers published in journals, monographs, and edited thematic collections must all have their place.

Acknowledgements This article is a revised and considerably enlarged version of author's earlier paper, published originally in Danish (Paasi, 2006). The first draft of the paper was prepared while the author was serving as an Academy Professor at the Academy of Finland. The institution's support is gratefully acknowledged. The author would also like to thank Peter Meusburger and an anonymous referee for their useful comments.

References

Ackerman, E. A. (1945). Geographic training, wartime research, and immediate professional objectives. *Annals of the Association of American Geographers, 35*, 121–143. doi:10.1080/00045604509357271.

Agnew, J. (2009). The impact factors. *AAG Newsletter, 44*, 3.

Agnew, J. A., & Duncan, J. S. (1981). The transfer of ideas into Anglo-American human geography. *Progress in Human Geography, 5*, 42–57.

Alasuutari, P. (2004). The globalization of qualitative research. In C. Seale, D. Silverman, J. Gubrium, & G. Gobo (Eds.), *Qualitative research practice* (pp. 595–608). London: Sage.

Albert, M. (2003). Universities and the market economy: The differential impact of knowledge production in sociology and economics. *Higher Education, 45*, 147–182. doi:10.1023/A:1022428802287.

Barnes, T. J. (2004). Placing ideas: Genius loci, heterotopia and geography's quantitative revolution. *Progress in Human Geography, 28*, 565–595. doi:10.1191/0309132504ph506oa.

Becher, T. (1989). *Academic tribes and territories*. New York: McGraw-Hill.

Bourdieu, P. (2004). *Science of science and reflexivity*. Chicago: University of Chicago Press.

Buttimer, A., Brunn, S., & Wardenga, U. (1999). *Text and image: Social construction of regional knowledge* (Beiträge zur regionalen Geographie, Vol. 49). Leipzig, Germany: Institut für Länderkunde.

Canagarajah, A. S. (2002). *A geopolitics of academic writing*. Pittsburgh, PA: University of Pittsburgh Press.

Castree, N., & Sparke, M. (2000). Professional geography and the corporatization of the university: Experiences, evaluations, and engagements. *Antipode, 32*, 222–229. doi:10.1111/1467-8330.00131.

Frijhoff, W., & Spies, M. (2004). *Dutch culture in a European perspective*. Basingstoke, UK: Palgrave Macmillan.

Fuller, S. (2002). *Knowledge management foundations*. Oxford, UK: Butterworth Heinemann.

Garcia-Ramon, M. D. (2003). Globalization and international geography: The questions of languages and scholarly traditions. *Progress in Human Geography, 27*, 1–5. doi:10.1191/0309132503ph409xx.

Gregson, N., Simonsen, K., & Vaiou, D. (2003). Writing (across) Europe: On writing spaces and writing practices. *European Urban and Regional Studies, 10*, 5–22. doi:10.1177/0969776403010001521.

Guetzkow, J., Lamonot, M., & Mallard, G. (2004). What is originality in the humanities and the social sciences. *American Sociological Review, 69*, 190–212. doi:10.1177/000312240406900203.

Hartshorne, R. (1939). *The nature of geography*. Lancaster, PA: Association of American Geographers.

Harvey, D. (1984). On the history and present condition of geography: An historical materialist manifesto. *The Professional Geographer, 36*, 1–11. doi:10.1111/j.0033-0124.1984.00001.x.

Harvey, D. (2006). Editorial: The geographies of critical geography. *Transactions of the Institute of British Geographers NS, 31*, 409–412. doi:10.1111/j.1475-5661.2006.00219.x.

Kallo, J. (2009). *OECD education policy: A comparative and historical study focusing on the thematic reviews of tertiary education* (Research in Educational Sciences, Vol. 45). Jyväskylä, Finland: Finish Educational Research Association.

King, R. (2004). *The university in the global age*. London: Palgrave Macmillan.

Kitchin, R. (2005). Commentary: Disrupting and destabilizing Anglo-American and English-language hegemony in geography. *Social & Cultural Geography, 6*, 1–15. doi:10.1080/146493 6052000335937.

Krieger, J. (2006). *American hegemony and the postwar reconstruction of science in Europe*. Cambridge, MA: MIT Press.

Kuhn, T. (1962/1970). *The structure of scientific revolutions*. Chicago: The University of Chicago Press.

Kyvik, S., & Larsen, I. M. (1997). The exchange of knowledge: A small country in the international research community. *Science Communication, 18*, 238–264. doi:10.1177/10755470970 18003004.

Livingstone, D. N. (2003). *Putting science in its place: Geographies of scientific knowledge*. Chicago: The University of Chicago Press.

Minca, C. (2000). Venetian geographical praxis. *Environment and Planning D: Society and Space, 18*, 285–289. doi:10.1068/d1803ed.

Nayak, A., & Jeffrey, A. (2011). *Geographical thought: An introduction to ideas in human geography*. Harlow, UK: Prentice Hall.

Paasi, A. (2005). Globalisation, academic capitalism and the uneven geographies of the international journal publishing spaces. *Environment and Planning A, 37*, 769–789. doi:10.1068/a3769.

Paasi, A. (2006). "Internationalismens" ulige geografi inden for den globale akademiske basar ["Internationalism": Unequal geography within the global academic market]. In K. Buciek, J.-O. Baerenholdt, M. Haldrup, & J. Ploger (Eds.), *Rumlig Praksis* (pp. 49–62). Roskilde, Denmark: Roskilde Universitetforlag.

Paasi, A. (2011). From region to space – Part II. In J. A. Agnew & J. S. Duncan (Eds.), *The Wiley-Blackwell companion to human geography* (pp. 161–175). Oxford, UK: Blackwell.

Paasi, A. (2013). *Fennia*: Positioning a 'peripheral' but an international journal under conditions of academic capitalism. *Fennia, 191*, 1–13. doi:10.11143/7787.

Paasi, A. (2015). Academic capitalism and the geopolitics of knowledge. In J. Agnew, V. Mamadouh, A. Secor & J. Sharp (Eds.), *The Wiley-Blackwell companion to political geography* (pp. 509–523). Oxford, UK: Blackwell.

Parmar, I. (2002). American foundations and the development of international knowledge networks. *Global Networks, 2*, 13–30. doi:10.1111/1471-0374.00024.

Parsons, T. (1947/1966). Introduction. In M. Weber (Ed.), *The theory of economic and social organization* (pp. 3–86). New York: The Free Press.

Rodriquez-Pose, A. (2004). On English as a vehicle to preserve geographical diversity. *Progress in Human Geography, 28*, 1–4. doi:10.1191/0309132504ph467xx.

Samers, M., & Sidaway, J. (2000). Exclusions, inclusions, and occlusions in 'Anglo-American geography': Reflections on Minca's 'Venetian geographical praxis'. *Environment and Planning D: Society and Space, 18*, 663–666. doi:10.1068/d1806ed.

Schoenberger, E. (2001). Interdisciplinarity and social power. *Progress in Human Geography, 25*, 365–382. doi:10.1191/030913201680191727.

Scholte, A. (2000). *Globalization: A critical introduction*. London: Palgrave Macmillan.

Shusterman, R. (2000, August 11). The perils of making philosophy a lingua Americana. *The Chronicle of Higher Education, 46*(49), B4–B5. Washington, D.C.

Tietze, S., & Dick, P. (2009). Hegemonic practices and knowledge production in the management academy: An English language perspective. *Scandinavian Journal of Management, 25*, 119–123. doi:10.1016/j.scaman.2008.11.010.

Ward, K., Johnston, R., Richards, K., Matthew, G., Taylor, Z., Paasi, A., et al. (2009). The future of research monographs: An international set of perspectives. *Progress in Human Geography, 33*, 101–126. doi:10.1177/0309132508100966.

Wright, J. K. (1966). *Human nature in geography*. Cambridge, MA: Harvard University Press.

Power/Knowledge/Geography: Speculation at the End of History

13

Richard Peet

Finance Capitalism

The late twentieth century saw the emergence of a new kind of society. A capitalism dominated by multinational corporations producing commodities and services was replaced by a capitalism dominated by multinational banks and investment corporations controlling access to capital. In this new "global finance capitalism," finance is the leading form of capital; finance and its persuasive apparatus are integral parts of the governance system; finance capitalism normally operates on a global scale; finance capitalism takes the total form of a political, economic, ethical, cultural, and spatial system; and the contradictions underlying specifically financial crises shape the ongoing dynamic of capitalism. The term "finance capital" was originally coined by the Austrian Marxist Rudolf Hilferding (1981) to describe an increasing concentration and centralization of capital, in the institutional form of corporations, cartels, trusts, and banks, that organized the export of surplus capital from the industrial countries, especially Britain, in search of higher rates of profit elsewhere. More recently, David Harvey (2005) has argued that in capitalist enterprises, ownership (shareholders) and management (CEOs) have been fused together, as upper management is paid with stock options. Increasing the price of the stock becomes the main motive in operating the corporation; moreover, productive corporations, diversifying into credit, insurance, and real estate, become increasingly financial in orientation—hence "the financialization of everything," meaning the control by finance of all other areas of the global economy. Thus, nation-states, individually (as with the United States) and collectively (as with the G7/8/20), have to support financial institutions and the integrity of the financial order, for that is what keeps economies going (witness the massive intervention of the central banks in the

R. Peet (✉)
Graduate School of Geography, Clark University, Worcester, MA 01610-1477, USA
e-mail: rpeet@clarku.edu

© Springer Netherlands 2015
P. Meusburger et al. (eds.), *Geographies of Knowledge and Power*,
Knowledge and Space 7, DOI 10.1007/978-94-017-9960-7_13

financial crisis of 2007–2010). Within this rearranged capitalist system, Harvey finds the power of shareholders declining, whereas that of CEOs, key members of corporate boards, and financiers is increasing. The tremendous economic power of this new entrepreneurial-financial class enables vast influence over the political process (Harvey, 2005, pp. 31–38).

The main difference between Hilferding's finance capitalism, or even Harvey's more recent version, and today's global finance capitalism, is the greater abstraction of capital from its original productive base into a more virtual realm; the faster speed at which money moves across wider spaces into more places; the level, intensity, and frequency of crises that take financial rather than productive forms; and, most important, the spread of speculation and gambling into every sphere of economic, social, and cultural life. We have seen the "democratization" of finance capital through the inclusion into the reserve army of the financiers of millions of people who benefit from home ownership, pension fund investments, mutual funds, and education savings. No longer do we just have fat cats manipulating share prices; now millions of quasi-capitalists worry all night about their retirement savings and their hopelessly inflated house prices. Nonetheless, contemporary finance capitalism remains a fat-cat world in terms of the market power exercised by the leading component of finance capital—the assets of a few hundred thousand superrich people put into the hands of "expertly run" wealth management companies. On the one hand, finance capitalism has developed massive, sophisticated powers of social and cultural control over governments, classes, and regional populations, so that critical, political response to widening inequalities and instabilities may be muted for long stretches of time, and restrained even during intense moments of system-threatening crises—we now live in a time of global co-optation. On the other hand, the level and depth of financial crisis have grown, the "space of crisis" has extended to include virtually all national economies, and the "space of victims" (direct and indirect) is now virtually universal. The intersection of these tendencies creates an air of unreality and distancing in which crises are dealt with superficially, even as their intensity deepens. Crises that are structural and endemic appear to burst onto the political-economic stage as spontaneous events, spectacular in their array of corrupt, star actors—the "economy of the spectacle" defines our times (cf. Debord, 2004). But in reality these crises accumulate, for they are neither understood nor controlled, nor even is there much popular will to control them, because many people combine the roles of perpetrator and victim. Moreover, the finance system is so big, amorphous, and almighty powerful, that it seems both unimaginable as a whole and impregnable as an ongoing system, even as it erupts. Inevitably, this vast system of subservient neglect tends toward catastrophe.

This new finance capitalism appeared on the global scene in a burst of cultural and economic exuberance that can only be admired as a sign of the new global times, by an awestricken public. At least that is what appears in the popular media, which are a crucial part of the very system on which they are supposed to objectively report. But then, global finance capitalism might be rewritten as global media capitalism, because mediatization is as powerful as financialization, and both share that air of fantastic unreality (the world as global reality show) that takes the place of what once passed for everyday life. So sports news gets more time and, alas, more attention, than news of war; the media do not carry the news on a Saturday; and nothing "happens" on a day with football games, even as the economic system is collapsing. Yet, as Karl Marx once almost said, social analysis means breaking the mesmerizing dazzle of the global spectacle by moving analysis toward the discovery of structural essence.

Institutional Analysis

With the term "geography of power," I am referring to the concentration of power in a few spaces that control a world of distant others. How should one analyze this geography? First, I wish to avoid the kind of structural analysis according to which vaguely defined "capitalists" or "state apparatuses" functionally make things happen. Yet the depth and reach of the present crisis necessitates talking in broad structural terms. The way out, surely, is to add an intermediate-level analysis that focuses on the specific agencies that act in broader systems—producing definite ideologies as specific discourses, for example. In other words, I aim to construct a critical institutional analysis embedded within structural terms and categories—the best of two analytical worlds.

In the past a few exceptional, and therefore critical, economic theorists were interested in institutions (Polanyi, 1944; Veblen, 1912). An interest in institutionalism has more recently been revived (Hodgson, 1988; Metcalfe, 1998; Samuels, 1995). Many institutional economists share with mainstream economists the dominant (and safe) convention that economics is the study of the efficient allocation of resources. But they no longer see *the market* as the economy's sole guiding mechanism. Instead, they argue, the structure of society organizes markets and other institutions (Ayres, 1957). For these institutionalists, the optimizing results of neoclassical economic theory could be realized only in an institution-free environment, where transaction costs are minimal. By contrast, an exchange process incurring transaction costs implies significant modifications in economic theory and has different implications for economic performance (North, 1990, 1995; Williamson, 1985). Institutional economics therefore engages a broader set of interests than does conventional economics, being more concerned with power, institutions, individual and collective psychologies, the formation of knowledge in a world of radical indeterminacy, and the relations among culture, income, and control in societies. In other words, it engages an economic universe that begins to look like the reality envisaged by social theorists such as Gramsci or Foucault, except that, unlike the originators of this line of thought, contemporary institutionalists are usually not social critics—they simply want to produce a more inclusive, conventional economic theory that better serves existing power.

In the discipline of geography, similar arguments have been made, for example, that conventional economic geography abstracts economic action from its contexts, whereas in reality economic activity is socially and institutionally situated (Martin, 1994, 2000; Scott, 1995; Sunley, 1996). Institutional economic geography examines the shaping of space economies within environments of institutions characterized by path dependency. One fertile strand of institutionalist interest focuses on social regulation, governance, and the effect of policies in shaping national, regional, and local economies (Amin, 1999). This kind of thinking comes easier to geographers because of the discipline's environ-*mentalism*. Even so, the problem with the term "institution" is that no one is sure what it means—is it an organization, such as a corporation or bureaucracy, or values, such as efficiency or benevolence?

I shall explore this question a bit further, using the case of economic action and policy. The institutions making policy might be conceptualized materially as organizations, located in buildings, with the less tangible conventions, norms, rules, discourses, mentalities, and imaginaries conceived as the *institutional products* made by organizations. Simply put, in what follows, *institution* initially means a physical-organizational entity, located in a space, with a mission and declared purpose, backed by command over some kind of resource (ideas, expertise, money, connections). But *institution* is also used in the Foucauldian sense of a *community of experts*, an elite group of highly connected individuals controlling an area of knowledge and expertise. This community of experts shares the same ideas and ideals. It takes the same things for granted—indeed that is the meaning of *consensus*. Although there may be interpersonal stresses, the members stick up for each other because they share a common interest that, if broken, might reveal too much about their biases and communal insecurities—expertise is always in part a front presented to the outside world. There are basic ideas and methods that do not have to be discussed, so debate focuses *productively* on slight differences within a meaning structure that is presumed within a set of institutions, such as government, governance, and elite academic institutions. Membership in such a community is the main source of an expert's power, status, and income. Saying something outside the accepted discourse can lead to banishment from the inner, institutional circle, with its status and prestige, to the outer fringes of quasi-responsibility, or even to the desert of nutcase irresponsibility! Institutions as communities of experts are self-policing. Of course, the main problem is that the institutionally taken-for-granted may be the main cause of the problem that the expert community is supposed to be addressing. Community discourse therefore takes the form of perpetual variations on the always irrelevant.

Power Centers

These communities of experts can be *mapped* in the sense of analyzing the arrangement of institutions in space—with "arrangement" being formative, rather than derivative, of power—and examining the power relations among institutional complexes. This concept of *mapping* is similar to the notion of "policy networks" used in political science to connote the "structural relationships, interdependencies, and dynamics between actors in politics and policy-making" (Schneider, 1988, p. 2; translation mine). In political science, policy networks are understood to be "webs of relatively stable and ongoing relationships which mobilize and pool dispersed resources so that collective (or parallel) action can be orchestrated toward the solution of a common policy" (Kenis & Schneider, 1991, p. 36).

Economic elites circulate within institutional complexes—from Harvard University to an investment bank, from bank to treasury, from treasury to think tank, with time spent at the IMF in between … it's a great life if you can get it. Well paid too. In effect, the institutional mapping I have in mind realizes the "networks" and "webs" of network analysis in a spatial matrix composed of points (institutions), clusters (power centers), and flows (power relations). In other words, I contend that mapping institutions as they are located in space is a productive, materialist approach to understanding the generation of power. That is, power is more forceful when it is concentrated, accumulated, given momentum by the fame of its place. Clusters of power-generating institutions make up a "power center." Each power center can be thought, in and of itself, as a place, in the sense of a cluster of interconnected institutions with an ambience (Wall Street in New York, K Street in Washington, DC, Cambridge in England and New England), or what might be termed an institutional complex, a congregation of grouped experts with its distinct culture and reputation—when Cambridge speaks, people listen.

Furthermore, centers of power can be classified as belonging to three main types, according to the dominant purpose of their leading institutions and the type of power they initiate; economic, meaning that their leading institutions deal primarily in money—such as financial markets, investment banks, and corporate headquarters—and transmit power as control over investment and financial expertise; ideological, meaning that institutions deal in ideas produced at the level of theory—such as universities, research institutes, and foundations—and transmit power as scientifically justified ideas, rationalities, and discourses; and political, meaning that institutions construct and enforce ideas in practical formats—such as government and governance centers—that transmit power as policy. (And the more I think about it, the greater my inclination to add a fourth type: Popular media centers of power transmitting power as entertainment.) In finance capitalism, economic centers of power predominate over the others, using the political centers to marshal collective power on their behalf, ideological centers to manufacture the myths that legitimate their actions, and media to keep people happy as they are manipulated into mindless compliance. In other words, there is a "central place hierarchy" conceived in terms of the centralization, accumulation, and exercise of powers that have structural inequalities.

Power centers, composed from complexes of institutions, can also be mapped in terms of the power relations they engage in across space and the discourses they transmit across space. To begin with, each power center concentrates resources (capital, ideas, expertise) from a broader field of power. This field may be physically contiguous, in the sense of the hinterland of a power center. Or it may have a "virtual" aspect in the sense of its position in a World Wide Web. Experts clustered in power centers, or highly connected via e-mail, process intellectual, theoretical, and practical resources drawn from fields of power by applying their concentrated knowledge and expertise. I contend that power centers formed by institutional complexes can be classified as hegemonic, meaning that they produce ideas and policies with enough theoretical depth and financial backing that they dominate thought over wide fields of power (New York, London, Frankfurt, etc.); sub-hegemonic, referring to peripheral centers of power that translate received discourses and practices, modify and add ideas, and exercise power in specialized regional ways (Delhi, Mumbai, Singapore, São Paulo, etc.); or counter-hegemonic, meaning centers, institutions, and movements founded on opposing political beliefs that contend against the conventional, exercise counter-power, and advocate policy alternatives (Havana, Caracas, etc.). In the modern world, ideas backed by political, cultural, and economic resources are transmitted among power centers in the specific ideological form of discourses legitimated by their backing in key resources, for example by claims to science. At the receiving end, in sub-hegemonic centers, discourses release their contents as "power effects," such as theoretical persuasion or replicated practices. Nonetheless, even power of the most apparently solid, indisputable kind is continually being destabilized by class, gender, ethnic, and regional differences in experience and interpretation (cf. O'Tuathail, 1997).

Analyzing the Exercise of Power

On the whole, the power centers of the modern capitalist world do not produce commodities in the form of physical objects. They mainly produce the ideas that order and control the production of objects. Hence, we need a set of concepts that link power with the ideas that create worlds. In the critical theoretical tradition there is only one place to begin this analysis: Karl Marx's concept of "ideology." The Marxist concept of ideology refers to the production and dissemination of ideas, primarily by the state and its bureaucratic apparatus, that support and legitimate the prevailing social order (Marx & Engels, 1932/1970). The ideas behind institutional practices, such as making investment decisions or framing policies, are not neutrally conceived, as science pretends, nor exercised in the interest of everyone, as modern humanitarianism hopes. Instead, power is conceived and exercised to serve dominant political-economic interests. In Marxist theory, dominant interests are those of the richest people in society, powerful because they possess capital, defined as ownership of productive wealth, and typified by the shareholders and upper management of companies and corporations. There are several different versions of the Marxist theory of ideology. Two seem germane here. In structural dependence versions of the theory of ideology, private owners of productive assets impose binding constraints on other institutions, such as governmental operations. Offe (1985) emphasizes the power of capitalist refusal, especially through "investment strikes"; for example, investors refusing to invest in a corporation or country whose policies they do not like. Power elite theory, by comparison, argues that all types of powerful institutions act on behalf of capital because their managers share similar interests and many of the same values with capitalists, often because the elite move back and forth between the corporate, financial, academic, and government worlds. For example, the government officials whom President Obama employed to "solve" the financial crisis of 2007–2010 had investment banking backgrounds. According to

Miliband (1969; cf. Poulantzas, 1978), capitalists are able to control state institutions, and use them to realize their interests, because capitalists, elected representatives, elite academics, and high state officials are all the same people, sharing the same political values, using the same clichés ("moving forward"), and encased within the same circle of responsibility—the culture of the power elite. (Who can forget that video clip in Michael Moore's [2004] film *Fahrenheit 9/11*, in which President George W. Bush, addressing what is clearly a very rich audience, says, "This is an impressive crowd—the haves and have-mores. Some people call you the elite. I call you my base.") In the Marxist tradition, then, power takes the form of persuasive ideologies, circulating through dominant clusters of highly interconnected institutions, before spreading over space, carried by an array of conventional media.

The Marxist notion of ideology establishes a critical analytics of power, but we can hardly stop there. Moving "up" the institutional hierarchy toward global power institutions seems to be paralleled by "ideological deepening," from the production of particular, persuasive ideologies to a broader, sociocultural construction of logics, or ways of thinking. One concept dealing with this sociocultural construction of power is "hegemony." This concept was derived from Marx's theory of ideology by the Italian Marxist Antonio Gramsci (1971), with important additions coming later from the French philosopher Louis Althusser (1971). Gramsci thought there were two levels of political control over people; *domination*, which he understood to mean direct physical coercion by the police, the army, and the courts—what Althusser would later term the "repressive state apparatus"; and *hegemony*, which referred to ideological control and the production of consent by non-physically coercive means and institutions—what Althusser would later call the "ideological state apparatus."

Domination we know only too well—torturing prisoners, clubbing demonstrators, detaining people at airports, making police visits in the middle of the night, and so on down a frightening list. By *hegemony* Gramsci meant the cultural production of systems of values, attitudes, beliefs, and morality so that people supported the existing social order and the proscribed way of life. Hegemony, for Gramsci, was an "organizing principle" diffused, through socialization, as common sense into every area of daily life. Or in Althusser's typically even stronger version, for he was a man of extremes, the ideological state apparatus instilled systems of meanings in people's minds which placed them in "imaginary relations" with reality—the social construction of the imagination precludes anything like a true understanding of the real. What these theorists had the audacity to suggest is that the philosophy, culture, and morality favored by the ruling elite are made to appear as the natural, normal way of thinking, believing, and creating for entire groups of people—national pride and prejudices, for example, or even, in advanced versions, a specification of the good of global humanity, as in the case of philanthropic neoliberal reform. More precisely for the present topic, hegemony is constituted by a set of related ideologies that include, in advanced liberal capitalism, the validation of competitive individualism and the fetish of expertise resulting from technological rationality (Boggs, 1976).

Furthermore, whereas Marx had stressed the role of the political, coercive superstructure (basically, the state apparatus) as the producer of legitimating ideologies, Gramsci looked more closely at the roles played by supposedly non-coercive social institutions, such as churches, schools, trade unions, and so on, institutions that he collectively designated "civil society." (Althusser argued that the dominant role in the construction of minds was played by the "educational ideological apparatus"—that is, the schools and universities.) Gramsci also paid more attention than Marx to the specific people actually thinking up, supporting, elaborating, and spreading hegemony—the ideological agents, so to speak. Each social group, Gramsci said, organically creates a stratum of intellectuals that lends meaning to that group's collective experience, binds the group together, represents it persuasively, and helps it function effectively—that is, without too much stress. Hegemony, for Gramsci, was produced for the ruling class by the civil servants, managers, priests, professionals, and scientists of his day. For us it is produced by movie directors, script writers, media personalities, talk-show hosts, investment analysts, think-tank experts, and superstar professors. In other words, there is a special class thinking up and spreading dominant modes of thought, and they congregate in power centers. The essence of power is the creation of hegemony in places that specialize in the making of minds.

This whole process of hegemony construction and translation is not a smooth operation whose end is known from the first instance. The great thing about ideas is that they can always be countered. The great thing about thought is that it can fall silent. Specifically, countering the hegemony of a broadly defined ruling class meant, for Gramsci, constructing a counter-hegemony as part of class struggle. Gramsci did not believe in structural contradictions playing themselves out automatically in social transformation. He thought instead that activists had to seize moments of structural crisis by employing powerful counter-ideas conceived in advance by people who thought differently. In all of these struggles over ideology and hegemony, intellectuals play leading roles.

Interpretive Power

My own analysis essentially adheres to this kind of Gramscian project (Peet, 2007). But adherence comes with a few words of criticism, and with something of a redirection of course. First, the criticism, though I venture it reluctantly. Although Gramsci points to the social construction of common sense, and Althusser takes this further (perhaps too far) with his notion of the sociocultural production of imaginary relations to the world, neither theorist quite gets to the heart of the matter, at least to my mind. The great puzzle, surely, is how society manages to produce a safe, system-supporting common sense when many people commonly experience a horrific, everyday world of poverty, hunger, and death? To get to the root of this deep social constructionism, we have to repeat Gramsci's original question: "What prevents miserable life experiences from forming mass critical consciousness?" The answer, I think, involves re-examining the relations between material reality, collective and individual experience, and the making of consciousness. A complete answer would go beyond the confines of the present chapter, so I will only briefly introduce my thoughts here.

Material events do not form experience directly, nor does experience flow into consciousness as photographic memories. Instead, experience passes through what might crudely be termed the filter of socialized beliefs. Beliefs, in turn, are best thought of as interpretive devices so interventional that what one person sees as mere coincidence, another perceives as definitive evidence of divine intervention. In other words, experience is lived belief, rather than material existence lived directly. Thus, the key to hegemonic domination is the social production of frames of interpretation, especially the collective beliefs through which people think their lives. Rephrasing Gramsci and Althusser with this kind of existential emphasis on interpretation in mind, I suggest that civil society institutions inhabited by expert intellectuals clustered in power centers create hegemony by inventing and recasting deeply embedding belief-structures that are projected into mentalities so that they filter and direct interpretations of experiences in organized, channeled ways. That is, I do not think that conventional rationality is a perfect reflection of the world, but rather that the rationality that prevails is a socially produced mode of careful thought based in beliefs conceived and perpetuated in the interests of elite power. (I am thinking in particular about economic rationality and economic theory here.) Control the belief system and you control the interpretive framework in which social life occurs. Control the interpretive framework and you control the passage of thought from experience into consciousness. Control interpretation, and you can let people "think for themselves." In other words, interpretive hegemony is the necessary basis for modern freedom. Power centers are masters of interpretation.

Expert Discourse

Each age has its version of hegemonic power and its cities of influence. In premodern times these cities were outstandingly religious in content and ambience—so the "great cities" of Europe with their physical trappings of churches and monuments that now form the touristy historical backdrop to our far more subtle environments of power. In modernity, by contrast, hegemonic power centers exude the aura of science. They are populated by highly intelligent, trained, and experienced individuals—"experts"—and well-established, abundantly financed institutions—universities, banks, government departments, think tanks, banking associations, etc.—that are more economic-institutional than civil-institutional in character. This high-level, economic-institutional thinking employs a kind of symbolic representation for which the Gramscian term "common sense" is insufficient; something else, perhaps "expert sense," is involved. Examining this expert sense requires grafting onto the Marxian-Gramscian-existential theory I have outlined Foucault's (1979, 1980) notions of discourse, discipline, and expert. Foucault claimed to discover a previously little noticed kind of linguistic function, the "serious speech act," or the statement, backed by validation procedures, made from the standpoint of experts, and developed within communities of experts (Dreyfus & Rabinow, 1983, pp. 45–47). For Foucault, serious speech acts exhibit regularities as "discursive formations" with internal systems of rules determining what statements are taken seriously and what objects included in discussions are deemed important or responsible. Foucault thought that these regularities of presence and absence could be analyzed archaeologically (identifying the relations that bound statements into whole arguments) and genealogically (analyzing how discourses were formed within institutions claiming power). Therefore, as Best and Kellner (1991) summarize, discourse theory analyzes "the institutional bases of discourse, the viewpoints and positions from which people speak, and the power relations these allow and presuppose … [as well as] … discourse as a site and object of struggle where different groups strive for hegemony" (p. 26; see also Rabinow, 1984).

Here again, however, I need to add more in the way of critical conceptualization. What is it that is being socially constructed? Behind a discourse we find a set of concepts with labels on them—theories with analytical terms that a discourse subsequently employs. Particularly important are the originators of a line of interpretation, thinking, and discussing—in what might be called the construction of a "theoretical memory." At the other end of a discourse, at points of influence, lies what Cornelius Castoriadis (1991, p. 41) calls a social imaginary, a system of significations that organizes the (presocial, biologically given) natural world, institutes a social order (articulations, rules, and purposes), establishes ways in which socialized and humanized individuals are fabricated, and saturates consciousness with the motives, values, and hierarchies of social life. Although this sounds interesting, I find it a bit vague. I like the term "social imaginary" because it places imagination at the creative edge of the hegemonization of culture, while still rooting the imaginary in the social—in other words, it combines the social construction of ways of thinking and believing with the mind's creativity. Social imaginaries, then, are collective forms of consciousness, structured by specific social environments, that make people not only think in similar ways, but imagine in similar pictures, words, and that subjective mixture of the two ("picture-words") that minds love to play with. Imaginaries take class and regional forms, that is, the imagination uses materials (images, memories, experiences) from the familiar to project imaginative versions of the already known. There is a limit placed on what we can legitimately think about, formed within the prison cell of our interpreted experiences.

However, despite such structuring, the word "imaginary" clearly implies imaginative interpretation and creativity—projecting interpretations into the scarcely known—so that social imaginaries are vital sources of transformational, as well as reproductive, dynamics. The imaginary realm must therefore be seen as tension-filled, between visionary and more grounded logics, between received wisdom and new interpretations, between fundamental beliefs and practical forms of consciousness, between alternative ways of knowing and different ways of envisioning. The connection between memory and imaginary is a set of ideas running through the entire ideological formation. The people who originally think these ideas up, and lend them terms to speak with, originate a system of discursive and imaginary power. One example is the discourse of classical-neoclassical-neoliberal economics—and, hence, the power of Adam Smith, David Ricardo, and others in forming the collective memory, the social imaginary of contemporary mainstream economics. Another example is the rapid spread of speculative fever from New York and London to every financial power center in the world during the 1990s and early 2000s.

For some time now, experts have been trained as intellectuals primarily in universities, where they learn to think theoretically—that is, employing theories to understand and change natural and social reality. Theory is the quintessential form taken by serious thinking in a deep level of contemplation, when the mind seeks the original, causal sources of events. Theory restructures the mind to think in deeper, more powerful ways. A persuasive general theory, learned at school or university, becomes the social-imaginary base for hundreds of expert discourses, which structure thousands of policies, which affect billions of people ("intellectual leveraging"). Theory structures the expert imaginary by forming the concepts through which even creative thinking reaches into the scarcely known. Thus, in modernity hegemony is produced as dominant theoretical imaginaries in disciplines claiming power by presuming the status of science. Put slightly differently, hegemony in its most basic sense means controlling what is taken to be "rational," with specialized power centers monopolizing particular types and styles of hege-monization—inventing and updating specific logics and rationalities. That is their economy. That is how they make a living. Cities specialize in different kinds of imagineering. If they fail to do this creatively, others take their place. The race to control the minds of the world leaves trails of second-rate places in its wake.

Speculation at the End of History

In the 1980s, as neoliberalism took control, income was deliberately redirected toward people who *could not spend* it, no matter how hard they tried (e.g., $20 million apartments in financial centers that became real-estate price multipliers); they could only save and invest it. So under neoliberalism, in the United States alone, $1 trillion a year flows into the investment accounts of a few hundred thousand already very wealthy people (Saez & Piketty, 2007). Financial institutions compete to use the investment funds over-accumulated by "high-net-worth individuals" and by workers' savings in pension funds, insurance contributions, and so on. Corporate capital experiences this competition for investment as an external compulsion originating in the dominant financial component of capital: CEOs who fail to deliver are subject to scrutiny by private equity firms that make their money by buying up "non-performing" corporations, ruthlessly restructuring them (i.e., firing workers), and then selling them to make a quick profit that yields high returns to investors. Thus the disciplining of General Motors by finance capital, new dinosaur versus old, was a sight to behold!

As this description suggests, the reach of financial power (in all of its aspects) has expanded outwards from its original, capitalist bases in the advanced industrial countries into a global playing field, where trillions of dollars range daily with ease and speed in search of high returns—four trillion a day passes through the currency markets, for example. Clearly, this global playground for capital is lined still with political and cultural boundaries. But increasingly, within global investment space, countries are assessed merely as risk–benefit ratios; by being included that way in the profit calculus, states are reduced in significance, except as they act as minders or cheerleaders to the profit-seeking actors behind global capital. This new version of finance capitalism is centered on the deployment of large accumulations of wealth by specialized institutions, such as investment banks and risk assessment firms, which are concentrated in a few centers of financial power—the upper rank of "world-class cities."

Yet even with ruthless profit-raiding of corporations as modus operandi, the securities market is a relatively safe, stable investment outlet. The stock market is regulated by the state, in the United States by a government agency—the Securities and Exchange Commission established in 1934, during a previous time of crisis. Investment management companies controlling collective assets in the form, for example, of mutual funds are also regulated under the Investment Company Act of 1940. However, under neoliberalism, the superrich have increasingly found ways of avoiding state regulation of investment. They do so partly by escaping national jurisdictions, as with phantom corporate headquarters in places like Liechtenstein and the Cayman Islands. And they escape regulation at home by creating exotic investment vehicles. In the United States, investment funds open to small numbers of "accredited investors" (fewer than a hundred), and funds made up of "qualified purchasers" (the qualification being over $5 million in investment assets), are not subject to governmental regulation other than the registration of traders. So temporary investing in the equity market (the stock market), making a quick profit, and then selling, competes with other, far more speculative and lightly regulated hedge funds, private equity deals, subprime mortgage bundlers, futures, derivatives, and currency trading. The situation is similar in many other countries because speculation pays and states bail you out in times of trouble. In the context of global-ization, "emerging markets," and exotic investment outlets, investment funds are expected to return at least 20 % a year, doubling elite wealth every 4–5 years. So we live in societies where the dynamic of the leading component of capital is the pursuit, by any means, of more money for those who already have too much. This reckless pursuit of money for the sake of more money is financial, societal madness. It can only result in disaster.

The price of high returns is … eternal risk. Any investment fund that does not take extreme risks and thus does not generate high returns suffers disinvestment in highly competitive markets, where money changes hands in computer-quickened moments. Hence, in the centers of financial power, there is a competitive compul-sion to take increasingly daring risks in search of higher returns that temporarily attract investment. Speculation, risk, and fear are structurally endemic to finance capitalism. Fear itself becomes the source of further speculation—buying gold, or futures, for example. Speculation and gambling have spread from Wall Street into all sectors of society—housing prices, state lotteries, casinos on Indian reservations, bingo in church halls, sweepstakes, Pokémon cards—everyone gambles, even little kids. (My favorite is Dubai, an entire city built on islands of speculation.) The interlocking of speculations is the source of their intractability and of the growing space of their effects. To take one example, the financial crisis of 2007–2010 had the following moments: Vastly overpriced housing, particularly near booming financial centers; competition among financial institutions to offer easy credit to anyone; the bundling of home mortgages into tradable paper; very high levels of leveraging; and the use of assets whose value can disappear in an instant to securitize other, even more risky, investments. It is not just that crisis spreads from one area to another. It's more that crisis in one (such as the inevitable end to the housing-price bubble) has exponential effects on the others (investment banks overextended into high-risk speculations) to the degree that losses accumulate that are potentially beyond the rescuing powers of states and governance institutions. Hence, the tendency toward catastrophe.

I shall conclude this tale of sorrow by peering into the minds of the speculators, clustered in financial power centers, making instantaneous decisions that burst economies apart, wrecking the lives of millions. This behavior is "distanced thinking" in terms of space (clustered versus spread) and in terms of effect (deliberate ignorance of social effects). It is "driven thinking" in terms of the single-minded pursuit of quick and nasty profits. It is greedy, "egotistical thinking" in terms not so much of neglect of others, but more of a kind of selfish hatred of others, a will to power made pathological by a desire to damage. It is a style of interpretation, thought, and discourse best produced in fantasy cities, in buildings that reach to the sky amid landscapes spectacularly displaying ostentatious overabundance—all power corrupts, but spatially concentrated power corrupts absolutely. It is a psychosis best understood by reading its psychotic advocates, authors such as Ayn Rand (1957), whose objectivist philosophy of "rational egoism" spurs speculation into what her student, Alan Greenspan, called "irrational exuberance"—the economy produced by the exercise of millions of intersecting, selfish acts, made without regard for their consequences, motivated only by the mad desire for ever more money. These people use their creative fantasies to make speculative super-profits that spectacularly destroy the social fabric, replacing it with a culture of interacting, escalating, creative selfishness. These are the cities of the mad. Only in this way can we begin to fathom the willful destruction of the environment in the last quarter of the twentieth century and the first quarter of the twenty-first century. Speculation at the centers of global power, where the future is dreamed. Speculation that brings about the end of history.

References

Althusser, L. (1971). *Lenin and philosophy and other essays*. London: New Left Books.
Amin, A. (1999). An institutionalist perspective on regional economic development. *International Journal of Urban and Regional Research, 23*, 365–378. doi:10.1111/1468-2427.00201.
Ayres, C. E. (1957). Institutional economics: Discussion. *American Economic Review, 47*, 26–27.
Best, S., & Kellner, D. (1991). *Postmodern theory: Critical interrogations*. New York: Guilford Press.
Boggs, C. (1976). *Gramsci's Marxism*. London: Pluto.

Castoriadis, C. (1991). *Philosophy, politics, autonomy: Essays in political philosophy*. New York: Oxford University Press.

Debord, G. (2004). *Society of the spectacle*. Wellington, New Zealand: Rebel Press.

Dreyfus, H. L., & Rabinow, P. (1983). *Michel Foucault: Beyond structuralism and hermeneutics*. Chicago: University of Chicago Press.

Foucault, M. (1979). *Discipline and punish: The birth of the prison*. New York: Vintage Books.

Foucault, M. (1980). *Power/knowledge: Selected interviews and other writings, 1972–1977*. New York: Pantheon.

Gramsci, A. (1971). *Selections from the prison notebooks*. London: Lawrence & Wishart.

Harvey, D. (2005). *A brief history of neoliberalism*. Oxford, UK: Oxford University Press.

Hilferding, R. (1981). *Finance capital: A study of the latest phase of capitalist development*. London: Routledge & Kegan Paul.

Hodgson, G. M. (1988). *Economics and institutions*. Cambridge, UK: Polity Press.

Kenis, P., & Schneider, V. (1991). Policy networks and policy analysis: Scrutinizing a new analytical toolbox. In B. Marinand & R. Mayntz (Eds.), *Policy network: Empirical evidence and theoretical considerations* (pp. 25–59). Frankfurt am Main, Germany and New York: Campus Verlag.

Martin, R. L. (1994). Economic theory and human geography. In D. Gregory, R. L. Martin, & G. E. Smith (Eds.), *Human geography: Society, space and social science* (pp. 21–53). Basingstoke, UK: Macmillan.

Martin, R. L. (2000). Institutional approaches in economic geography. In E. Sheppard & T. Barnes (Eds.), *A companion to economic geography* (pp. 77–94). Oxford, UK: Blackwell.

Marx, K., & Engels, F. (1970). *The German ideology. Part one, with selections from parts two and three, together with Marx's "Introduction to a critique of political economy"* (Lawrence & Wishart, Trans.). New York: International Publishers. (Original work published 1932)

Metcalfe, S. (1998). *Evolutionary economies and creative destruction*. London: Routledge.

Miliband, R. (1969). *The state in capitalist society*. New York: Basic Books.

North, D. (1990). *Institutions, institutional change and economic performance*. Cambridge, UK: Cambridge University Press. doi:10.1017/CBO9780511808678.

North, D. (1995). The new institutional economics and third world development. In J. Harriss, J. Hunter, & C. M. Lewis (Eds.), *The new institutional economics and third world development* (pp. 17–26). London: Routledge.

O'Tuathail, G. (1997). Emerging markets and other simulations: Mexico, the Chiapas revolt, and the geofinancial panopticon. *Ecumene, 4*, 300–317.

Offe, C. (1985). *Disorganized capitalism*. Cambridge, MA: MIT.

Peet, R. (2007). *Geography of power*. London: Zed Books.

Polanyi, K. (1944). *The great transformation*. New York: Farrar and Rinehart.

Poulantzas, N. (1978). *State, power, socialism*. London: Pluto.

Rabinow, P. (Ed.). (1984). *The Foucault reader*. New York: Pantheon.

Rand, A. (1957). *Atlas shrugged*. New York: Dutton.

Saez, E., & Piketty, T. (2007). Income inequality in the United States, 1913–1998. *Quarterly Journal of Economics, 118*, 1–39. Retrieved August 16, 2012, from http://elsa.berkeley.edu/~saez/

Samuels, W. (1995). The present state of institutional economics. *Cambridge Journal of Economics, 19*, 569–590.

Schneider, V. (1988). *Politiknetzwerke der Chemikalienkontrolle: Eine Analyse einer transnationalen Politikentwicklung* [Political networks of chemical substance control: An analysis of transnational policy development]. Berlin: De Gruyter.

Scott, A. (1995). The geographic foundations of industrial performance. *Competition and Change, 1*, 51–66.

Sunley, P. (1996). Context in economic geography: The relevance of pragmatism. *Progress in Human Geography, 20*, 338–355. doi:10.1177/030913259602000303.

Veblen, T. (1912). *The theory of the leisure class: An economic study of institutions*. New York: Macmillan.

Williamson, O. E. (1985). *The economic institutions of capitalism: Firms, markets, relational contracting*. New York: The Free Press.

Jürgen Wilke

The Beginnings of Media Control

In modern times, the production and especially the distribution of knowledge is due to—and is driven by—the availability of methods of duplication and distribution of statements and messages to a large number of receivers. We call these methods "media." In the middle of the fifteenth century, Johannes Gutenberg invented the first technology that was able to fulfill this function: printing with movable type. It was not long before measures were taken to control the use of this technology and thus to supervise the availability of information to mankind. Media control is nearly as old as modern media itself.

In the early days of the printing press, mainly theological books and books for ecclesiastical use were produced. Measures for control of the new medium were first taken by the Catholic Church, aimed at the purification of the faith and the protection of Christian morals and conventions. Soon, however, the state joined in these efforts. The principles of control thus widened. The intention was to avoid disturbances of public peace and order and to safeguard state secrets and personality rights.

In the sixteenth century, a system of control was gradually built up in the German Empire. Not only books but also journalistic publications were subject to it; first broadsheets and pamphlets, later periodical newspapers and magazines. The core element of this system was censorship (Wilke, 2007a, 2013). In its preventive form, it was in principle the most efficient means to inhibit the production of undesired printed work in general. However, other measures were added: the censorship and confiscation of already printed works, as well as regulations regarding the

J. Wilke (✉)
Department of Communication, Mainz University,
Jakob Welder-Weg 12, 55099 Mainz, Germany
e-mail: wilke@uni-mainz.de

© Springer Netherlands 2015
P. Meusburger et al. (eds.), *Geographies of Knowledge and Power*,
Knowledge and Space 7, DOI 10.1007/978-94-017-9960-7_14

settlement of printers only in free municipal, residential and university cities. Furthermore, printers were obliged to swear an oath that they would observe the rules of censorship. Violation of the rules incurred severe punishment. Unannounced visitations to printing shops by the authorities served the purpose of control.

The continuous repetition and expansion of control measures suggests that they did not have the desired effects. There were two crucial aspects to this: on the one side were the economic interests of the printers. Initially this was more important than the idea of something like freedom of the press, which was entirely lacking in the beginning. On the other side, the early modern state lacked the capacity to assert its will. Of course there was an apparatus for media control that, in the case of Germany, was headed by an official of the imperial court at the seat of the emperor in Vienna. Enforcement of the rules, however, depended on local authorities. Therefore the German territorial structure moderated the effects of media control, as is apparent when compared with centralized states such as France.

In early modern times, a system of media control was not exclusive to Germany, where printing technology had been invented and public media developed in especially varied ways. As this technology was rapidly introduced in other countries, control measures were also established. This happened, for example, under the Tudor monarchy in England, where a form of self-regulation by the printers' guild (the Stationers' Company) was also put into practice at an early date (Siebert, 1965).

The Decline and Fall of Media Control

The media control that developed in early modern times persisted over centuries. In the European countries, it diminished over longer or shorter periods. England led the way in this process, thus becoming the "motherland" of press freedom. There were several reasons for England being distinctive in this. First there was a long legal tradition that, since the Middle Ages, had recognized the protection of the personal sphere (the principle of *habeas corpus*). Second, a parliament existed that, although it was not yet constituted by general free elections, was a representative organ of (partial) social interests opposing the absolute sovereign, the king. Thirdly, due to the Reformation, England had a very strong pluralization of different religious groups.

Conflicts over power and religious freedom escalated in the 1640s with the Puritan revolution (Siebert, 1965, pp. 165–233). The Long Parliament (1640–1660) dismissed press control. However, when this resulted in anarchically escalating publication activity, Parliament itself re-introduced control. This was no longer accepted without protest, however. In 1644, John Milton wrote his fictitious speech "Areopagitica," the first major essay in the Western world to claim and to support press freedom. In this discourse, Milton summarized all the possible reasons against censorship and in favor of freedom of the press. At the center of his argumentation were two reasons: an individual-anthropological and a collective-sociological one (Wilke, 1983). The first one was normative, the second one utilitarian.

The "Areopagitica" had no direct impact, however; in fact, the *Printing Act*, making the licensing of printing works obligatory, was repeatedly extended. In 1695, however, Parliament relinquished another continuance. Thus pre-censorship (licensing) was abolished in England. For the first time there was virtual press freedom. It was then in the former British colonies in America where, after the Declaration of Independence, press freedom was constitutionally guaranteed for the first time (Virginia in 1776; Pennsylvania in 1790; the "First Amendment" in 1791).

Developments in mainland Europe lagged behind. In France, the 1789 Revolution brought a breakthrough for the acceptance of freedom of opinion and the press, but this did not last long. A delay in the development was especially typical for Germany. There was a succession of occasional signs of progress and new setbacks, depending on the respective understanding and exercise of power. In the nineteenth century, breaks were marked by the Carlsbad Decrees in 1819, the Revolution of 1848 and the founding of the Empire in 1871, leading to the imperial press law (*Reichspreßgesetz*) in 1874, which for the first time guaranteed press freedom all over Germany (Wilke, 2008, pp. 253–255). Where the state was forced to renounce pre-censorship, it normally looked for other means to control the press or at least to make life difficult for it: by enforcing penal laws, by financial means (taxation, bail enforcement) or by restrictions in distribution.

Making Media Control Totalitarian in the Twentieth Century

At the end of the nineteenth century, pre-censorship had generally been abolished in the European countries. Even in Russia liberalization came after the death of Tsar Nicholas I in 1855. This was described in the press with terms familiar to us today, such as "*glasnost*" and "*perestrojka*" (Bljum, 1999, p. 26).

However, the twentieth century did not become the era of media freedom that could have been expected at its beginning. On the contrary, media control achieved a level of totality as never before. There were two principal kinds of reasons for that: on the one hand, political-ideological reasons, on the other hand, technical and media-related reasons.

Political-Ideological Reasons

In the twentieth century, the re-institutionalization of media control was above all a matter of totalitarian states. Emerging from roots in the nineteenth-century, ideologies developed that aimed at changing human society according to certain principles. On the one hand, these principles were economic ones—they aimed to abolish exploitation. On the other hand, they had racial roots and aimed at the predominance of a particular race or ancestry. On the one side was communism; on the other was National Socialism and fascism. Despite the vast differences between these ideologies, and although they fought each other fiercely, they were similar in a certain way: They allowed for only one center of power, practiced a monopolization of information and formation of a collective will and aimed at total social representation.

Although the seizure of power by the Bolsheviks in Russia in 1917 did not immediately lead to the reintroduction of pre-censorship, the press decree of 9 November 1917 prohibited those newspapers that, in the opinion of the new leaders, openly appealed for resistance against the government (Roth, 1982, p. 36). The constitution passed on 10 July 1918 guaranteed in Article 14 the free distribution of all print products exclusively to "the working people … in order to guarantee their free expression of opinion" (Roisko, 2015, p. 35, translation by the author). To concentrate the Russian editorial and publishing activities, the state publishing house of the Russian Soviet Federative Socialist Republic, known under the acronym Gosizdat, was founded (20 May 1918) (Bljum, 1999, p. 50). The implementation of a department for military censorship followed (23 December 1918). Censorship activities were also transferred to the *Main Administration for Political Education* (Glavpolitprosvet), which was put under the control of the *People's Education Ministry* (Ermolaev, 1997). In 1922, Glavlit emerged as a new central control institution that from then on was responsible for licensing and pre-censorship (Bljum, 1999). Furthermore it published regulations for press organs and publishers. Already in December 1917, a press revolutionary tribunal had been installed that punished violations of the press decree.

The Bolshevization of the Russian press was promoted by all these measures. For three-quarters of a century, Glavlit exerted the control of the rulers over printing and press in the Soviet Union. It was the state institution used to suppress the free distribution of knowledge, information and opinions. Glavlit officially existed until August 1990 and was then superseded by an institution operating under the abbreviation GUOT (Roisko, 2015, pp. 330–336; Trepper, 1991). Apart from the omission of the pre-censorship of Soviet media content, which had been obligatory until then, its field of activity did not decisively differ from that of the precursor organization.

From the 1920s, the Nazis appeared with a similarly totalitarian claim to power. They wanted to re-educate the whole German nation according to their ideology. For this purpose, they created an extensive control and propaganda apparatus (Abel, 1968). After 30 January 1933, the NSDAP had the political power to subjugate the media to state and party supervision and control. Although there was no formal pre-censorship, several measures were taken that in principle fulfilled the same purpose. They stretched over several levels. To organize control, a separate ministry was established in Germany, this had never happened before. The Ministry of Popular Enlightenment and Propaganda (Ministerium für Volksaufklärung und Propaganda) was "responsible for all questions of mental influence on the nation, the propaganda for the state, culture and economy, the information of the national and international public about it and the administration of all institutions serving these purposes" (Wilke, 2007b, p. 116, translation by the author). With the help of the *Reichskulturkammer* (Reich Cultural Chamber), a professional organization of a compulsory nature consisting of seven individual chambers, all persons working in the different cultural sectors were registered. Among the seven chambers were the *Reichspressekammer* (Reich Press Chamber), the *Reichsrundfunkkammer* (Reich Broadcasting Chamber) and the *Reichsfilmkammer* (Reich Film Chamber) On the juridical level, laws and orders were issued, above all the *Schriftleitergesetz* (Editor's Law), which came into force on 1 January 1934 and secured control over admission to the journalist's profession. Those who were not registered in the professional list (or had been removed) were not allowed to work as journalists. The orders issued in 1935 by Max Amann, the president of the Reich Press Chamber and head of the NSDAP central publishing house (Franz Eher Nachf. = Franz Eher and Successors), were economically significant. The remaining bourgeois press was to be eliminated by the Nazi press. Newspapers were prohibited and there were further orders to close down newspaper companies, above all during World War II. This led to a clear divide in the German press landscape.

Furthermore there was a pervasive attempt to gain influence over media content. Above all the press instructions, issued at the Berlin press conferences and forwarded even to the provincial press via the Reich Propaganda Offices, served this purpose (Wilke, 2007b). Over the course of the years, the number of these press instructions increased considerably. A whole range of more than 20 instruction types were used for this: Not only bans were imposed, but also regulations, permissions and requests regarding the publication, the commentaries, and the layout of press articles. Furthermore the use of language (terminology) was prescribed, objections were expressed, and in rare cases compliments given. At first these press instructions were only given verbally and the journalists could reproduce them in their own words. Later on, when World War II had begun, the most important instructions were dictated and printed precisely (*Tagesparole*). Besides the political there were specific instructions in economical and cultural matters and even in sports.

In Italy, where the fascists came to power in 1924, press freedom was also abolished in the same year. At the same time, authorities and ministries to supervise the media were established (Cannistraro, 1975; Galasso, 1998; Murialdi, 1986). For the production of official press releases, the *Ufficio Stampa* was created. This institution was closed in 1934 and replaced by a central *Sottosegretario di Stato per la Stampa e la Propaganda*, following the German model. In 1935, it was converted into a ministry unifying all supervisory powers over newspapers, books, radio, cinema, theatres and tourism. In May 1937 it became the *Ministero della Cultura popolare* ("*MinCulpop*"). Under its guidance, significant propaganda campaigns were realized.

Technical and Media-Related Reasons

Besides the political reasons, technical and media-related reasons were decisive for the expansion and totality of media control in the twentieth century. For several centuries, until the end of the nineteenth century, the printed press had been the only medium of mass communication, although in diverse forms like books, newspapers, magazines, brochures etc. Their significance, however, had increased considerably, both through the number of titles and their circulation. While the print-run of the German periodical political press amounted to approximately 30,000 copies at the end of the seventeenth century, one century later it had increased tenfold to approximately 300,000 (Welke, 1977). At the end of the nineteenth century, it was around 10 million (Wilke, 2008, p. 274). The daily newspaper played a primary role for information on current affairs, and later also for the formation of opinion. Magazines diversified in numerous types and fulfilled both the function of special information and knowledge as the function of entertainment.

From the end of the nineteenth century, new mass communication media appeared. 1895 is considered as the year of birth of the film, the medium of moving images that found its social place in the cinema. At the beginning of the 1920s, the auditory medium of radio followed. In the mid-1930s, television technology was also developed to the point that regular broadcasts of programs could start. However, due to World War II, the development of this audio-visual medium was delayed.

These "new media" updated the problem of media control. Film censorship was established in many countries for almost as long as films have been shown (for France: Jamelot, 1937; for Great Britain: Mathews, 1999; for Germany: Binz, 2006). Initially and for obvious reasons the control followed the rules of theatre censorship. In Germany, it was first exerted by local police authorities. To achieve a nationally consistent system of film censorship, a central assessment became necessary here. After World War I, Article 118 of the Weimar Constitution guaranteed freedom of opinion and prohibited censorship, but for the medium of film (in the "fight against pulp and smut literature"), it allowed for differing regulations. That is why in 1920, a special cinema law was passed (Binz, pp. 121–254). According to this, the presentation of films was to be forbidden if a film was able to threaten "public security and order," "to hurt religious feelings" or if a "brutalizing and vulgarizing" effect was expected. Another reason for prohibition could furthermore be the "threatening of the German reputation or of Germany's relations to foreign countries" (Wilke, 2008, p. 321, translation by the author). The Nazis tightened these regulations in the Cinema Law of 1934. For content control, a *Reichsfilmdramaturg* (Reich film dramatic adviser) was appointed. For financial control; a film credit bank was established. The gradual nationalization of the film industry was done surreptitiously.

The control of broadcasting was even easier, at least in Germany. This was due to technology. The broadcasting sovereignty of the Reich gave state authorities the possibility to organize this medium according to their ideas. In Germany, this was mainly done by Hans Bredow. In the 1920s, he secured significant influence in the organization of broadcasting for the national postal service although, initially, private-sector investors were sought to finance radio programs (Lerg, 1980). To control the programs, surveillance committees were implemented (Bausch, 1956). The Reich sent one representative to these committees, and the respective state in which the broadcasting association was based sent two. A second committee, the program advisory board (*Programmbeirat*), had mere consulting functions.

In the twentieth century, broadcasting was organized under decisive state influence in many other countries for two reasons: the necessity to organize the technology centrally and the fear of negative consequences of its uncontrolled use. In the United States, on the contrary, things proceeded differently, as radio stations developed in "rank growth." Only belatedly, in 1927, was the Federal Radio Commission created as an independent regulatory authority for a broadcasting system which organized almost exclusively as a private-sector industry. In 1934, it turned into the Federal Communications Commission (Barnouw, 1970; Emery & Emery, 1984). Great Britain also followed its own way. There, the British Broadcasting Corporation (BBC) was established in 1927 on the basis of a Royal Charter which was designed to secure independence both from the state and from owners (Briggs, 1995).

The totalitarian states especially used the possibilities of the radio (and later of television) in the twentieth century to indoctrinate the population and control its information. The advantage was that this medium could be controlled centrally and effectively, and consolidated by staff policy. In the Soviet Union, this was the task of a state committee (Gostelradio). In Germany, a state broadcasting reform had already been effected in 1932 (Lerg, 1980, pp. 438–536). This made it easy for the Nazis to take over broadcasting after Hitler's rise to power. In 1933, this medium consequently also fell into the hands of the Ministry of Popular Enlightenment and Propaganda. Broadcasting was not only exposed to the Nazi influence with regard to structure and contents. By producing cheap "*Volksempfänger*" (people's receivers, a table-top radio), the regime aimed for the nationwide distribution of radio sets.

Since the nineteenth century, work had proceeded on the development of another medium that would not only enable the transmission of sounds, but also of images— namely television (Abramson, 1987). By the time the technical problems were solved and a practical use became apparent, the Nazis had already seized power in Germany. Ambitiously, the first presentation was arranged in Berlin on 22 March 1935 to demonstrate the precedence of Nazi Germany to the world (Winker, 1994). Television was also under political control, again inevitably, due to its technology and organization, but those involved did not yet recognize the use of this medium for propaganda purposes, and further development was interrupted by the beginning of World War II.

Media Control After 1945

In the twentieth century the two world wars were periods with pervasive and strong media control (as wars have ever been). Even democracies could not avoid imposing such means on the media. With the end of World War II, at least the fascist regimes disappeared (although not all authoritarian ones) and with them the media control they had established. At first the Allied powers in the defeated Germany exerted control over the reorganization of the media in the respective parts of the country they occupied. In the Federal Republic of Germany, established in 1949, the Basic Constitutional Law restored full freedom of opinion and the press. With the help of licensing, the Western Allies had already, before this, tried to establish a pluralist press system. With regard to broadcasting, however, it was not possible to allow for a larger number of stations, and therefore regulation was needed.

To withdraw broadcasting from state control, the Allies pressed for the implementation of a public service model following the example of the BBC. With this, a new form of media control was introduced, the so-called public or societal control. It was exerted by committees that consisted of representatives of relevant groups in society. They were responsible for supervising the radio and television stations and their programs. Representatives of the political system were also involved.

In other places, however, World War II led to a re-establishment or consolidation of media control. The Soviet Union as a victorious power was able to extend its political governance to Eastern and Central Europe, where political systems according to the Soviet model developed, including measures for media control. This was also true of the German Democratic Republic where a multi-faceted control and propaganda apparatus was created. The GDR Constitution included a guarantee of press freedom as a civic right, but this right was restricted by certain constitutional principles and a tight political penal law. The Constitution initially declared a prohibition of censorship which, however, was eliminated. In any case, pre-censorship was not needed as control could also be achieved with other measures: by licensing, by the allocation of printing paper, by personal instructions etc. A system very similar to that of the Third Reich served for instructing the press (Wilke, 2007b, pp. 256–309), and again the state orchestration of radio and television was even easier.

Media Control at the End of the Twentieth Century

A review of media control at the end of the twentieth century shows a rather mixed pattern, especially when the whole world is taken into consideration. On the one hand, media control has been reduced or its enforcement has become more difficult. On the other hand, there are still great discrepancies regarding media freedom on the international level—and new communication technologies have also produced new control measures.

In the former socialist countries of the Eastern bloc, the political changes at the beginning of the 1990s (widely) led to the abolition of the usual measures of media control. Indications of a certain liberalization can also be witnessed in some other countries, even in Africa. However, there is also opposite evidence, for example, in Russia, where authoritarian trends have again increased, or in Latin America, in the course of the sharp swing to the left that has occurred in Venezuela, Bolivia and Ecuador. The commercial broadcasting introduced in Western Europe since the 1980s, on the other hand, implies a lower degree of control than in the case of public broadcasting services. This bonus of freedom is bought with a greater dependence on advertisements, which can be seen as a type of economic constraint, but this reaches only as far as viewers and listeners tune in.

On an international basis, there is still a vast discrepancy between the normative level and actual conditions. There are hardly any countries that do not verbally avow freedom of opinion and of the press. This is embodied in the General Declaration of

Human Rights of the United Nations. A survey at the beginning of the 1990s showed that of the constitutions of 169 states, in 143 freedom of opinion, press and/or information was somehow guaranteed (Breunig, 1994); 43 constitutions guaranteed press freedom alone, 37 guaranteed freedom of opinion *and* freedom of the press, 42 guaranteed freedom of opinion *and* freedom of information. In 18 constitutions, like in the Basic Constitutional Law (*Grundgesetz*) of the Federal Republic of Germany, freedom of opinion, press freedom and freedom of information were included. Before the political changes in Eastern Europe, only half this number of countries had declared all three freedoms together in their respective constitution.

Naturally, the wording of the constitution is no sufficient guarantee for the actual validity of basic rights. The fact that freedom of opinion and press freedom is declared does not speak for the total absence of media control. In many cases, this guarantee only exists on paper, but is restricted in real life. This restriction can be achieved by more or less restrictive laws (documented by Breunig, 1994 and updated partially and regionally by Mendel, 2003, 2009), but also by practical obstacles and measures of the state or by threats emerging from society itself.

Surveys that have been conducted for decades give an empirical impression of the worldwide state of press freedom (or the state of media control). These surveys are all conducted by several non-governmental organizations: *Reporters Without Borders* (RWB; www.rsf.org), the *International Press Institute* (IPI, 2009; www.freemedia.at), the *American Freedom House* (www.freedomhouse.org), the *International Federation of Journalists* (IFJ; www.ifj.org), and the *Committee to Protect Journalists* (CPJ, 2009a, 2009b; www.cpj.org) (Becker, Vlad, & Nusser, 2007; Behmer, 2009; Holtz-Bacha, 2003; Löwstedt & Hahsler, 2003; Valentin, 2008). All these organizations are of Western origin and imply as their normative basis primarily the Western liberal idea of press freedom and of media independence from (state) control. Other non-legal forms of media control are more or less included. The "Western concept" is often criticized by representatives of other countries or regions of the world who plead for other concepts of the role of the media, such as in nation building, economic development, overcoming illiteracy and poverty (Becker et al., 2007, p. 6).

The value of the surveys of the state of press freedom depends on data collection. The organization *Reporters Without Borders* (originally the French *Reportères sans Frontières*) draws on reports by 120 correspondents, journalists, jurists and other experts. They have to assess the state of press freedom in their respective country, filling in a questionnaire with 42 questions (RWB, 2009c). On this basis, an index is compiled for each country. A score and a position is assigned to each country in the final ranking (RWB, 2009a). *Freedom House*, the US-American foundation, examines the level of press freedom on the basis of 23 questions divided into three broad categories: the legal environment (0–30 points), the political environment (0–40 points), and the economic environment (0–30 points). The countries are then classified in three categories: "Free" (0–30 points), "Partly Free" (31–60 points) and "Not Free" (61–100 points). Certainly this is a simplification and reduces the differences in its annual *Map of global press freedom* to only three colors (grades).

The International Federation of Journalists (IFJ) publishes regional *Press Freedom Reports* (i.e., for the Arab World and Iran, see IFJ, 2009a), documents additionally the cases of journalists who are killed in the pursuit of their profession (IFJ, 2009b), and compiles a "watch list" of countries in which press freedom is particularly endangered and must therefore be watched carefully (IFJ, 2009c). The CPJ publishes an updated list of the "Ten worst enemies of press freedom," a list that has been called a "hall of shame" (Löwstedt & Hahsler, 2003, p. 388).

Looking to previous surveys of *Freedom House* we see that the degree of media freedom in the world obviously has become greater (www.freedomhouse.org). In 1980, 42.2 % of the countries were classified as "Not Free"; in 2008, it was only 33 %. In 1980, 23.4 % of the countries were regarded as "Partly Free"; in 2008 it was 30 %. 34.4% were regarded as "Free" in 1980; in 2008 it was 37% of the countries. Thus the number of "Partly Free" countries rather than the number of really "Free" countries has increased.

To give a more differentiated picture we will use the data of *Reporters Without Borders*. The questionnaire for compiling its Press Freedom Index contains 40 criteria for each country. These are grouped into eight sections:

1. Physical attacks, imprisonment and direct threats
2. Indirect threats, pressures and access to information
3. Censorship and self-censorship
4. Public media (i.e., monopoly or not)
5. Economic, legal and administrative pressure
6. Internet and new media
7. Number of journalists murdered, detained, physically attacked or threatened, and government's role in this
8. Country media data (figures of national media, opposition news media, journalists)
   Any relevant points not included in this list may be added.

This measurement instrument of course has its problems. The replies to the questions have to be quantified with points between 0.5 and 5 (Becker et al., 2007, pp. 24–25). Some of the replies are themselves quantitative (i.e., number of journalists murdered), but others (and most) are nominal. In transforming the latter into figures, the decisions and discriminations made may be arbitrary. Physical attacks on journalists or media companies are assessed with 5 points, improper use of legal action or summonses against journalists with 0.5 point. So the RWB index necessarily confounds different criteria.

On the whole the survey is carried out from the journalist's perspective and implies their claim for press freedom with few if any limitations. The inquiry does not take into account that there may be legitimate limits of press freedom, i.e., in other human or social rights of equal significance (for example, the protection of one's reputation or private life). In 2009 the annual report of *Reporters Without Borders* mentioned that Slovakia fell in the ranking list from place 7 to 44 and explained this with a law that was passed in 2008, granting everybody a right of corrective response against the media.

Nevertheless the questionnaire of *Reporters Without Borders* is a helpful instrument, particularly for comparisons between different countries (measured in the same way) and over time. Based on the results an index for each country is constructed that ranges between 0 and 100 (if not more). The results have been published annually since 2002 in a rank order and in groups for the different regions in the world (RWB, 2009a, 2009b, 2009c). We present these results for 175 countries in 2009 (1 September 2008 – 31 August 2009) for the first time as a world map.

For this purpose the rank order has been classified into nine classes, namely between 0–4.99 and 100–115.50. This yields a more multi-colored picture of the state of press freedom (and the absence of media control) in the world nowadays, less simplified than the three-color map that *Freedom House* offers. Generally the lighter the color, the higher the degree of press freedom, and the darker the color, the lower the degree of press freedom (Fig. 14.1).

The countries in which the state of press freedom is assessed to be the best are those in central and northern Europe (Denmark, Finland, Ireland, Norway and Sweden are jointly ranked 1st; the Netherlands and Switzerland 7th; Belgium 11th; Austria 13th; Germany 18th and the United Kingdom 20th), and also Canada (19th) and the United States (2009 back in 20th place thanks to an "Obama effect"). From Asia and the Pacific only Japan (17th), New Zealand (13th) and Australia (16th) are among the top 20. Most of the other European countries in 2008/09 showed some (minor) deficits in press freedom (Greece 35th; Poland 37th; France 43rd; Spain 44th; Italy 49th) for different reasons. Italy got its score of 12.14 (which means rank position 49) due to the harassing influence of Silvio Berlusconi and due to attacks on journalists by the Mafia. Bulgaria brings up the rear of the countries of the European Union (68th). Still much more suffering under press control and violent acts is experienced by the media in Russia (153rd), which in 2009 ranked two places behind even Belarus (151st). (Ten out of twelve former Soviet republics are rated by *Freedom House* as "Non Free.")

In South America, Uruguay is classified as the best (29th in the world), followed by Chile (39th), Argentina (47th), and Paraguay (54th). Brazil takes 71st place. Some Caribbean states are ranked better (Jamaica 23rd; Trinidad and Tobago 28th). Stronger means of control and problems for the media do exist (as already mentioned) in Ecuador (84th) and Bolivia (95th), and even more in Venezuela (124th), Columbia (126th) and Mexico (137th). Elsewhere in Central America violent crimes are reported targeting the press, and Honduras fell in the rank order (from 99th to 128th) because of a coup d'état in June 2009. Cuba is the only totally "black" country in South America (170th).

Comparatively "dark" in general is the situation in Africa. A high degree of freedom is guaranteed in Ghana (27th), Mali (30th), South Africa (33rd) and Namibia (35th), whereas the Horn is the region with the most violations of press freedom: Sudan (148th), Somalia (164th) and Eritrea (175th) where no independent media are tolerated. Other countries that are assessed low in press freedom (or have

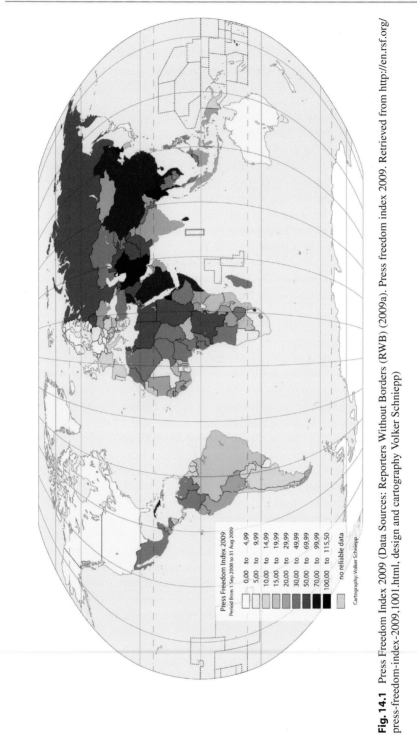

Fig. 14.1 Press Freedom Index 2009 (Data Sources: Reporters Without Borders (RWB) (2009a). Press freedom index 2009. Retrieved from http://en.rsf.org/press-freedom-index-2009,1001.html, design and cartography Volker Schniepp)

plummeted recently in the list) are Chad (132nd), Gabon (129th), Madagascar (134th), Nigeria (135th), Zimbabwe (136th), Democratic Republic of Congo (146th) and Rwanda (157th).

Countries with a lot of media control and hardly any press freedom are in general the Arab and Muslim states. While some—Kuwait (60th), Lebanon (61st) and the United Arab Emirates (86th)—are ranked in the middle of the list, there are others in which the conditions for the media are (much) worse: Morocco (127th), Egypt (143rd), Tunisia (147th), Libya (156th), Syria (165th), Yemen (167th). In 2009, Israel sank 47 places in the index from 46th to 93rd position. The reasons for that were military censorship and arrests and imprisonment of journalists. In the Muslim world, Iran (172th) stands in 2009 at the threshold of what RWB calls the "infernal trio": Turkmenistan (173rd), North Korea (174th), and Eritrea (175th).

Mostly dark red to black are the colors for Asia in the World Press Freedom Map. Japan is the only Asian country among the top 20. Even democratic countries like Taiwan (58th) and South Korea (69th) fell far in the rank order in 2009 because of arrests of journalists or bloggers and interference by governments. Even worse is the situation for free media in Malaysia (131st), Singapore (133rd), Afghanistan (149th), Pakistan (159th), Sri Lanka (162nd), Vietnam (166th) and China (168th). India, proud to be the most populous democracy in the world, is assessed somewhat better (105th), but it stands surprisingly behind Mongolia (91st). It should be said, however, that the differences in the scores of the *Press Freedom Index* between positions 1 and 108 range only from 0 to 30, whereas for the positions 109–175 the range of scores is from 30.5 to 115.5. The gaps between the countries that are dark colored on the map thus exceed these between the light colored countries.

From year to year the countries' positions in the rank order of the index may change because of relief or trouble in the area of press freedom that recently emerged. But the overall rank order is rather stable over time, particularly at the top and at the bottom. While several countries lifted their position from 2009 to 2013 (i.e. Lybia, Burkina Faso, Malawi, Ivory Coast etc.), the position of others more or less worsened (i.e. Hungary, Japan, Tanzania, Mali) (RWB, 2013).

Do the surveys of media control in the world lead to the same results? If they do, it would confirm the validity of the data. Becker et al. (2007) statistically compared the ratings of *Reporters Without Borders* and *Freedom House* and found that they were quite similar: "Despite the differences in measurement techniques and between the countries of the organizers, the two groups mostly agree on the classification of the media systems of the world" (p. 13). Valentin (2009) compared RWB, IPI and IFJ. Only 22 % of the cases that they documented were ranked in identical ways. Valentin found little difference in the assessment of the cases, particularly again between RWB and IPI. They differed most widely in documenting physical attacks on media personnel. IFJ documented the most cases of labor legislation that affected freedom in journalism. This indicates different professional interests of the organizations.

Besides that, the NGOs differ in their strategies. *Reporters Without Borders* is the most active agent, publicizing news, protesting publicly, organizing campaigns etc. The strategy is to accuse and blame governments and countries in front of the world audience, hoping that this might improve the situation of the journalists and the media under threat. The IFJ prefers to address the journalistic profession and people responsible for media control. Additionally events like days of action are organized to mobilize people. IPI's target group is primarily the top media personnel, not the general public.

New Challenges for Media Control

Problems for the relationship between media freedom and media control at the turn of the millennium result once more from technological development. In the course of only a few years, a new communication technology has conquered the world: the Internet. This technology has changed, even revolutionized, social communication in many respects. One important aspect is that this kind of communication, which runs across computer networks without borders, defies individual state control and even controllability as no other technology before. The characteristics of the technology make it difficult to react to its possibilities of misuse, which range from (child) pornography to terrorism. Above all those states that already control the conventional media feel especially challenged. They try to block websites with contents they want to withhold from their citizens. *Reporters Without Borders* now collects material for a report on "Internet Enemies" (RWB, 2009e) and *Freedom House* has published individual national surveys on this subject (Freedom House, 2009d).

According to *Freedom House* (2009a, 2009b, 2009c), low "web control" and high "net freedom" exist, for example, in Estonia, Great Britain, South Africa and Brazil, four countries representing the status "Free." Kenya, India, Turkey and Egypt are assessed as "Partly Free"; China, Tunisia and Cuba as "Not Free." In recent times, after the disputed presidential election in 2009, one could observe in Iran the attempts of a government to maintain power via the control of the media, on the one hand, and the possibilities of the new technologies to undermine these attempts on the other. In its report "Internet Enemies" (2009), Reporters Without Borders accuses as such Saudi Arabia, China, Cuba, Egypt, Iran, North Korea, Syria, Tunisia, Turkmenistan, Uzbekistan and Vietnam. They

> have all transformed the network into an intranet, preventing Internet users from obtaining news seen as 'undesirable'. All of these countries mark themselves out not just for their capacity to censor news and information online but also for their almost systematic repression of Internet users. (RWB, 2009f, p. 2)

Australia and South Korea, two countries that ordinarily protect press freedom, have been put "under surveillance" because of certain draft laws or "some disproportionate measures to regulate the Net" (ibid.). With regard to the Internet, even liberal constitutional democracies face problems. In general, they allow for media freedom, but want their borders drawn by general laws to be protected also on the Internet. However, it is disputed how this could be done. At the time of writing, in 2009, there is a debate in Germany: a law passed by the federal government provides for the blocking of websites with child pornography. This encounters the Internet society's resistance, as they consider it as a general attack on the freedom of the Internet. Is this, as it is believed, only a solution for individual criminal cases (for absolutely justified reasons) or is it about an infrastructure for an extensive control of the web? Much more than in earlier times, the people in society are provided with means to counteract restrictions being imposed on the use of these means. Nevertheless, the question of media control thus remains on the agenda.

References

Abel, K.-D. (1968). *Presselenkung im NS-Staat* [Press control in the National Socialist state]. Berlin: Colloquium.

Abramson, A. (1987). *The history of television, 1880–1941*. Jefferson, NC: McFarland.

Barnouw, E. (1970). *A history of broadcasting in the United States*. 3 vols. New York: Oxford University Press. (Original work published 1966)

Bausch, H. (1956). *Der Rundfunk im politischen Kräftespiel der Weimarer Republik* [Broadcasting during the political power struggle in the Weimar Republic]. Tübingen, Germany: C.B. Mohr.

Becker, L. B., Vlad, T., & Nusser, N. (2007). An evaluation of press freedom indicators. *International Communication Gazette, 69*, 5–28.

Behmer, M. (2009). Measuring media freedom: Approaches of international comparison. In A. Czepek, M. Hellwig, & E. Novak (Eds.), *Press freedom and pluralism in Europe: Concepts & conditions* (pp. 23–36). Bristol, UK: Intellect Books.

Binz, G. (2006). *Filmzensur in der deutschen Demokratie* [Movie censorship in the German democracy]. Trier, Germany: Kliomedia.

Bljum, A. V. (1999). *Zensur in der UdSSR: Hinter den Kulissen des "Wahrheitsministeriums": 1917–1929* [Censorship in the USSR: Behind the scenes of the "Ministry of Truth": 1917–1929]. Bochum, Germany: Projekt-Verlag.

Breunig, C. (1994). *Kommunikationsfreiheiten:. Ein internationaler Vergleich* [Freedom of Communication: An international comparison]. Konstanz, Germany: Universitätsverlag Konstanz.

Briggs, A. (1995). *The history of broadcasting in the United Kingdom. 4 vols*. Oxford, UK: Oxford University Press. (Original work published 1961–1979, Rev. ed. 1995)

Cannistraro, P. V. (1975). *La fabbrica del consenso: Fascismo e mass media* [Manufacturing consent: Fascism and mass media]. Rome: Laterza.

Committee for the Protection of Journalists (CPJ). (2009a). *Attacks on the Press in 2008*. Retrieved September 10, 2012, from http://www.cpj.org/attacks/

Committee for the Protection of Journalists (CPJ). (2009b). *2009 prison census: 136 journalists jailed worldwide*. Retrieved September 10, 2012, from http://www.cpj.org/imprisoned/2009.php

Emery, E., & Emery, M. (1984). *The press and America: An interpretive history of the mass media* (5th ed.). Englewood Cliffs, NJ: Prentice Hall.

Ermolaev, H. (1997). *Censorship in Soviet literature 1917–1991*. Lanham, MD: Rowan & Littlefield.

Freedom House. (2009a). *Map of press freedom*. Retrieved July 31, 2009 from http://www. freedomhouse.org/template.cfm?page=251&year=2008Freedom

Freedom House. (2009b). *Country reports*. Retrieved July 31, 2013 from http://www.freedom-house.org/template.cfm?page=107&year=2008

Freedom House. (2009c). *Press freedom ranking by region*. Retrieved July 31, 2009 from http://www.freedomhouse.org/uploads/fop08/FOTP2008_RegionRankings.pdf

Freedom House. (2009d). *Freedom of the net: A global assessment of internet and digital media*. Retrieved October 20, 2009, from http://www.freedomhouse.org/template. cfm?page=383&report=79

Galasso, G. (1998). Die Umgestaltung der Institutionen: Das faschistische Regime in der Machtergreifungsphase [The process of institutional reorganization: The fascist regime during the phase of seizure of power]. In J. Petersen & W. Schieder (Eds.), *Faschismus und Gesellschaft in Italien: Staat, Wirtschaft, Kultur* (pp. 19–47). Cologne, Germany: Böhlau.

Holtz-Bacha, C. (2003). Wie die Freiheit messen? Wege und Probleme der empirischen Bewertung von Pressefreiheit [How to measure freedom? Ways and problems of empirical evaluation of press freedom]. In W. R. Langenbucher (Ed.), *Die Kommunikationsfreiheit der Gesellschaft: Die demokratischen Funktionen eines Grundrechts (Publizistik Sonderheft 4/2003)* (pp. 403–412). Konstanz, Germany: Universitätsverlag Konstanz.

International Federation of Journalists (IFJ). (2009a). *Perilous assignments: Journalists and media staff killed in 2008*. Retrieved September 10, 2012, from http://www.ifj.org/assets/docs/051/004/ eb26233-0f25804.pdf

International Federation of Journalists (IFJ). (2009b). *Breaking the chains: The Arab world and Iran press freedom report 2009*. Retrieved September 10, 2012, from http://www.ifj.org/assets/ docs/051/237/f086033-fa2f3ed.pdf

International Federation of Journalists (IFJ). (2009c). *Watch list*. Retrieved September 10, 2012, from http://www.freemedia.at/our-activities/watch-list/

International Press Institute (IFI). (2009). *World press freedom review*. Retrieved September 10, 2012, from http://www.freemedia.at/publications/world-press-freedom-review/singleview/325 fede2a7/688/

Jamelot, Y. (1937). *La Censure des Spectacles: Théatre, Cinèma* [Censorship of performances: theater, cinema]. Paris: Editions Jel.

Lerg, W. B. (1980). *Rundfunkpolitik in der Weimarer Republik* [Broadcasting policy in the Weimar Republic]. Munich, Germany: dtv.

Löwstedt, A., & Hahsler, K. (2003). Global guardians of the freedom of expression. In W. R. Langenbucher (Ed.), *Die Kommunikationsfreiheit der Gesellschaft: Die demokratischen Funktionen eines Grundrechts (Publizistik Sonderheft 4/2003)* (pp. 385–402). Konstanz, Germany: Universitätsverlag Konstanz.

Mathews, T. D. (1999). *Censored: What they didn't allow you to see, and why—Story of film censorship in Britain*. London: Chatto and Windus.

Mendel, T. (2003). *Freedom of information: A comparative legal survey*. Retrieved September 10, 2012, from http://portal.unesco.org/ci/en/file_download.php/fa422efc11c9f9b15f9374a5eac-31c7efreedom_info_laws.pdf

Mendel, T. (2009). *The right to information in Latin America: A comparative legal survey*. Retrieved September 10, 2012, from http://unesdoc.unesco.org/images/0018/001832/183273e.pdf

Murialdi, P. (1986). *La stampa del regime fascista* [The press of the Fascist regime]. Rome: Universale Laterza.

Reporters Without Borders (RWB). (2009a). *Press freedom index 2009*. Retrieved September 10, 2012, from http://www.rsf.org/en-classement1003-2009.html

Reporters Without Borders (RWB). (2009b). Retrieved September 10, 2012, from http://www.rsf. org/spip.php?page=impression&id_rubrique=1003

Reporters Without Borders (RWB). (2009c). *Worldwide press freedom index 2009: How the index was compiled*. Retrieved September 10, 2012, from http://www.rsf.org/IMG/pdf/ note_methodo_en.pdf

Reporters Without Borders (RWB). (2009d). *Questionnaire for compiling the 2009 Press freedom index*. Retrieved September 10, 2012, from http://www.rsf.org/IMG/pdf/quest_en.pdf

Reporters Without Borders (RWB). (2009e). Retrieved September 10, 2012, from http://www.rsf. org/spip.php?page=impression&id_article=26126

Reporters Without Borders (RWB). (2009f). *Internet enemies*. Retrieved September 10, 2012, from http://www.rsf.org

Reporters Without Borders (RWB). (2013). *Press freedom index 2013*. Retrieved July 2, 2013 from http://en.rsf.org/press-freedom-index-2013,1054.html

Roisko, P. (2015). *Gralshüter eines untergehenden Systems. Zensur der Massenmedien in der UdSSR 1981–199* [Grail Keeper of a Declining System. Censorship of Mass Media in the USSR 1981–1991. Cologne, Germany: Böhlau (meanwhile published), Johannes Gutenberg-University of Mainz.

Roth, P. (1982). *Die kommandierte öffentliche Meinung: Sowjetische Medienpolitik* [Public opinion under command: Soviet media policy]. Stuttgart, Germany: Seewald.

Siebert, F. S. (1965). *Freedom of the press in England 1467–1776: The rise and decline of government control*. Urbana, IL: University of Illinois Press.

Trepper, H. (1991). Kulturbetrieb [Cultural Industry]. In Forschungsstelle Osteuropa (Ed.), *Kultur im Umbruch*. Bremen, Germany: Forschungsstelle Osteuropa.

Valentin, S. (2008). *Konzepte von Medienfreiheit und ihre Umsetzung durch NGOs: Reporters sans frontières, International Press Institute und International Federation of Journalists im Vergleich* [Concepts of media freedom and their implementation by NGOs: Reporters sans frontières, International Press Institute and International Federation of Journalists in comparison]. Marburg, Germany: Tectum.

Valentin, S. (2009). Konzepte von Medienfreiheit und ihre Umsetzung durch NGOs [Concepts of media freedom and their application]. Marburg, Germany: Tectum

Welke, M. (1977). Zeitung und Öffentlichkeit im 18. Jahrhundert: Betrachtungen zur Reichweite und Funktion der periodischen deutschen Tagespublizistik [Newspaper and the public in the 18th Century: Reflections on the scope and function of periodic German daily journalism]. *Presse und Geschichte: Beiträge zur historischen Kommunikationsforschung* (pp. 71–99). Munich, Germany: Verlag Dokumentation.

Wilke, J. (1983). Leitideen in der Begründung der Pressefreiheit [Guiding principles in the justification of freedom of the press]. *Publizistik, 28*, 512–524.

Wilke, J. (2007a). Pressezensur im Alten Reich [Press censorship in the Old Reich]. In W. Haefs & Y.-G. Mix (Eds.), *Zensur im Jahrhundert der Aufklärung: Geschichte – Theorie – Praxis* (pp. 27–44). Göttingen, Germany: Wallstein.

Wilke, J. (2007b). *Presseanweisungen im zwanzigsten Jahrhundert: Erster Weltkrieg – Drittes Reich – DDR* [Press instructions in the twentieth century: World War I—Third Reich—GDR]. Cologne, Germany: Böhlau.

Wilke, J. (2008). *Grundzüge der Medien- und Kommunikationsgeschichte* [Essential features of media and communication history] (2nd ed.). Cologne, Germany: Böhlau. (First ed. published 2000)

Wilke, J. (2013). Censorship and Freedom of the Press http://ieg-ego.eu/en/threads/european-media/censorship-and-freedom-of-the-press

Winker, K. (1994). *Fernsehen unterm Hakenkreuz: Organisation, Programm, Personal* [Television under the Swastika: Organization, program, staff]. Cologne, Germany: Böhlau.

Tau(gh)t Subjects: Geographies of Residential Schooling, Colonial Power, and the Failures of Resistance Theory

15

Sarah de Leeuw

Introduction

John Henry knows a thing or two about grappling with powerful forces and then putting them to rest. He is a hunter and a fisherman and a sometime logger living in northern British Columbia (B.C.), Canada. Once he told me about a wrongly aimed gunshot and one of the most powerful animals he knows, a moose. He told me how that wrongly aimed shot meant chasing down an angry thousand-pound animal for over five hours along remote logging roads. Eventually the powerful animal died, and John was able to get on with the job of skinning and gutting it. Of readying it for the back of his truck. Of preparing meat for its journey home to freezers and stews, skin for drums made of stretched hide, and bones for dogs to chew on.

John told me about that wrongly aimed gunshot the same evening he showed me the scars that run up and down the insides of both of his arms. It was during the Annual General Assembly of Carrier Sekani Family Services, up on the Grassy Plains Indian Reserve on the south side of François Lake in northern B.C.[1]

---

[1] The term Indian is problematic, highly dated, and offensive because it legitimizes and formalizes the conflation and homogenization of peoples on the basis of exclusion from the category European and/or White. Nomenclature, however, continues to confound many who write about Indigenous issues in Canada. This confusion is evidence of colonialism's ongoing influence in twenty-first-century geographies. Since 1982, the terms Aboriginal and Indigenous are used in Section 35 of Canada's Constitution Act to inclusively denote First Nations, Inuit, and Métis people and to reflect the ongoing research and contemporary shifts in colonial languages toward more accurate and respectful descriptors of the territory's First Peoples. The term First Nations replaces the term. I use the term "Indian" in this chapter to denote historical conceptualizing of First Nations; wherever I am able, I use the specific names of peoples and Nations to whom I am referring (e.g., the Sekani peoples).

S. de Leeuw (✉)
Northern Medical Program, University of Northern British Columbia, Faculty of Medicine, UBC, 3333 University Way, Prince George, BC V2N 4Z9, Canada
e-mail: deleeuws@unbc.ca

© Springer Netherlands 2015
P. Meusburger et al. (eds.), *Geographies of Knowledge and Power*,
Knowledge and Space 7, DOI 10.1007/978-94-017-9960-7_15

When John showed me his scars, he also told me about the other most powerful force at which he has ever wanted to take aim: Indian schools. These schools were built at the behest of colonial governments and churches intent on both settling Indian lands and saving Indian souls: In line with what were unquestionably benevolent intents of the day, they were also focused on educating Aboriginal children to prepare them for life in modern times.[2] John's experience with Indian schools began when he was young. At about the age of six, he began attending Lejac Indian Residential School. Then, during the 1960s and as a teenager, he attended Prince George College, a Catholic school attended primarily by First Nations children from across northern B.C. John got the scars that run up and down his forearms for "talking Indian" and for "talking back" to nuns even after they had warned him not to speak the devil's tongue. He got the scars from being hit with straps of leather sliced from long strips of the turbine belt that turned the engine on the wood-planer behind the school.

When John Henry tells me about his schooling and his scars, he is reflective about the very act of talking about them or about his schooling experience. Indian schooling is not a topic he enjoys speaking about, yet it is one he cannot seem to stop thinking about. It is also, as he acknowledges, a complex subject; every person who spent time in one of the province's 18 residential schools, which operated between 1861 and 1984, had a different experience of their schooling. Thus, although the schooling project is now roundly understood as a "national crime" for which a national apology was offered, testimonial literature about the experience underscores that there has never been a single or uniform experience of the residential schooling project (see, e.g., Nuu-chah-nulth Tribal Council, 1996; Secwepemc Cultural Education Society, 2000). The lesson that John takes from all of this, however, remains relatively simple. He has spent most of his life, both inside and then beyond the Indian schools, being taught a lesson. The lesson is that colonial power "just won't quit." From John's perspective, the trouble with colonial power is that no matter how you shoot at it, it seems to transform and rise up. By talking about the schooling, memories of it are unearthed and experiences are relived. Both the experiences and the memories are also opened up to (re)interpretations. At the same time, John is loath to stop speaking about colonial education, particularly because it has implications that reach into the present day and that continue to touch him, his family, and his community. Thinking about talking or not talking, thinking about trying to make sense of his experiences or simply trying to forget them, all involve a seemingly endless battle that John Henry would, on many days, simply like to sidestep.

---

[2] Although residential schooling in Canada is now clearly considered a deeply problematic and often violent practice, and although a national apology to all Indigenous peoples in the country has been offered (Waterstone & de Leeuw, 2010), it cannot fairly be asserted that *all* people at the helm of the residential schooling project were monstrous or behaving with malevolent intent. Nor can it fairly be asserted that *all* Indigenous peoples uniformly experienced residential schooling as a bad thing. Indeed, some former residential school students have attested to enjoying and benefiting from their time in the schools. For further discussion, see Edwards (2009) and Raibmon (1996).

While knowing and acknowledging that John Henry's experiences are not universal or representative of all former residential school students (see, e.g., Edwards, 2009/10), I, too, am interested in considering, albeit from a very different perspective, the conundrum of how to speak about and how to conceptualize the complexities of the residential schooling project. In this chapter, John's story forms the cornerstone of how I approach questions about the geographies of residential schooling. With his story in mind, my aim is to contribute to broader discussions in geography about power, knowledge and education, Indigenous peoples, and theories of resistance—theories that I argue have been overused in considerations about colonial relations and geographies of power. More specifically, I am interested in thinking about Aboriginal students' responses, both at the time and then after the fact, to the residential schooling project.

I approach this topic from two perspectives. On the one hand, given that this chapter attempts to broaden what I argue is a prevalent tendency in geography to understand relationships involving power imbalances through theories of resistance, the chapter is in great part theoretical in nature. Taking John's lead, namely, that there are no clear-cut or straightforward ways of thinking about how to make sense of his experience, I suggest that in order to fully conceptualize residential schooling, theories of resistance will likely need to be re-evaluated. On the other hand, and again with John's story in mind, this essay attempts to respect the complexity of residential schooling by approaching it from a humanistic tradition, one that looks to story and narrative as a means of conveying difficult and emotionally powerful events. In this way, the chapter is also about colonial education and a story about sidestepping colonial power. It is about tau(gh)t subjects and, in its structure and approach, attempts to embody the tautness of relationships that continue to permeate the geographies of British Columbia. It is about colonial knowledge-systems embedded in the teachable subjects (math, history, social studies) taught to Indigenous peoples, about the spaces where these subjects were and are taught, about the people (or subjects) who teach and taught the lessons, and the students (or subjects) who learn and learned the lessons. This chapter is also about tense or *taut* relations. After all, none of the lessons taught were uniformly delivered or unself-consciously accepted at face value by Aboriginal children. I am fundamentally interested in untangling and theorizing what John Henry is alluding to: How might we understand efforts to lay a powerful beast (colonial education) to rest and simultaneously understand that lives, particularly those of Indigenous people in Canada, continue to be vibrantly lived while continuously (re)forming around the scars that colonialism inflicted?

Resistance

One of the first scholarly texts written about residential schooling in British Columbia met with significant criticism just prior to publication and release. *Resistance and Renewal: Surviving the Indian Residential School* (Haig-Brown, 1988) was, according to the author, criticized as unwanted interference by a non-Indigenous academic into the lives and histories of First Nations. In an effort to ameliorate interpretations of the text as a simplistic recounting of the abuses and violence inflicted by residential schooling, and with the desire to understand and document students' diverse experiences in residential school, Haig-Brown's analysis of interviews with former students focused on resistance:

> Throughout [the] onslaughts described in the stories [in the book], the people resisted and found strength with that resistance … Some of the determination First Nations people now exhibit found its roots in the resistance to the invasive culture of the schools designed to annihilate First Nations cultures. (p. 11)

Resistance, in other words, was used to theorize the residential schooling project in line with relationships in other colonial contact zones: Places where "disparate cultures meet, clash, and grapple with each other in highly asymmetrical relations of domination and subordination" (Pratt, 1991, p. 4), which are always marked by uneven, splintered, messy, shifting, and varied interactions between subjects who play multiple roles with often competing and unclear agendas. Is, however, "resistance" an accurate rendition of what Aboriginal students did in residential schools? Can contemporary dealings with British Columbia's colonial present (Gregory, 2004) by Aboriginal peoples be usefully theorized through the language of resistance?

For some time now, geographers concerned with systems of power have relied on the concept of resistance when accounting for the ways in which dispossessed subjects subvert, undermine, transform, or, perhaps more plainly but still very strategically, survive in the face of dominant hegemonic forces engaged in attempts to actively or passively control or suppress them (Rose, 2002; Sparke, 2008). Others have undertaken similar critical analyses of power as monolithic or unopposed (see, e.g., Scott, 1990), and theories about resistance have deeply entrenched and normalized the idea that powerful or dominant forces never go unchallenged. There are always countervailing practices and strategies levied by those whom dominant forces attempt to subordinate. Resistance theory in geography gained prominence during the so-called cultural turn, a transformation of the discipline from "a period when few geographers used or thought about issues of resistance—instead analyzing and criticizing structural relationships of power—to one where everyone seem[ed] to be talking about resistance and domination" (Pile & Keith, 1997, p. xi). Cited as an important lens for use by radical and critical geographers, resistance is expressly about opposing power: It is "people fighting back in defense of freedom, democracy and humanity" (Pile, 1997, p. 1). Indeed, geographies of resistance are conceptualized as so universal that "resistance can be found in everything" and "potentially, the list of acts of resistance is endless" (p. 14).

Even to its proponents, finding resistance everywhere and in everything risks rendering the concept problematically broad and potentially conflating it with something as mundane as "washing your car on a Sunday" (Thrift, 1997, p. 124). In efforts to ensure that the concept does not slip into such banality, and at the same time recognizing that everyday and ordinary acts can be and often are acts of resistance, proponents of the concept further clarify that neither the concept nor the practice can be properly understood "simply [as] the underside of the map of domination" (Pile, 1997, p. 23). Instead, resistance should be conceptualized as outside the given or expected: "If there is a beaten track, a track laid down through the spatial technologies of power configurations, then resistance will stray from the track, find new ways, elaborate new spatialities, new futures" (p. 30). Precisely because of its prevalence, and thus the potential of it becoming invisible through normalization, proponents of the concept argue that there must be vigilance in uncovering it and making it visible, in recognizing its dynamic and multiscalar nature and reinforcing that, just as there is never one geography of power, there is no singular geography of resistance: "Resistance is resistance to both fixity and to fluidity" (p. 30). Despite careful unpacking of the concept, however, practices of resistance remain slippery and difficult to document or explain. The concept is equally elusive.

The risk of broad conceptualizations is that a term loses any constraints: It comes to mean everything and thus, potentially, nothing. The dilemma becomes how to document the concept in practice and action. Some resistance theorists have attended to these challenges by proposing that consciousness is a key element of any act of resistance: Resistance, it seems, implies some type of consciousness and self-awareness on the part of the person(s) enacting a response to power, no matter how passive the consciousness may be. Attributing consciousness to an act of resistance ultimately signals intent, which—as some resistance theorists argue (see, e.g., Rose, 2002)—may be too narrow an understanding of both the concept and the practice. The difficulty remains, then, that

> resistance studies tend to conceptualize agents as responding to a dominant system. Thus, in an endeavour to recognize, describe, and theorize various forms of response, resistance theory necessarily establishes the system (that which is responded to) as a preestablished force. (Rose, 2002, p. 384)

In other words, theorizing events/knowledges/subjects as "off the beaten track" does not allow one to imagine what is possible if the track were not there at all. The production of new spaces espoused by an "off the beaten track" theory of resistance geography always risks validating, if not reproducing, perhaps in somewhat retooled forms, existing spaces that never entirely allow us to imagine the still unimaginable spaces of something (and somewhere) utterly new.

How to theorize new spaces and politics that are entirely decoupled from preexisting forces remains a pressing—and mostly unanswered—question when attempting to conceptualize, and potentially mobilize, counter-hegemonic or anticolonial possibilities. Posed succinctly, the conundrum becomes, "How do we pass from the politics of 'resistance' [which equates to] 'protestation,' which parasitizes upon what it negates, to a politics which opens up a new space outside [a] hegemonic position and its negation?" (Žižek, 2006, p. 382). Indeed, the significant dilemma for studies of Indigenous geographies is the following: If resistance can be found everywhere and in everything, and if there is always, in its practice of negation, an element of intention (even when, paradoxically, it seems unintended), does it not become impossible for actors—particularly those without power or with lessened access to power—to ever achieve a semblance of undetermined or independent being in the world? Those with less power remain perpetual resisters, distanced from and outside that which continually (re)produces them. In such an equation, John Henry will never be able to deal in his own way with the scars and memories that define him, nor will he ever be at liberty to make sense of his experience of residential schooling, an experience that included "talking Indian" and "talking back," in any way other than as an experience that focused on resistance and dominance. This equation, as others have suggested, results in some potentially problematic politics.

Refuting Resistance

Captured poignantly by Sparke (2008), a kind of "romance" permeates resistance. This romanticization of the resister and the act of resisting risks obfuscating the pain and violence experienced by resisters who, as noted above, seem perpetually positioned as having no undetermined or independent possibility other than that of resister (see also Thrift, 1997). Because resistance turns on ideas of combativeness, on acts of opposition, both passive and active, and on practices of enduring tremendous and powerful outside forces, the work of resistance extracts a great deal from those undertaking it. Resistance theorists offer little thought about either the toll that resistance work takes on resisters or what other work might be done when resistance work is finished and/or just put aside at the end of a long day. Indeed, if resistance is understood as potentially everywhere and in everything, it seems unlikely that resistance work will ever be finished. Additionally, as noted by others (see Ortner, 1995), resistance demands a constant refusal of that which is being resisted, a refusal that prevents any potential for the equalizing of actors. With respect in particular to Indigenous and non-Indigenous relationships in neocolonial and contemporary colonial landscapes, resistance studies and its consequent refusal of the colonial are always at risk of (re)positioning Indigenous peoples as monochrome heroes dedicated to struggles against colonial power:

> Resistance studies are thin because they are ethnographically thin: Thin on the internal politics of dominated groups, thin on the cultural richness of those groups, thin on the subjectivity—the intentions, desires, fears, projects—of the actors involved in these dramas. (Ortner, 1995, p. 190)

For those who are skeptical of the relatively uncritical deployment of the concept of resistance, Ortner's observation confirms the idea has not yet been properly problematized and does not account for the complexity of lives lived outside of or with minimal reference to power.

Efforts have been made to move away from the potentially exhausting, and inherently aggressive and combative, connotations embedded in the lexicon of "resistance."

Katz (2004), a feminist geographer attuned to the eminently real, fleshy, and messy implications and exhaustions associated with social work, and social (re)production in a globalized world, offers concepts such as "resilience," "reworking," and even "engagement" as antidotes to languages and theories of resistance. These terms leave room to understand the ways in which people more creatively and less combatively navigate—and ultimately survive—the multiple systems of power within which we all exist.

Because resistance calls upon and solidifies dualisms, it is difficult to apply the concept to those times and spaces in which many actors grapple asymmetrically and chaotically with one another. The idea of resistance imposes logic and structure upon human efforts and risks erasing the nuance and pluralities of multiple actors who make up the complex interactions that produce the times and spaces in which we live. In his "talking Indian and talking back," John Henry may not have been resisting. He may have been, more simply, being himself. He may have been taking a few "wrongly aimed shots" at a system he was trying to make sense of. If resistance is the only way to conceptualize or describe the actions for which John Henry was severely punished during residential schooling, his abilities to operate in a self-determining way, entirely outside the colonial geographies of power that contain him, are negated. He is, paradoxically through a language of "resistance," transformed from an at least somewhat autonomous subject—with the ability to fully refuse through a complete lack of recognizing a system that subjugates him—to a subject who is always tethered to residential schooling.

Lefebvre (1991) argues that rebelling against, opposing, or decrying systems of power ends up "playing into the hands of the bourgeoisie" (p. 233). For Lefebvre, resistance is not an effective oppositional act or even a dialectical effort that reforms the systems it negates. Instead, resistance according to Lefebvre is always an act that fortifies that which is being decried, opposed, or resisted:

> This is why the rebel and the anarchic protester who decries all of history and all the works of past centuries because he sees in them only the skills and the threat of domination is making a mistake.... There is a kind of revolt, a kind of criticism of life, that implies and results in an acceptance of this life as *the only one possible*. As a direct consequence this attitude precludes any understanding of *what is humanly possible*. (pp. 232–233)

At the heart of Lefebvre's argument is the idea that opposition and resistance require distance. Distance, he suggests, results in a feeling of alienation. A feeling of alienation risks a refusal of the dialectic, an inability (or lack of desire) to imagine what else is possible, and, ultimately, an acquiescence (albeit frustrated and alienated) to what is. In addition to reinforcing power, solidifying dualities, and obfuscating human suffering, resistance theory also exaggerates the distance between people. This aspect is of particular import to (post)colonial studies and understandings of Indigenous geographies, including those lived in northern B.C. by John Henry. Without distance, a central tenet of resistance, categories of otherness—categories upon which colonialism rests—might begin to crumble. Ultimately, then, if resistance propagates concepts of distance, and if distance (re)produces difference and otherness, resistance studies may not be productive means of understanding, reconciling, or undoing neocolonial relationships, including the relationships that John Henry had and continues to have to residential schooling and the scars that mark him.

Indigenous Geographies, Residential Schools, and the Failures of Resistance Theory

Thinking about the colonial present, and in particular Indigenous geographies in Canada and other neocolonial nations, increasingly means understanding "colonial discourses without denying agency to colonized peoples or overlooking practices of resistance" (Nash, 2002, p. 221). Indeed, resistance theory is deployed almost axiomatically to any mention of hegemony, power, or the subjugation, dispossession, or deterritorialization of othered subjects, including Indigenous peoples in B.C. (see Harris, 2002, 2004). Discussions about the multiple spaces and times in which Indigenous peoples actively, aggressively, and intentionally fought (or are fighting) back against colonial powers are growing, and much of this work deploys theories of resistance (see, e.g., Harris, 2002; Pualani-Louis, 2007; Radcliffe, 2000; Watson & Huntington, 2008). The concept of resistance, however, does not receive much critical attention. It is rarely examined critically, implicated in colonial assumptions about "the Indigenous other," or theorized as possibly (re)producing homogenous understandings of both Indigenous and colonial geographies.

In British Columbia, colonial landscapes were settled and worked differently by racialized peoples whose access to power was mediated in part by where they were or with whom they were interacting (Kobayashi & de Leeuw, 2010; Mawani, 2009). It would also be an error to understand Indigenous peoples in what is now B.C. in any homogenizing way; the lands were—and are—home to 198 First Nations, each with unique sociocultural protocols and many of which occupy overlapping and competing territories (Sterritt, Marsden, Galois, Grant, & Overstall, 1998). In many cases, even the categories of Indigenous and non-Indigenous peoples in B.C. must be understood as a fiction, a conceit of governments (and sometimes churches) intent on stabilizing identities for the purpose of legal and fiduciary management (Lawrence, 2003, 2004). At the very literal and embodied levels, people crisscrossed boundaries by migrating, moving, falling in love, and having children. As in other colonial spaces, colonialism in B.C. was very much about diminutive and intimate relations, or what historian Stoler (2006) calls "tense and tender ties" (p. 6). These ties, including ties that the federal and provincial governments tried to monitor and restrict, included intermarriage (and not just between First Nations people and settlers from Europe), the production of alternative (e.g., non-heterosexual and extended) families, and intimate and caring connections between people in domestic or diminutive spaces such as boarding schools, work sites, and foster homes (Bednasek & Godlewska, 2009; de Leeuw, 2007).

Given the need to account for the utterly unstable, pluralistic, diminutive, intimate, and tender nature of colonialism, several conceptual problems arise when using resistance theory to understand the relationships between Indigenous peoples and the various non-Indigenous peoples who settled British Columbia. Distance between subjects is shrinking all the time, particularly as strict delineations between Indigenous and non-Indigenous people become less and less clearly articulated. Studies that default to resistance theory as the primary means of understanding Indigenous lives in the colonial present do not address this development. Resistance, and its attendant distancing, poses other challenges when trying to understand the complexities of colonial settlement within what were First Nations geographies. Despite references to dynamic adaptive practices being everywhere and in everything, a feature of resistance theory is that of singularity with reference to that which is being responded to (e.g., "*a* beaten track" or "*the* track"). Resistance theory in studies of colonialism seems unable to fully account for the complexity of those whom the theory conceptualizes (and thus positions) as resisters: Can a "resister" ever switch sides? What happens when a "resisting" subject becomes a member of the group(s) against which he or she rebelled? To be sure, colonialism is about forcibly establishing and maintaining power. But this establishment and maintenance is undertaken in such complicated and multiple ways, and by such diverse actors, that conceptualizing it—even metaphorically—as "a" track, "the" track, or "a" force (as opposed to multiple, competing, and diverging tracks and forces) risks misunderstanding its complex and pluralistic nature, a nature that arguably affords colonialism the very resiliency and power that resistance theorists want to combat.

Similar plurality—and its corresponding challenges—applies to the Indigenous peoples who navigated and engaged colonialism. Vesting colonialism or any of the peoples it impacts with any kind of singularity, which resistance theory necessarily does in its inability to account for multiplicity and simultaneity, risks not conceptualizing possibilities for decolonization. If colonialism indeed turns (even in part) on very intimate and tender relationships, does a theoretical lens that demands opposition and combativeness not risk masking or even missing the nuances of human interactions, particularly as those interactions unfold(ed) in microscale places? Recalling Žižek's question about how to achieve a politics which opens up new spaces outside hegemonic positions and their negation, the challenge becomes understanding Indigenous geographies in British Columbia in ways that are deeply anticolonial and that account for contemporary geographies inhabited today by complex and pluralistic configurations of peoples. Here, lessons offered by John Henry have some bearing on the discussion. What lessons might John Henry be conveying when, within the same story, he speaks of wrongly aimed gunshots and scars inflicted upon him during his time at Indian schools? Answering this question requires some understanding of the geographies of education writ large and, more specifically, geographies of colonial education in British Columbia.

Schooling children within state-sanctioned curricula and educational spaces has always been a vital means by which governments, or other powerful systems, impart certain kinds of knowledge to produce citizens and subjects who conform to and embody particular norms, protocols, and sociocultural expectations (Giroux, 1981). In many of the earliest colonial strategies concerning "the Indian problem" in Canada, education and schooling were proposed as one of the most effective means to civilize, settle, and quell Indigenous people and the challenges to state power that they posed (Milloy, 1999). Residential schooling, which peaked during the early and middle decades of the twentieth century, evolved from the earlier and somewhat less invasive day-school educational efforts of the eighteenth century. Residential schooling relied on pedagogical visions of children transforming through sustained spatial separation from families and communities and full immersion into "circles of civilized care" (Davin, 1879, n.p.; see also Miller, 1997; Milloy, 1999). Residential schools were, according to ecumenical and governmental leaders who oversaw their operations, a form of aggressive civilization, and a spatial intervention into barbarism. The broadest intent of the schooling, admittedly with variations over time and across geography, was to de-Indigenize the Canadian landscape (Razack, 2002) by educating, civilizing, Christianizing, and modernizing Indigenous children so as to kill the Indian in the child and save the man (Churchill, 2004).

Eighteen residential schools operated across B.C. between 1861 and 1984. They operated with the objective, as stated in 1947 by Federal Minster J. A. Glen, of educating Indians "capable of meeting the exacting demands of modern society with all it complexities" (Glen, 1947, n.p.). "To produce Indians of such capacity," Glen argued, "… is not an easy task … It may mean 100 or 200 years of the keenest kind of insight and understanding. Education of every type must be utilized. This should include schools … and all available forces, both positive and negative" (n.p.). Glen's objectives relied on space; material sites became active forces in the sociocultural transformation of Aboriginal children undertaken by churches and colonial governments. The schools, constructed at the behest of those who held social power, functioned as material realizations of that power and imposed order, constraint, and discipline onto the lives and bodies of Aboriginal children. In 1924, British Columbia's longest-serving residential school principal, Reverend George Raley, described Coqueleza School as "a monument to the advancing policy of the Department of Indian Affairs" (n.p.). For Duncan Campbell Scott, Deputy Superintendent-General of Indian Affairs Canada, built spaces *were* pedagogy:

[This first brick marks] an epoch in the education of the Indian of British Columbia. Every effort will be made to impress [on] the native mind that the occasion [of the school building's completion] is one when the standard of ideas is raised to a higher plane than ever before. (quoted in Raley, 1924, n.p.)

The schools were looming structures, bifurcated in design so as to separate children by gender, with the result of children being separated from family members. They were structured so as to ensure that students were under the gaze of teachers and staff at all times, were generally terribly overcrowded, and were rife with disease (Kelm, 1998; Titley, 1986). Students understood school spaces, including dormitories and bathrooms, as violent and inseparable from pedagogic efforts to transform them into clean, civilized, White citizens. Former St. Mary's student Mary Anne Roberts remembers:

There were five showers … and they would put us in and [a] senior would scrub us down. Then the nun was standing at the door … and she would check to see if we were clean, and with me, I am naturally dark, so I would always get sent back. I always got sent back because to her I was, not that I was dirty, it is just because I was naturally dark. So I would get sent back and they would scrub the heck out of me and that had a really, really bad effect on me, and through the years, even up to now, I would feel myself washing and washing and never feeling clean. It had a lifelong effect on me. (quoted in Glavin and former students of St. Mary's, 2002, p. 47)

Curricula and the subjects delivered to Indians in Indian schools differed from that delivered to non-Aboriginal children across the province. Moreover, Aboriginal children's experiences of those curricula and subjects differed from those of White children. Subjects taught to Aboriginal children were less about imparting a knowledge base—of mathematical skills, for instance—and more about erasing characteristics associated with savage Indianness. Consequently, Aboriginal children understood a subject such as history not as knowledge per se, but as an active force designed to sublimate Indigeneity (de Leeuw, 2009). Schooling of all children rests on a vision of shaping children into future adults (see, e.g., Gagen, 2004; Gleason, 1999), but Aboriginal children were always understood by their non-Aboriginal educators as markedly different from White children because, by virtue of being non-Aboriginal, White children embodied the possibility of a colonialist future. Aboriginal children were thus not schooled by educators who recognized in their students qualities of future White colonial subjects, qualities that required appropriate nurturing in order to mature. Instead, childness in an Indian was a particular *Indian-childness* (see de Leeuw, 2009). It was not something merely to shape into adultness, because that would assume Indian adulthood. Rather, Indian-childness was something to do away with entirely, thus preventing the child from maturing into an Indian adult. Knowledge became a power by which to eradicate identities.

In efforts to expunge any characteristic of Indianness in Aboriginal children, teachers and educators in residential and other Indian schools severely punished children for expressions indicative of Indigeneity. Rules and school policies changed and shifted, which was utterly confounding for students, but in general what was deemed Indian, and thus punishable, tended to be communication in Indigenous languages, interactions with family members, and, most often, non-adherence to curriculum and ways of thinking set by non-Aboriginal teachers or school staff (de Leeuw, 2007, 2009). John Henry bears the scars of such practices.

What might replace the dualistic antagonisms of resistance theory when trying to understand colonial projects and Indigenous peoples' responses to them in British Columbia? How did Aboriginal children, within historical residential schools and in very non-confrontational and non-resisting ways, find ways to survive? What are the linkages between that survival and Indigenous peoples' ongoing engagement of contemporary educational systems, systems from which Indigenous peoples across B.C. continue to be alienated (Richardson & Blanchet-Cohen, 2000)? Resistance theory, as I suspect John Henry's story hints at, may not be the best means by which to understand the long-standing articulation by many Aboriginal peoples that they want neither to relinquish characteristics such as language and cultural expressions tied to identifying as Aboriginal nor to forgo skills and knowledges that will allow them to prosper in times and spaces altered by colonial presences.

Although by no means in a universal or homogenous way, testimonial literature
on the residential school experience does suggest that many Aboriginal children in
residential schools during the mid-twentieth century were mindful of being vio-
lently punished, and aware of a changing world to which they were bound to require
adjusting, but were loath to relinquish connections to cultures and communities
(Nuu-chah-nulth Tribal Council, 1996; Secwepemc Cultural Education Society,
2000). They thus found ways of creatively and artistically engaging the schooling
systems and spaces in which they lived. They imagined alternatives and seemed
unafraid of making even wrongly aimed shots at the spaces in which they were
confined. Such practices and expressions were described by Clutesi (1967), a prominent Aboriginal artist and writer who attended residential school in British
Columbia, not as resistances but as approaches "from the backdoor" (p. 9). In a
widely publicized book that documented his First Nation's rich cultural histories
and criticized colonial interventions, Clutesi argued, with wit and wryness, for the
need for a mediated middle ground between Indigenous and non-Indigenous
peoples: "What can be done to really help the Indian at this time? One way would
be to look for his better qualities. He has some. Meet him halfway" (p. 11). Almost
as if anticipating being romantically and stagnantly positioned as a resisting agent
against colonial power, Clutesi went on to write: "You cannot fool an Indian with
the gushing displays of hypocritical prying of a would-be do-gooder" (p. 13).

Within the tightly monitored and strictly enforced (yet always shifting and contradictory) boundaries of residential schools, Indigenous children found ingenious
and "backdoor" means to express Indigeneity and grapple with colonial education.
Unique evidence of such strategies comes in the form of student-produced visual
and literary arts. For instance, in a poem published in a student-produced school
newsletter, a student from St. Joseph Indian School clearly disavows the lessons and
texts of residential school and states her intent to return to her community after she
finishes school; she poetically writes about returning to her mother's arms, burning
books, and once again walking the gravel roads of her home community (see
Fig. 15.1). During the 1940s, students at St. Michael's School produced, within
their classes in carpentry and domestic science, a plethora of objects featuring First
Nations iconography and imagery (see Fig. 15.2).

These and many other student-produced art objects can be conceptualized as
material records of Indigenous adaptability and resilience, characteristics that
allowed for a navigation of two worlds. The work is an imagining of something
utterly new. The art allowed students to remain connected to their Indigenous
communities and lineages while surviving within colonial education structures.
These records suggest the prodigious nature of children's efforts to sidestep at least
some of colonialism's lessons. The work manifests all of the "tautness" in colonial
school relationships. Even though there exists little student-produced explanation
about the objects and texts, it is difficult not to read the materials as students
maintaining ties to community, holding fast to Indigeneity, and making efforts to
decolonize educational space.

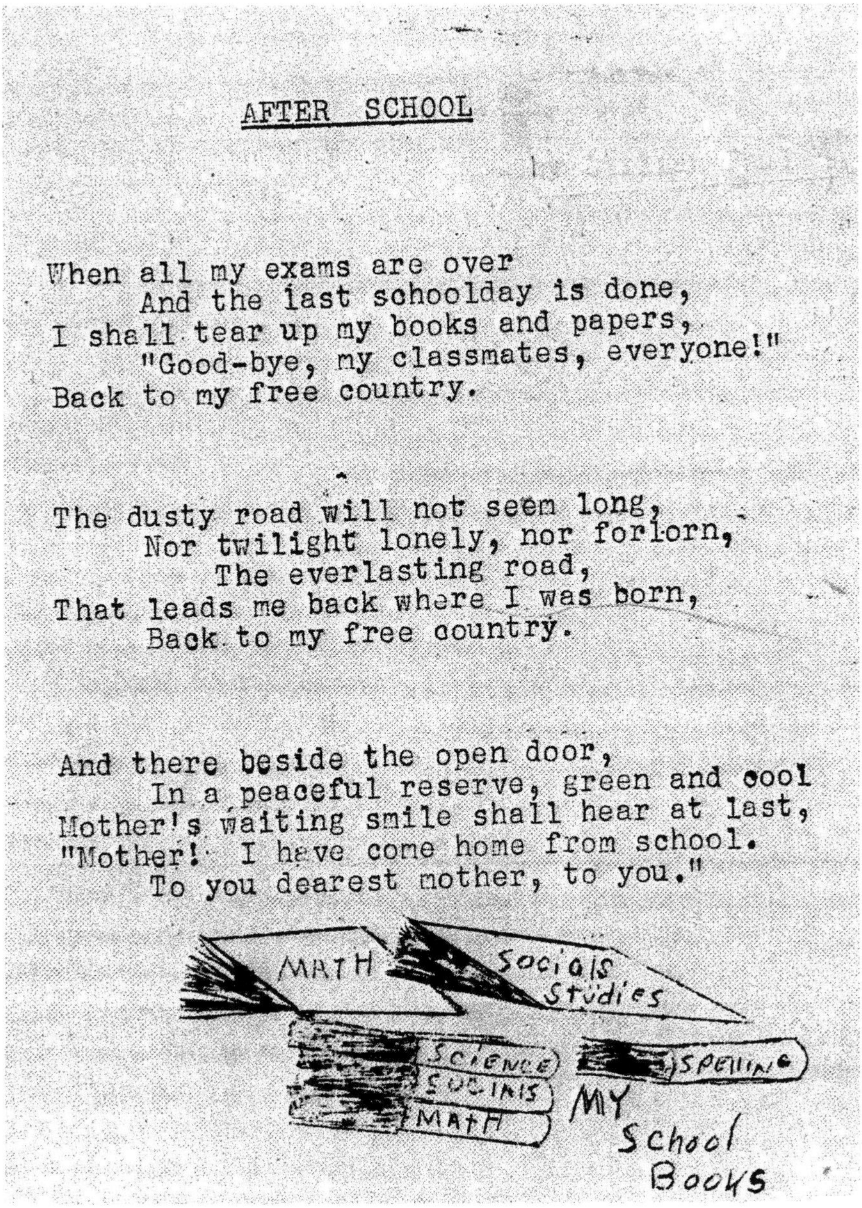

AFTER SCHOOL

When all my exams are over
    And the last schoolday is done,
I shall tear up my books and papers,
    "Good-bye, my classmates, everyone!"
Back to my free country.

The dusty road will not seem long,
    Nor twilight lonely, nor forlorn,
        The everlasting road,
That leads me back where I was born,
    Back to my free country.

And there beside the open door,
    In a peaceful reserve, green and cool
Mother's waiting smile shall hear at last,
"Mother! I have come home from school.
    To you dearest mother, to you."

Fig. 15.1 Student poetry from Williams Lake Indian Residential School (Source: Published in the school newspaper *My Heart Is Glad*, 1965)

Fig. 15.2 St. Michael's Indian Residential School Art Display, ca. 1940 (Source: Original image courtesy of the Alert Bay Community Museum, Library, and Archives)

Some Tentative Conclusions

Let's return now to living with scars and John Henry's story. John Henry has children, and those children attend twenty-first-century schools in British Columbia. John wants them to do well (and by all accounts they are). Still, in part because of an educational lineage that did not anticipate Aboriginal children's success and in part because of the history of educational violence toward Indigenous children, educational success is still not easily attained for Aboriginal people in B.C. When compared with non-Indigenous peoples across Canada, Indigenous peoples have some of the highest rates of school dropouts and some of the lowest rates of post-secondary education. Education is a social determinant of health (Richmond & Ross, 2009), and a lack of education has an impact on the overall health, prosperity, and strength of Aboriginal peoples. This arguably results in the ongoing socioeconomic marginalization of Aboriginal peoples. Thought of slightly differently, marginalization from education translates into the ongoing maintenance of colonial power in the landscapes of British Columbia. Indigenous peoples in B.C. want to address this problem, and not necessarily by means that lend themselves to being conceptualized as "resistance."

In 2006, a task force was created in the northern B.C. community of Prince George, a city in which 25 % of the school-age population is Aboriginal. The task force, led by local First Nations, including members of the Carrier Sekani First Nations, voiced frustration about the marginalization faced by Aboriginal children in schools and articulated a desire for Aboriginal children to achieve educational success so they can lead healthier lives. Compared with a high-school completion rate of almost 100 % for non-Indigenous students, the provincial rate of graduation for Aboriginal students is 48 %, and for children in the care of the government that rate falls to 15 %. Research on Aboriginal students' experience of schools in the province documents pervasive feelings of marginalization and alienation at all levels of schooling (Richmond & Ross, 2009). In efforts to improve these rates, the task force proposed a quite unique approach: Develop an "Aboriginal Choice" school which, while open to anyone and fulfilling all provincial curricular standards of other schools, would focus predominantly on local First Nations languages, protocols, and customs and eventually be developed within an architectural space designed in consultation with Indigenous peoples in order to mirror First Nations' orientation to space and art. The vision of these Aboriginal Choice schools has never been articulated as one of resistance. Indeed, Charlotte Henay, the former Principal of Aboriginal Education in the region, spoke publicly about curricula that embrace storytelling and methods of alternative dispute resolution (Henay quoted in Bruner, 2009, p. 18). She quietly imagines utterly new spaces, without any reference to resistance.

The response to Henay and the schools by non-Aboriginal peoples across northern B.C. was swift and vitriolic, with much of the rhetoric based on lines of logic that differ little from those of nineteenth- and twentieth-century sentiments about Indigenous peoples and the need to assimilate them through colonial education. Bloggers opined that "the only way for the native situation to get better is for them to become fully integrated into society. One law, one system, period." Or that "everyone [is] equal … an eye for an eye … yes sir, no sir, yes mame [sic], no mame [sic] … There ain't no free ride … I know lots of natives that are the products of OUR education system and are successful and well respected. The majority of them, however, should quit whining and do something to help themselves." In other media outlets, sentiments included, "I know several Native people on a personal level who left the rez. [to go to mainstream schools] and they are far better off and are the first ones to admit that!" Finally, one person observed that "setting up schools that cater to one culture just creates bitter tax payers in our society: It's time they [Natives] either go back in the bush and want nothing from us or join the rest of us Immigrants and be as one." If responses by non-Aboriginal peoples in northern B.C. were this violent to efforts which so clearly avoided confrontational language and strategies, it seems understandable that First Nations groups might feel unsafe employing strategies of resistance with reference to education for their children. If, for reasons of safety, First Nations are avoiding narratives of resistance in the educational strategies for their children and communities, it seems disingenuous for others to overlay theoretical frameworks of resistance upon their work.

At the same time, and here I return to Žižek's question, it is clear that Indigenous peoples continue to make efforts at decolonizing educational spaces and systems. To paraphrase John Henry, they keep shooting at the colonial beast. But the ways in which they are doing this do not, I think, constitute a simple negation or a parasitization. To conceptualize their work in this way would be, I think, an injustice to the complexity of their efforts. It also cannot account for "wrongly aimed" shots. When the Principal of Aboriginal Education in Prince George speaks to the media about the Aboriginal Choice school, she does not do so antagonistically or with any narratives that resemble combativeness or resistance. Instead, she confidently speaks as if the school is a given, without directly referring to the racism or naysayers who speak out against it. In her silence, she seems to be clearly indicating that even acknowledging such (neo)colonial voices would only give the perspectives credence. Not unlike the children who quietly yet assuredly make creative works within residential schools, Charlotte Henay is going about the business of ensuring new spaces within which to express aspects of Indigeneity in order to educate Indigenous (and potentially non-Indigenous) children, to keep alive aspects of different Indigenous cultures, and in this way to circumvent the impositions of colonial curricula. She is in the process of imagining new spaces tied to nothing other than what she and her communities believe is best for their children.

Again, the stories of John Henry are worth returning to. There is a sense of something complex unfolding in the lands where John still hunts, a relationship beyond resistance. Certainly First Nations people in northern British Columbia, as in the rest of Canada, are with increasing force demanding recognition for historic and extant colonial wrongdoings. But do these define every moment of a First Nations person's life? Of John Henry's life? Probably not. After all, John Henry does many things, things like hunting for moose and talking about making stew. Despite John embodying many marks of colonial projects, it seems disingenuous to impose over his every action, his every story, a narrative or conceptual framework of resistance. Would John Henry call what he does "resistance"? Again, probably not. Neocolonial voices, and outright racist voices, are still strong in B.C. Those voices still want Indigenous people to assimilate quickly and quietly into settler-colonial landscapes. They still believe that education, and more particularly schooling, is a means of achieving this assimilation.

John Henry lives with the scars of his educational past, and he thinks about how to hunt and chase animals for hours along remote roads. He hopes for a better education for his children, and Charlotte Henay works toward making that hope a reality. They imagine what is possible, they tell stories, and they do not seek permission. They imagine without asking. So, to answer Žižek's question, perhaps part of the answer is to think about relationships of imagining, of constant and not always oppositional or antagonistic engagements between various peoples with different experiences and expectations. Perhaps this approach is what John Henry was alluding to one summer during a conversation on the shores of Francois Lake in northern B.C. Perhaps he was suggesting that we should always keep in mind imagining something utterly new, even if it starts with wrongly aimed shots at the powerful beast we are chasing.

References

Bednasek, C. D., & Godlewska, A. M. C. (2009). The influence of betterment discourses on Canadian Aboriginal peoples in the late nineteenth and early twentieth centuries. *Canadian Geographer, 53*, 444–461. doi:10.1111/j.1541-0064.2009.00281.x.

Bruner, T. J. (2009). Aboriginal choice school concept growing. *Windspeaker, 26*(10), 18.

Churchill, W. (2004). *Kill the Indian, save the child: The genocidal impact of American Indian residential schools*. San Francisco: City Lights Publishing.

Clutesi, G. (1967). *Son of raven, son of deer: Fables of the Tse-Shaht people*. Sidney, Canada: Gray's Publishing.

Davin, N. F. (1879). *Report on industrial schools for Indians and half-breeds*. Report presented to the Government of Canada, Ottawa.

de Leeuw, S. (2007). Intimate colonialisms: The material and experienced places of British Columbia's residential schools. *Canadian Geographer, 51*, 339–359.

de Leeuw, S. (2009). "If anything is to be done with the Indian, we must catch him very young": Colonial constructions of Aboriginal children and the geographies of Indian residential schooling in British Columbia, Canada. *Children's Geographies, 7*, 123–140. doi:10.1080/14733280902798837.

Edwards, B. F. R. (2009/10, Winter). "I have lots of help behind me, lots of books, to convince you": Andrew Paull and the value of literacy in English. *B.C. Studies, 164*, 7–50.

Gagen, E. (2004). Making America flesh: Physicality and nationhood in early twentieth-century physical education reform. *Cultural Geographies, 11*, 417–442. doi:10.1191/1474474004eu321oa.

Giroux, H. A. (1981). *Ideology, culture, and the process of schooling*. Philadelphia: Temple University Press.

Glavin, T., & former students of St. Mary's. (2002). *Amongst God's own: The enduring legacy of St. Mary's Mission*. Mission, Canada: Longhouse Publishing.

Gleason, M. (1999). Embodied negotiations: Children's bodies and historical change in Canada, 1930–1960. *Journal of Canadian Studies, 34*, 112–138.

Glen, J. A. (1947). Statements. *The Indian School Bulletin*. Canada, Department of Indian Affairs Branch, Department of Mines and Resources, and Department of Citizenship and Immigration. 2 vols. Les Archives Deschâtelets.

Gregory, D. (2004). *The colonial present: Afghanistan, Palestine, Iraq*. Oxford, UK: Blackwell Publishing.

Haig-Brown, C. (1988). *Resistance and renewal: Surviving the Indian residential schools*. Vancouver, Canada: Arsenal Pulp Press.

Harris, C. (2002). *Making native space: Colonialism, resistance, and reserves in British Columbia*. Vancouver, Canada: UBC Press.

Harris, C. (2004). How did colonialism dispossess? Comments from an edge of Empire. *Annals of the Association of American Geographers, 94*, 165–182. doi:10.1111/j.1467-8306.2004.09401009.x.

Katz, C. (2004). *Growing up global: Economic restructuring and children's everyday lives*. Minneapolis, MN: University of Minnesota Press.

Kelm, M. (1998). *Colonizing bodies: Aboriginal health and healing in British Columbia, 1900–1950*. Vancouver, Canada: UBC Press.

Kobayashi, A., & de Leeuw, S. (2010). Colonialism and the tensioned landscapes of Indigeneity. In S. Smith, R. Pain, S. Marston, & J. P. Jones III (Eds.), *The handbook of social geography* (pp. 118–139). London: Sage. doi:10.4135/9780857021113.n5.

Lawrence, B. (2003). Gender, race, and the regulation of Native identity in Canada and the United States: An overview. *Hypatia, 18*(2), 3–31. doi:10.1111/j.1527-2001.2003.tb00799.x.

Lawrence, B. (2004). *"Real" Indians and others: Mixed-blood urban Native peoples and indigenous nationhood*. Vancouver, Canada: UBC Press.

Lefebvre, H. (1991). *Critique of everyday life: Volume 1* (J. Moore, Trans.). London: Verso Books.

Mawani, R. (2009). *Colonial proximities: Crossracial encounters and juridical truths in British Columbia, 1871–1921*. Vancouver, Canada: UBC Press.

Miller, J. R. (1997). *Shingwauk's vision: A history of Native residential schools*. Toronto, Canada: University of Toronto Press.

Milloy, J. (1999). *A national crime: The Canadian government and the residential school system, 1879–1986*. Winnipeg, Canada: University of Manitoba Press.

Nash, C. (2002). Cultural geography: Postcolonial cultural geographies. *Progress in Human Geography, 26*, 219–230. doi:10.1191/0309132502ph365pr.

Nuu-chah-nulth Tribal Council. (1996). *Indian residential schools: The Nuu-chah-nulth experience*. Port Alberni, British Columbia: Nuu-chah-nulth Tribal Council.

Ortner, S. B. (1995). Resistance and the problem of refusal. *Comparative Studies in Society and History, 37*, 173–191. doi:10.1017/S0010417500019587.

Pile, S. (1997). Opposition, political identities, and spaces of resistance. In S. Pile & M. Keith (Eds.), *Geographies of resistance* (pp. 1–32). London: Routledge.

Pile, S., & Keith, M. (Eds.). (1997). *Geographies of resistance*. London: Routledge.

Pratt, M. (1991). *Imperial eyes: Travel writing and transculturation*. London/New York: Routledge.

Pualani-Louis, R. (2007). Can you hear us now? Voices from the margin: Using indigenous methodologies in geographic research. *Geographical Research, 45*, 130–139. doi:10.1111/j.1745-5871.2007.00443.x.

Radcliffe, S. H. (2000). Entangling resistance, ethnicity, gender and nation in Ecuador. In J. P. Sharpe, P. Routledge, C. Philom, & R. Paddison (Eds.), *Entanglements of power: Geographies of domination/resistance* (pp. 164–181). London: Routledge.

Raibmon, P. (1996). "A new understanding of things Indian": George Raley's negotiation of the residential school experience. *BC Studies, 110*, 69–96.

Raley, G. H. (1924, June 30). *Coqualeetza Institute commencement exercises*. Sardis, BC: Coqualeetza School. Chilliwack Community Archives.

Razack, S. (2002). When place becomes race: Introduction. In S. Razack (Ed.), *Race, space and the law: Unmapping a white settler society* (pp. 1–21). Toronto, Canada: Between the Lines Press.

Richardson, C., & Blanchet-Cohen, N. (2000). Postsecondary education programs for Aboriginal peoples: Achievements and issues. *Canadian Journal of Native Education, 24*, 169–184.

Richmond, C. A. M., & Ross, N. A. (2009). The determinants of First Nation and Inuit health: A critical population approach. *Health and Place, 15*, 403–411. doi:10.1016/j.healthplace.2008.07.004.

Rose, M. (2002). The seductions of resistance: Power, politics, and a performative style of systems. *Environment and Planning D: Society and Space, 20*, 383–400. doi:10.1068/d262t.

Scott, J. (1990). *Domination and the arts of resistance: Hidden transcripts*. New Haven, CT: Yale University Press.

Secwepemc Cultural Education Society. (2000). *Behind closed doors: Stories from the Kamloops Indian Residential School*. Penticton, BC: Theytus Books.

Sparke, M. (2008). Political geography – Political geographies of globalization III: Resistance. *Progress in Human Geography, 32*, 423–440. doi:10.1177/0309132507086878.

Sterritt, N., Marsden, S., Galois, R., Grant, P., & Overstall, R. (1998). *Tribal boundaries in the Nass watershed*. Vancouver, Canada: UBC Press.

Stoler, A. (2006). Tense and tender ties: The politics of comparison in North American history and (post)colonial studies. In A. L. Stoler (Ed.), *Haunted by empire: Geographies of intimacy in North American history* (pp. 23–70). Durham, UK: Duke University Press. doi:10.1215/9780822387992-002.

Thrift, N. (1997). The still point: Resistance, expressive embodiment and dance. In S. Pile & M. Keith (Eds.), *Geographies of resistance* (pp. 124–151). London: Routledge.

Titley, B. (1986). *A narrow vision: Duncan Campbell Scott and the administration of Indian affairs in Canada*. Vancouver, Canada: UBC Press.

Waterstone, M., & de Leeuw, S. (2010). A sorry state: Apology excepted. *Human Geography: A New Radical Journal, 3*(3), 1–28.

Watson, A., & Huntington, O. H. (2008). They're here—I can feel them: The epistemic spaces of Indigenous and Western knowledges. *Social and Cultural Geography, 9*, 257–281. doi:10.1080/14649360801990488.

Žižek, S. (2006). *The parallax view*. Cambridge, MA: MIT Press.

Communication, Identity, and Power   16

Jo Reichertz

Nothing Is as Persuasive as ...[1]

Sometimes, even American gangsters have something to say to researchers. Asked about the power of communication, John Dillinger (1903–1934) apparently uttered the following words of wisdom: "Nothing is as persuasive as a good argument. Except perhaps ...". Whereupon he paused for a moment's thought: "Except perhaps for a good argument backed up by a loaded gun." This combination of argument and loaded gun was to subsequently go down in media history and then enter into society's stock of common sayings, reformulated as "an offer you can't refuse." It conveys an underlying willingness and ability to use violence and that this violence will ultimately constitute the power of the argument.

Dillinger was not the first, and certainly won't be the last, to point out that nice-sounding, reasonable and well-structured statements do not always assert their effect on their own but that, behind those words, stands a speaker with muscle power, a club, a sword or a gun, and that should his words prove ineffective, he is prepared to inflict pain on his counterpart, or even take his life. *This* power of words is thus based on a propensity towards violence and the ability to use violence on the part of the person doing the communicating.

History teaches us that arguments do not have any force or power of their own. If they are to have an effect, arguments need to be voiced in surroundings and in a society in which arguments count for something. The argument requires a specific political and intellectual climate in order to grow, prosper and have an effect—a climate in which one appreciates and respects the argument. Clearly enough, Dillinger did not live in such a society, but in one in which first and foremost the

---

[1] See also (Reichertz, 2011).

J. Reichertz (✉)
Institute for Communication Science, University of Duisburg-Essen,
Universitätstraße 12, 45117 Essen, Germany
e-mail: jo.reichertz@t-online.de

© Springer Netherlands 2015
P. Meusburger et al. (eds.), *Geographies of Knowledge and Power*,
Knowledge and Space 7, DOI 10.1007/978-94-017-9960-7_16

theatrical depiction of power formed the basis of communication. The power of his word was not so much backed up by a *social* guarantee but by a *personal* guarantee. It was not a social institution standing behind the words but a tangibly woundable person with a limited lifespan. His power was not anchored on a sustainable basis, which is why it could not persist. On the contrary: because it was bound up with his life and his potential for violence, it was constantly in danger of waning or meeting an abrupt end.

Such power is therefore structurally fragile. It has to be continuously revisualised for all to see, to be theatricalized—it must be made to prove itself anew. Also, the power must be constantly demonstrated. Wealth (financial power) acquired, terror dealt out to enemies and renegades (the power to take someone's life), the demonstrative indulgence shown to children and the repentant (the power to bestow life), a large number of relatives and friends (the power of the large group) and so on were always the sources of the power of communication, sources to be performatively displayed again and again. That is exactly why a good argument backed up by a loaded gun is so powerful.

What Exactly Is Effect?

There is hardly any other human perception that has been met over time with so much approval as the following: words, once articulated, can reveal considerable power, and communication is in a position not only to change people but also to change the course of the world. Yet despite the certainty that communication is powerful, surprisingly little is known about the sources of this power.

Certainly, we are well aware of specific powerful sources of the spoken word. Our everyday common sense tells us that words accompanied by the threat of substantial violence will easily motivate others to do what they are told. For its part, sociology teaches us that the words of the master will inspire the servant to perform, the latter clearly knowing that non-compliance will lead to the loss of employment. Sociology also teaches us that those who hold someone as special—indeed who ascribe charisma to that person—will follow the words of the esteemed one even when they don't fully grasp the message.

It is hardly surprising that violence or domination or charisma lend power to words. On the contrary, it would be surprising if violence, domination, and charisma were to remain without effect. In fact, violence, domination, and charisma undoubtedly explain quite a lot—in some areas, in corporations for example, a great deal. The crucial point is that communication can also be (very) powerful even if neither violence nor domination nor charisma underlies it. More to the point, this kind of communicative power is frequently the norm and not an infrequent borderline case. *Normal* communication in everyday situations manages quite well without violence, domination, and charisma, but does involve power nevertheless. It is precisely this power that is of interest to me.

I am interested in why people, when asked by others to pass the salt or close the window, will do exactly that. Indeed, people will do much more than that for one another. Because they have communicated with one another, they (often) change their behavior, their attitudes, and even their lives. And much of all this is due to the everyday power of communication.

I am not interested in why a secretary carries out a boss's instructions when she (the latter) has requested him (the former) to do so. That is self-explanatory. I am interested in why the secretary carries out that request with particular care and even thinks of certain details that his boss may have overlooked. I am also interested in why the boss will respond to the secretary's reminder—that detail X is still missing, for instance—to supply that information and thank the secretary for the reminder.

I am interested in why communicative action among those present usually leads to the desired results and effects. For, as a rule, adults can achieve with other adults that which they wish to achieve—even if they know what effect words can achieve with others in the first place.

Of course there are situations in which communication no longer achieves anything: when silence suffices, when everyone is quiet, or if communicative action is used to hurt or belittle others. Even then, communication has power, lots of power in fact. Communication can dominate without domination, be hurtful, be belittling, and even render speechless.

Of course there are other situations in which communication ends and violence begins. But that is not the topic I wish to address. My focus is on the communication that has power among those present, those in situ. Therefore this article primarily addresses the everyday power of communication on this side of violence, domination, and charisma.

No academic discipline, however, seems to address the everyday power of communication among those present—in other words, with the power that first emerges during the interplay of communicative actions and subsequently evolves. Considerable research still needs to be done in this area.

Such a consideration—i.e. one that removes the subject from the center of communication, as well as the deep structure that engenders the communicative action—must also ask whether greater significance should be afforded the social practice of communication than hitherto generally assumed. If one agrees with that, then the function of communication might be defined from another perspective. Communication would then not be about the transmission of messages on the one hand and understanding on the other; communication would instead be about effect—and that can only be achieved via power. Power is what leads actors to do what is communicatively asked of them. This is one reason why it would make sense to switch the focus of communication theory away from "understanding" to "power."

I would furthermore like to garner support for the hypothesis that the power of communication must first establish itself within communicative interplay in order to be effective at all. The communicative construction of reality (Keller, Knoblauch, & Reichertz, 2012) that not only generates the world in which we deal with one another, but also the identities of the persons so communicating with one another, is also necessary for the power of communication to emerge. Communicating with one another implies the constant negotiating of identities with one another.

Action is prompted by problems. Since problems do not solve themselves, one has to undertake something to effect a solution. The traditional means thereof: communicative action. Communication is intended to ensure that what occurs is what one wants to occur. Communication is intended to close the gap between making the wish and its realization. Communicative action must therefore give a reason for taking the action without this reason becoming identical with classical causality. Here *power* takes the place of causality. But if causality compels and so permits no choice, power affords latitude, power merely suggests one course of action, provides reasons for it, formulates hopes.

Power enters the game when the interlocutor accepts what is expected and turns the expected into a deed—especially if that very interlocutor wants something different. Communicative action is therefore not really powerful when the participants share congruent goals. The proposal, "Come on, let's go to the cinema," is not really difficult to implement when it is precisely what the interlocutor wants anyway. The really interesting question is this: why does the interlocutor accept the other's expectations of action if, at that moment, he or she has other interests and plans and doesn't want to go to the cinema?

The Power of Communication Assumes Actions

Those who act through signs announce; they inform—not to trigger an inner *experience* but to trigger an *action*. The person evidencing the signs wants to influence the action of his or her counterpart. The question is, however, why would the announcee (the recipient of the announcement) want to accept being influenced by the signs? The announcee does not have to do what he or she understood is expected to do, but could also do something different. After all, in human interaction, objection and contradiction cannot be simply put to rest. The possibility of contradiction is constitutive to communication between people. In fact, contradiction is what creates a framework for communication in the first place. If one couldn't do differently, then the other wouldn't have to communicate. Communication is meant to give a reason for doing something. If no reason were needed, then we would have a case of stimulus and response, a case of causality. Yet human action does not underlie causality; rather, it requires reasons and motives. These then induce actions.

Language on its own, though, is not sufficient to induce others (as Habermas, 1981 assumes). Something needs to be added, something supplementary (Luhmann, 2003, pp. 6–7) that is not linguistic, something that induces or—to be more exact—something that gives reason to be induced in the desired direction. The choice between consequences and non-consequences can "not be guided by language alone because it indeed offers both possibilities" (p. 6). The question is then what is it that induces us to follow the wishes of the announcer. An initial answer might be, "Power, that's what it is." In the process, power is a kind of placeholder for all the reasons that offer the chance of inducing the other to action. Or, in the words of Max Weber: "Power means the chance of imposing one's own will on a social relationship even against resistance, no matter what this chance is based on," (Weber, 1972, pp. 28, 531). According to Weber, power is only a chance, not a certainty. Similarly, power can wane or be augmented. Plus, it's not as though there is just the *one* power; power can also feed off and arise from many different sources.

Hence I can follow the communicative request of the other—i.e. be acquiescent because *everyone* follows, because everyone is acquiescent. I can imitate what others do because, perhaps, I believe they had good enough reasons for doing so. Then the power of the person communicating would lie in the readiness of the counterpart to *imitate* others who have let themselves be influenced by him or her. This argument is not very convincing. But I can also follow a communicative request because others also follow it and have always followed it because it is *custom* and *tradition*. In this case, I would be associating myself with a socially established *practice*. This does happen.

If one looks from a higher vantage point at the *motives* that induce people to follow communicative impositions—i.e. look at the phenomenon of *power*—then it is possible to pinpoint three basically different motivations: *violence, domination* and *relationship* (even if they do overlap in the everyday and are only clearly separable from one another in the analytic field). All three sources of power consist of practices, namely, the practice of exerting violence, the practice of using domination and the practice of building up relationships. These practices are aimed at exerting power, that is imposing one's will on other participants—even against resistance. Power is the *hypernym*; violence, domination and relationship are the *hyponyms*. Or put another way: where there is compulsion, there is power; where there is an order, there is power; where there is love, there is power; and even where there is truth, there is power. Nevertheless, it does make a difference what the source of the power is.

The first reason to follow communicated acts of imposition has already been suggested in the foregoing: it is the readiness and the ability of the announcer, in the event of non-fulfillment of an expectation of action, to cause the other more or less serious bodily harm—in a word, pain. In brief, the reason for acceptance lies in the readiness and the ability of the announcer to exert violence and, of course, in the wish of the announcee to avoid pain and bodily harm.

The second reason to follow communicative acts, i.e. comply with acts of imposition, is to be found in the readiness and the possibility of the announcer, and in certain circumstances legally so, to cause the announcee some damage or to allow him certain advantages. Thus the speaker may utter commands, instructions or orders—in short, sentences that obligate and as a result of utterance are imposed. Because the speaker is acting within a certain capacity and because this right is not only associated with that capacity but is also a set right in some form, the consequence is that either compliance with the requirement is asserted, by means of force if need be, or the non-complier is excluded from the system. These reasons, citing Max Weber, fall under *domination* (cf. Weber, 1972, p. 28).

One can, in this way, legitimize domination in that the relevant dominators are merely the personal implementers of an entity recognized to be more powerful—God or a people, for example. Here individuals submit themselves to an organized power or institution, of which it is assumed that this entity possesses a much higher rationality or that it would, if necessary, be able to impose the wish as communicated by exerting violence. This kind of power makes actions predictable or at least more predictable—not certain as such, merely more expectable to a greater degree of probability. That being so, one can compute the acquiescence and count on the acquiescence, which makes the life of and living in organizations very much simpler because, on the basis of the legal situation, one knows what is expectable, what one may hope and what one has to fear.

The third general reason for reacting with acquiescence to the communicated expectations of an announcer can be found in the social relationship that emerges between announcer and announcee. This relationship must, however, be of a special kind.

> The term 'social relationship' will be used to denote the behavior of a plurality of actors .... The social relationship thus *consists* ... of a *probability* that there will be, in some meaningfully understandable sense, a course of social action. For purposes of definition there is no attempt to specify the basis of this probability. (Weber, 1972, p. 13)[2]

The relationship can be consciously brought about by the announcer, it can come about via negotiation, or it can simply emerge—perhaps against the participants' will. Due to this social relationship, the participants have become *relevant* to one another, which means that they can sanction the behavior, the person and thus the identity of the counterpart, sustainably so. To achieve a positive sanction—e.g., praise and recognition—someone is acquiescent. Or someone is acquiescent because he or she wishes to avoid a negative sanction, here a reprimand or derecognition. The decisive thing with this kind of power is that it is based on the voluntary recognition of the power of the other(s) and that in essence it is not rooted in violence and domination, but rather in the situation and in the common history of the participants.

If one now turns from general considerations to the particular forms of power and examines the theories and the sciences that over the course of the last few centuries have been preoccupied with *communicative power* and with the power of communication to achieve effects, then one can point to an array of different explanatory approaches, the majority of which (with the usual pinch of salt) can be assigned to two groups. The first group of theories and concepts explaining communicative power ascribes power to language itself, to the forms of speaking or to the forms of articulation. The second group sees power in the communicating actor, or to be more precise, in the social situation that the persons communicating jointly construct, if not always with the same interests—and not in the language.

---

[2] Talcott Parsons p. 118 "The theory of social and economic organisation", being a translation of Part I of Weber's *Wirtschaft und Gesellschaft*. Glencoe, Ill: Free Press 1964 paperback, first published 1947.

Translation note: This standard English translation is perhaps overly mathematical in that Weber did not use the term *Wahrscheinlichkeit* [probability] but *Chance* [chance, opportunity].

The first group of theories, i.e. the group that perceives the source of power in language itself, covers: (a) the notion of language as a magical force; (b) the notion that a specially processed (*rhetorical*) form of language spontaneously unfolds power by virtue of its 'truth'; and (c) the notion that a specific part of the speech act, namely the *illocutionary act,* has the force (*illocutionary force*) to trigger specific reactions in the counterpart almost compulsively (Habermas, 1981; Searle, 1979). I will not be going into these approaches here (for greater detail, please see Reichertz, 2009, pp. 202–204).

To make my argument even clearer, let me summarize the two positions, beginning with Bourdieu. For Bourdieu, language is not merely a means of communication but also primarily an indicator of social status and the social background of the person communicating—and therefore a means of domination. It is obvious to Bourdieu that the force effected via words does not lie in what is spoken; rather it lies with the speakers or, to be more precise, in their societal status (cf. Bourdieu, 2005). Only the communicative action that is authorized by the social status of the speaker has power. The speaker is thus empowered by society and/or its organizations or institutions while society also vouches for this authorization, meaning that it can also justify and initiate sanctions, if necessary.

People who succeed in being heard must have an office that is recognized (Bourdieu, 2005, p. 79). The persons communicating therefore do not (only) communicate in their name but also (and always) in the name of their group. It is through them that the social group to which they belong speaks. The importance of the social group determines the importance of communicative action. The possible elegance of the speech then belongs only to the symbolism of the power, not to the power itself.

Yet Bourdieu's reflections take only a specific kind of communication into account, namely the "official" communication between actors who are tied to one another by means of a power relationship. In short, Bourdieu explains the power of communication via domination. Although this is a perfectly feasible approach, such a restriction covers only that part of communication in which domination is effective because it is recognized as domination. Bourdieu explains (only) why the servant listens when the master says something. But that is really not surprising—even if this view of things is much closer to the power of communication than that espoused in many language philosophy debates about success conditions and conventions. Nevertheless, the specific problem as to why communicative action and communicative doing can involve power without violence and domination remains unresolved.

Something similar also applies to the power of communication that lies in charisma, which undoubtedly exists. Be that as it may, charisma lies in a non-everyday relationship in a crisis situation. Hence one cannot explain why communication can unfold power in the everyday. All the same, the effect of charisma shows emphatically that the power of communication essentially feeds off the personal relationship and the identity work associated with it. I will follow this train of thought below and also ask the question whether and for what reason the social relationship can build up the power of communication.

Into the second group of theories fall those that place the power of communication in the actor and in the situation in which communication between people takes place. One of these theories focuses on the *authorized speaker* (Bourdieu, 2005) and another on the concept of *charisma* (Weber, 1972). I will not be going into these approaches either (please see Reichertz, 2009, pp. 211–213 for greater detail).

Instead, I would like now to deal with the most important element relating to communicative power, an element that also places the speaker in the foreground and that is often overlooked: the intra-relationship of the persons communicating.

Social Relationship as the Prerequisite of Power

The everyday power of communication is, in this view, predicated on the relevance of the persons communicating to the identity of the participants, relevance built up in the course of the communicative interaction. Actors gain relevance to one another if they are reliable (Brandom, 1994, pp. 206–208). Accordingly, a reliable co-actor in the communication process is the one whose communicative action and conduct, to a high degree of probability, always has the (mostly) implicitly stated reasons and consequences. One trusts a reliable co-actor to ensure that deeds will follow his or her words. This trust can be brought into the communication process by virtue of the shared interaction history. However, it can also be built up via communication.

> In producing assertions, performers are doing two sorts of things. They are first authorizing further assertions (and the commitments they express), both concomitant commitments on their part (inferential consequences) and claims on the part of their audience (communicational consequences). In doing so, they become responsible in the sense of answerable for their claims. That is, they are also undertaking a specific task responsibility, namely the responsibility to show that they are entitled to the commitment expressed by their assertions, should that entitlement be brought into question. This is the responsibility to do something, and it may be fulfilled for instance by issuing other assertions that justify the original claim. (Brandom, 1994, p. 173)

The understanding and unfolding of communicative power are therefore only possible if words and deeds correspond with one another, if words are "true." After all, if every act of speaking were untrue, then it would be meaningless and without effect. The attribute of *true* does not refer to whether we actually *mean* what we say (i.e. it does *not* refer to an inner doing) but to whether we do what we say (i.e. it refers to the consequences). The problem is therefore not *authenticity* (I say what I really mean); the problem is the certainty of action, reliability. The crucial question is: do I let the deeds as announced follow my words? Less important is whether what I do in the performance of my words is actually what I want in my deepest inner being —or whether I do something only because I have said it. It is not the inner attitude that is decisive, but the deed.

For specific groups, communicative action must—at least to a certain extent—have a specific form of commitment. Otherwise, we could and should ignore it. Committed and, as a result, consequential acts of speech thus form an essential basis for human community and human coordination. No wonder there is a ban on lies in all societies—even if the truth need not be always be told to every single person and in every single situation. The social norm does not intend words to blow in the wind; it intends spoken words to be actions that abide. Or to put it better, they are indeed meant to be actions and can be effective only if action and word (normative) are *linked* with one another. Seen from this perspective, the history of mankind can be read as the ongoing attempt to regulate and keep stable the relationship between word and action.

In this view, the power of words and communication is predicated on the powerful implementation of specific forms of sociation that are aimed at the creation of reliability. Goffman has in fact named this process the "standardization" of communication (Goffman, 2005). Berger and Luckmann (1969, pp. 58–59) would speak here of *institutionalization*. Or, to borrow a term from Foucault's discourse, one can say *disciplinization* (Foucault, 2004, but see also White, 2008, pp. 63–65).

The disciplinization of communication is older than the invention of military and economic discipline at the end of the sixteenth and the beginning of the seventeenth century. Discipline existed long before the military, long before work cycles. Yet although discipline can go hand in hand with domination, it does not necessarily have to. Discipline can operate without domination and can also emerge from social relationships. What is meant by *discipline* here is a reliable process of advancement and progress that is oriented to others and that is directed at a specific social cycle. It's a matter of reliability that does not feed off violence and domination.

Communicative discipline is a bidding, a ruling that is aimed at the predictability of further action. It thus creates a structure of enabling. The disciplinization of communication is not repressive and forbidding, nor does it rule out specific actions. On the contrary, it assures a horizon of expectation; it sets up a framework; it creates reliability. Thus discipline empowers, making many things possible. Yes, it does on occasion forbid; forbidding is, after all, the converse of bidding. Non-adherence to what was said may be forbidden, for example. Nevertheless, this discipline belongs to the enabling of actions and *not* primarily to the forbiddance of actions.

Communication Makes Identity

The will not to be ordinary but to be someone special or at least appreciated by others is constitutive to the life of humans. This will to identity does not need any self-reflexive re-assurance. Rather, the will to identity is as certain as pain suffered, of which one is also certain. We not need to re-assure ourselves that we have an identity; we are certain that we do. This will is not in the sense of a person wanting one thing and not the other. It is the will to survive. It is fundamental. Without that will, you have nothing.

Identity wills recognition, i.e. feedback from the other. This need not necessarily be a positive reply, a consent to the being-thus (as understood by Honneth, 1994). Recognition means seeing that which there is to see. And when what is seen is the other, the alien, the disruptive, the misshapen, the sick, then even this reply is one that goes to create identity, recognize identity. In the struggle for recognition, there are not just merely winners and losers; it is more a matter of allocation across the entire field. Not everyone, not even most of us, become masters; most of us end up as servants or indeed as merchants, purveying to the masters and the servants what they need.

Communication always creates the identity of the persons communicating with one another (see Cooley, 1998; Mead, 1973, 1999). Initially, communication does this in a structural way, which means that all those who participate in the communication process are generally understood to be the owners of identity. This assumption, this doing as-if, consequently leads to the identity being created at the same time. Furthermore—and with a much greater role in this connection—the communication process always ascribes a particular identity to the counterpart as well as to the communicating person (Strauss, 1959). To the particularity of this identity belongs, on the one hand, the categorization of which social group we belong to (to the group of fathers, for instance) and, on the other, which ranked place we take in that group (or what kind of father one actually is—essential reading here Durkheim & Mauss, 1987).

We are evaluated and rated during the communication process, and above all, via communication. Communication therefore does not just say that we are somebody but also what we are to others and what we are to ourselves. Hence communication does not just say that we are a person, but also what person we are. Communication allocates us in our field of interaction. Some move up, some move down—with the others, namely the majority, somewhere in between. It is from this fundamental task of communication that the power of communication grows. For just as communication can ascribe a specific identity, so too can it deny that identity or re-interpret it and bring an entirely new identity to the light of day. To that extent, in the course of every communicative action, all the participants involved in the communication process commit their entire identity in order to reach their goals (Tomasello, 2008).

Each communication process has consequences—not just for the imminent problem that is to be solved with the help of communication. No, each communication process also has consequences for the subsequent communication process. With each act of communication, the persons involved contribute to the writing of a future open-ended history that will never really be deleted. Indeed, it will always influence the following communication processes (cf. Goffman, 2005). No communication process starts at point zero; each one picks up from the preceding one, continues it, modifies it or transforms it, even if the persons so involved have previously never communicated with one another.

Relationship as the Basis of Communicative Power

Power induces people to accede to impositions communicatively made, according to the argument advanced above. One form of power—the one that is of particular interest here, since it is the most frequent kind of everyday communication—is the power the arises from the special relationship that the persons communicating enter into with one other, create with one another. This form of power grows from the social relationship created *in and with the communication process* and from the motives constituted by the relationship. With the motives so validated, the action of each counterpart can be referred to the other's, because only specific reasons were permitted for this action. However, as Marcel Mauss asserts, a social relationship only comes about through exchange (Mauss, 1978, pp. 11–13). First of all, looks are given, taken and reciprocated; then (at which point the relationship is further consolidated) communicative actions are given, taken and reciprocated—before, finally, reasons for action are given, taken and reciprocated. If, at the beginning, the individuals' bodies 'speak' to one another and recognize whether they want to have something further to do with one another, the voices of the actors and their culture will also come into effect. A social relationship so originated binds all the participants, because a relationship does not just connect the consciousness of the participants with one another but also their identities (see also the concept of figuration—Elias, 2004, pp. 139–145).

This power does not precede the relationship; rather, a relationship such as power is successively built up in and with the communication process. This power only arises from the actual communication process. For communication is not just an interplay relating to the coordination of action but also, as we play the game, we learn what we think of the other. And the other learns what he or she thinks of us. Thus the relationship is built and if the participants succeed in becoming important to one another, then this will be a special social relationship, to which Brandom gives *deontic status* (Brandom, 1994, pp. 201–203).

This relationship results from the fact that the participants in the communication process voluntarily commit themselves to the validity of specific norms via their communicative action and conduct. In a kind of self-commitment, the persons communicating with one another take on rules during the communication process. And whether one is prepared to follow these rules will be revealed in the communication process itself. The communicative action thus creates *reliability*.

> It is our attitude toward a rule, our acknowledgment or recognition of moral necessity alone, that gives it a grip on us—not just in terms of its effect on our actual behavior, but in terms of our liability to assessment according to the rule that expresses that necessity. In this sense the norms that bind us rational creatures are instituted by our practical attitudes and activity. They are what we bring to the party (Brandom, 1994, p. 52).

If the persons communicating agree on what they should "bring to the party" as Brandom puts it, then they have reached a common status, the *deontic status*.[3] The deontic status, and this is an essential point here, cannot be established by one of the speakers alone: he or she is not able to set up a norm as obligatory. Both speakers must commit to following the norm since the commitment of the one does not necessarily mean that the other will follow suit. All the participants must play the same game of giving and demanding reasons. Otherwise, that deontic status

---

[3] *Deontic* (from greek *deon* = obligation, necessity) means concerning duties or obligations. By 'deontic status' Brandom means a communicatively acquired set of commitments and entitlements in a specific figuration. "Social practices are games in which each participant exhibits various deontic statuses—that is, commitments and entitlements—and each practically significant performance alters those statuses in some way" (Brandom, 1994, p. 166).

will not be achieved. When both do that, when both have become relevant to one another in their action, then they share a deontic status. The deontic status is therefore a specific kind of relationship, a special social relationship (as similarly argued by Taylor, 1994). The effectiveness is therefore based on voluntary recognition on the part of the actors concerned (Searle, 1997). To power belongs the consent to power by the counterpart.

Communicative Power Is Power Over Identity

It does exist—an everyday communicative power that can operate without commanding, threatening and corrupting. Indeed, communication usually succeeds in the everyday, namely without force or threat and corruption, but never without power. But it is a power that grows from the *relationship* of the actors to one another and from the significance of the other for one's own definition of identity. This power is ultimately based on recognition, on voluntariness.

As mentioned several times, communication creates identity and because identity is never really fixed, communication can define identity anew, damage it or—in the worst case—destroy it. "This vulnerability cannot simply be wished away" (Butler, 2006, p. 260). Or put more positively, the configurability of identity can never be put to rest. Identity is not something that one receives forever thanks to social interaction; rather, identity is assigned to you until further notice. For this reason, each identity always needs to be communicatively renewed via recognition, communication and exchange. However, this also implies that identity can at any time be attacked, injured and damaged via insult, belittlement and disrespect. Identity is never fixed—despite all attempts to make it so. Identity is always a provisional result if the current expression of societal communication processes also always have a history and create history in which everyone has his or her own place.

The world in which we live is symbolically structured, uncircumventably so, precisely because it produces by means of communication, that is symbolically, and precisely because it is also conveyed symbolically. Hence this world consists of a complicated, unevenly woven net of different senses, a net that is displayed in non-lingual and lingual signs and that embraces the entire world—i.e. the actor's exterior and interior.

The actor unfurls himself in and with communication and in it becomes visible to all and, as a result, also configurable. As he or she learns the techniques of competent communication (which always consist of a combination of words and deeds), the space of reasons is also conveyed, a space that makes it possible within the language and interaction community to separate the legitimate from the illegitimate reasons. Furthermore—and this is connected with the space of reasons—the actor is also provided with typical motives for symbolic as well as non-symbolic action. These motives are articulated in the action situation as typical *intentions* that drive actions on. Seen from this viewpoint, intentions are also of a social origin: they are internalized forms of the socially wishable, the expectable and the fearable. Intentions are socially configured and socially fixed ways of channeling biologically rooted desires into acceptable and recognizable forms. What we wish, what we feel, what we reject and what we recognize, it all has its social base and social roots.

Power—at least the power meant here, i.e. the power of relationship—does not arise from the relationship between words and humans, but rather always from the (social, not private) relationship from human to human—i.e., from social relationships and from the significance that relationships possess for the construction and maintenance of identity. It is always humans whose words have power, not words that have power. Of course, this involves a kind of *control* (cf. White, 2008, pp. 280–282), control over what something is worth to us because we are something or want to be something for ourselves and for others. It is always humans who in the course of communication processes commit to norms via communicative action and conduct. Communicative power is therefore not made by the word (see also Bischop, 2009) but made by humans—or to be more precise: by the interplay of humans, by their relationship. Without humans to back them up, words would have no power at all.

This definition of power binds power—that is, the ability to give others a motive for their actions—to the actors, even if the power over which the actors dispose through communication is in essence the power of sociality. This social power, though, always requires an actor if it is to be effective. Without actors, power is inevitably empty. And since this kind of power is bound to a certain degree to the actors, they can dispose of their acquired power only within limits. It can be stored and multiplied (one has, after all, a "reputation") and one can pass it on, again within limits—when one recommends someone and thus vouches for the person so recommended, for example. Nevertheless, communicative power is not the characteristic feature of a person as such, but rather it emerges again and again from the relationship that persons always enter into with one another.

This "unforced force" of communication is vouchsafed by social recognition. The closer the relationship between the persons communicating is—i.e. the more relevant they are to one another, the more power unfolds communicative action. One assigns identity, reliability and social competence to those who ensure that their words are followed by matching deeds. One knows with whom one is dealing; one shares the same world with them; one trusts them. One likes to be in their company, do business with them, and possibly even build a life together. But those whose words do not mean anything, since no-one takes heed of them, must first reminded and warned. If that remains ineffective, one soon begins to avoid them, deny them their identity, mark them out to others and exclude them.

Reliability and Good Reasons

The starting point for all is therefore the special, the relevant relationship that I, borrowing from Brandom, have called the "deontic relationship." If one wishes to unfold power, then it is the first thing that has to be created (please see Schröer, 2002, 2007, essential reading). It is the starting point from which a common space of good reasons can be set up or, alternatively, from which an already existing space of good reasons can be put into force. If that succeeds, then it is no longer just the *significant other*—sitting opposite us, with us—who communicates within the *relevant relationship*; but rather, within this relevant relationship, it is the generalized other, to whom both are committed, who speaks out of the significant other. For constitutive to the creation of a relationship is the establishment of a common generalized other—or formulated differently, the establishment of a space of good reasons that all participants consider meaningful, to which they therefore feel themselves voluntarily committed and to which they also expect others will voluntarily commit.

However, there are many spaces of good reasons in this world and there are equally many good reasons for preferring one space of good reasons over another. To that extent, the 'good reasons' do not really get us further. What is crucial is which space of good reasons can be established as the valid space for the participants. Yet even if there are many good reasons for (almost) everything, even if nobody really knows what the truth is (cf. Reinhard, 2006, p. 125), at least everybody knows what reliability is. For the meaning of truth—and perhaps less so, the meaning of truthfulness—is co-extensive with the meaning of reliability. One can perhaps no longer determine what is true, but one can surely determine what reliability is—if only because anyone can determine and test it and verify it within the common history.

Reliability, i.e. the certainty that the person communicating will have his or her words followed by deeds, is key to achieving communicative power. Reliability does *not* arise from a moral imperative or a philanthropic ethic but from a calculation: persons who are reliable are predictable and can be counted on. What they say has substance and we can orient ourselves to them (one way or another). For us, they have an identity because their words are identical with their deeds. However, we speak with those deemed unreliable increasingly less, or only about what is necessary, or not at all—at least about anything of relevance. Those denied reliability lose their "linguistic ability to act" (Kuch & Herrmann, 2007, p. 193). First what they say becomes meaningless and then they themselves: they are excluded, marginalized. Their identity suffers damage—but admittedly only in the eyes of the excluders, who still hear them but they no longer listen and pay even less heed to what they say. They may have words but their words lack the power to move others and provide others with a motive for their actions. To a certain degree, people who are denied reliability die a "communicative death" before the real one. What they say no longer induces anything.

The Social Significance of Communicative Power

The idea of communicative power conveyed here is not concerned with morals, let alone with improving the world through more honesty or more sincerity. Of the latter, sincerity, it may be doubted, and with good reason, whether it would lead to a better world anyway. Instead, it is a matter of improving communicative power in the sense of *making effective* communicative power, a power that in essence is a relationship power. Such effectivity means increasing the virtuosity of (non-violent and non-dominant) dealings between humans with one another and with oneself. After all, this communicative power helps people to better coordinate their behavior.

Communication is neither the place of reason and self-determination nor the means of bringing about reason and self-determination. Nor is it a tool for depicting the world in such a way that all reasonable people (have to) consent. Communication is in equal measure the reason and the unreason of utility. It is open to everything that can communicatively express itself and be reached—and is also open to naming the unreasonable reasonable.

What can communication do? On this side of violence and domination, it can suggest and provide other human actors with motives for their actions. It can do so because communicative action and conduct can create identity—one way or another. Which is why laying claim to communicative power for one's self always means the risk of compromising one's identity—if, admittedly, only on condition that the participants have entered into a relationship that all the participants consider important to one another.

Communicative power makes easier the coordination of human behavior, renders it *sustainable*. If one dispenses with the power of relationship in the communication process or if it becomes ineffective, then domination (and ultimately violence as well) must close the gap between the wish for action as communicated and its fulfillment. However, domination and violence do not just generate considerably greater social costs; they are also much more ineffective, since they always sow the seeds of resistance and revolt instead of consent and emulation.

How strong and how forceful is the power of communication? The answer to this question is not an easy one since it is often the case that the power of communication achieves little to nothing. Sometimes, though, its power is almost limitless. It always depends on how much communicative action counts among the participants. If, on the one hand, communicative action among the participants counts for little to nothing, then one cannot induce the other to clean up his or her room or close that window. If, on the other hand, communicative action is rated high, then one can indeed motivate others to skyjack planes or confess to crimes. Communicative power is able to inspire someone a whole life long or to bind him or her to the past a whole life long. Communicative power can set free and it can enchain. What is decisive for the power of communication are the relationship and the ensuing significance of the person communicating for the identity negotiation by the counterpart. If this significance is high enough, then the power of communication is stronger than domination and violence.

References

Berger, P., & Luckmann, T. (1969). *Die gesellschaftliche Konstruktion der Wirklichkeit: Eine Theorie der Wissenssoziologie* [The social construction of reality: A treatise on the sociology of knowledge]. Frankfurt am Main, Germany: Fischer.

Bischop, U. (2009). *Wortmacht – vom Wort gemacht? Wie gelingt es Kommunikation, Kraft zu entfalten?* [Power of word–made by the word? How can communication develop strength?]. Aachen, Germany: Shaker Media.

Bourdieu, P. (2005). *Was heißt Sprechen? Zur Ökonomie des sprachlichen Tausches* [What is speech? The economics of linguistic exchange]. Vienna, Austria: Braunmüller.

Brandom, R. B. (1994). *Making it explicit: Reasoning, representing, and discursive commitment.* Cambridge, MA: Harvard University Press.

Butler, J. (2006). *Hass spricht: Zur Politik des Performativen* [Hate speaks: Politics of the performative]. Berlin, Germany: Berlin Verlag.

Cooley, C. H. (1998). *On self and social organization.* Chicago: University of Chicago Press.

Durkheim, E., & Mauss, M. (1987). Über einige primitive Formen der Klassifikation [About some primitive forms of classification]. In E. Durkheim (Ed.), *Schriften zur Soziologie der Erkenntnis* (pp. 169–256). Frankfurt am Main, Germany: Suhrkamp.

Elias, N. (2004). *Was ist Soziologie?* [What is sociology?]. Weinheim, Germany: Juventa.

Foucault, M. (2004). *Geschichte der Gouvernementalität* [History of governmentality] Vol. 2. Frankfurt am Main, Germany: Suhrkamp.

Goffman, E. (2005). *Rede-Weisen: Formen der Kommunikation in sozialen Situationen* [Ways of Speech: Forms of communication in social situations]. Konstanz, Germany: UVK.

Habermas, J. (1981). *Theorie kommunikativen Handelns: Band 1* [Theory of communicative action: Vol. 1 Reason and the rationalization of society]. Frankfurt am Main, Germany: Suhrkamp.

Honneth, A. (1994). *Kampf um Anerkennung: Zur moralischen Grammatik sozialer Konflikte* [Struggle for recognition: The moral grammar of social conflicts]. Frankfurt am Main, Germany: Suhrkamp.

Keller, R., Knoblauch, H., & Reichertz, J. (Eds.). (2012). *Kommunikativer Konstruktivismus* [Communicative Contructivism]. Wiesbaden, Germany: Springer.

Kuch, H., & Herrmann, S. K. (2007). Symbolische Verletzbarkeit und sprachliche Gewalt [Symbolic vulnerability and linguistic violence]. In S. K. Herrmann, S. Krämer, & H. Kuch (Eds.), *Verletzende Worte: Die Grammatik sprachlicher Missachtung* (pp. 179–210). Bielefeld, Germany: transcript.

Luhmann, N. (2003). *Macht* [Power]. Stuttgart, Germany: Lucius und Lucius.

Mauss, M. (1978). *Soziologie und Anthropologie: Band 2* (E. Moldenhauer, H. Ritter, & A. Schmalfuß, Trans.) [Sociology and anthropology: Vol. 2.]. Berlin: Ullstein. Original version *Sociologie et anthropologie précédé d'une introduction à l'oevre de Marcel Mauss par Claude Lévi-Strauss.* Paris: Presses Universitaires de France, 1950.

Mead, G. H. (1973). *Geist, Identität und Gesellschaft* [Mind, self and society] Frankfurt am Main, Germany: Suhrkamp.

Mead, G. H. (1999). *Play, school, and society.* New York: Peter Lang.

Reichertz, J. (2009). *Kommunikationsmacht: Was ist Kommunikation und was vermag sie? Und weshalb vermag sie das?* [Power of communication: What is communication and what can it do? And why can it do it?]. Wiesbaden, Germany: VS-Verlag.

Reichertz, J. (2011). Commmunicative power is power over identity. *Communications: The European Journal of Communication Research, 36,* 147–168.

Reinhard, W. (2006). *Unsere Lügengesellschaft: Warum wir nicht bei der Wahrheit bleiben* [Our society of lies: Why we do not tell the truth]. Hamburg, Germany: Murmann.

Schröer, N. (2002). Was heißt hier „Sprechen"? Lässt sich Bourdieus "Ökonomie des sprachlichen Tausches" für eine Theorie kommunikativer Verständigung nutzen? [What means "speaking"? Can Bourdieu's "economics of linguistic exchange" be utilized for a theory of communicative understanding?]. *Österreichische Zeitschrift für Soziologie, 27*(3), 37–52.

Schröer, N. (2007). Geständnis gegen Beziehung [Confession against relationship]. In J. Reichertz & M. Schneider (Eds.), *Sozialgeschichte des Geständnisses: Zum Wandel der Geständniskultur* (pp. 195–228). Wiesbaden, Germany: VS Verlag.

Searle, J. R. (1979). *Sprechakte: Ein sprachphilosophischer Essay* [Speech Acts: An Essay in the Philosophy of Language]. Frankfurt am Main, Germany: Suhrkamp.

Searle, J. R. (1997). *Die Konstruktion der gesellschaftlichen Wirklichkeit* [The Construction of Social Reality]. Reinbek, Germany: Rowohlt.

Strauss, A. L. (1959). *Mirrors and masks. The search for identity*. Glencoe, IL: Free Press.

Taylor, C. (1994). *Quellen des Selbst: Die Entstehung der neuzeitlichen Identität* [Sources of the self: The making of the modern identity]. Frankfurt am Main, Germany: Suhrkamp.

Tomasello, M. (2008). *Origins of human communication*. Cambridge, MA: MIT Press.

Weber, M. (1972). *Wirtschaft und Gesellschaft* [Economy and society]. Tübingen, Germany: Mohr.

White, H. C. (2008). *Identity and control: A structural theory of social action*. Princeton, NJ: Princeton University Press.

The Klaus Tschira Stiftung

Physicist Dr. h.c. Dr.-Ing. E. h. Klaus Tschira established the German foundation Klaus Tschira Stiftung in 1995. The Klaus Tschira Stiftung is one of Europe's largest privately funded non-profit foundations. It promotes the advancement of natural sciences, mathematics, and computer science and strives to raise appreciation for these fields. The focal points of the Foundation are "Natural Science – Right from the Beginning", "Research" and "Science Communication". The Klaus Tschira Stiftung commitments begin in the kindergartens and continue at primary and secondary schools, universities and research facilities. The Foundation champions new methods of scientific knowledge transfer, and supports both development and intelligible presentation of research findings. The Klaus Tschira Stiftung pursues its objectives by conducting projects of its own, but also awards subsidies upon application and positive assessment. The Foundation has also founded its own affiliations that promote sustainability among the selected topics. Klaus Tschira's commitment to this objective was honored in 1999 with the "Deutscher Stifterpreis", the prize awarded by the National Association of German Foundations.

The Klaus Tschira Stiftung is located in Heidelberg with its head office in the Villa Bosch (Fig. 2), which used to be the residence of the Nobel laureate in chemistry Carl Bosch. www.klaus-tschira-stiftung.de

Fig. 1   Klaus Tschira (Photo and copyright Klaus Tschira Stiftung)

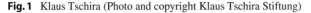

© Springer Netherlands 2015
P. Meusburger et al. (eds.), *Geographies of Knowledge and Power*,
Knowledge and Space 7, DOI 10.1007/978-94-017-9960-7

Fig. 2 The Villa Bosch (© Peter Meusburger)

Fig. 3 Participants of the symposium "Knowledge and Power" at the studio of the Villa Bosch in Heidelberg (© Thomas Bonn, Heidelberg)

Index

A

Ability, 14, 25, 29–32, 42, 52, 55, 56, 61, 76, 79, 80, 118, 126, 301, 315, 319, 327, 328
Aboriginal, 295–298, 304–306, 308–310
Absolute truth, 39–41
Absorptive capacity, 37
Abu Ghraib, 50
Abu Ishaq al-Shatibi, 137
Academic credibility, 10, 199
Academic disciplines, 11, 257
Academic fraud, 12
Academics, 10–12, 32, 35, 57, 58, 60, 61, 180, 189–195, 199, 209, 227, 235–237, 239, 240, 242, 243, 247–260, 266, 268, 269, 298
Academy, 10, 59, 207, 208, 210, 211, 215, 217, 222, 225, 230
Acquisition of power, 6, 24, 28, 33, 61, 190, 191, 236
Action-guiding norms, 27
Adams, J., 48, 155
Administration, 20, 23, 31, 50, 52, 55, 117, 178, 197, 223, 280
Aerial observation, 93, 94
Aerial reconnaissance, 7, 91, 94, 96, 97, 102, 105, 116
Afghanistan, 117, 118, 141, 289
Agency, 29, 33, 54, 79, 110, 163, 211, 224, 226, 274, 302
Aircraft, 7, 91, 92, 94, 96, 99–101, 104
Allied powers, 13, 283
Al-maslaha al-'amma, 8, 137
Althusser, L., 269–271

Note: If a term appears more than 100 times, only the first page number of its appearance in each chapter is listed.

American Association for the Advancement of Science, 166
American civil religion, 9, 155, 159
American expeditionary forces, 96
American Freedom House, 285–287, 289, 290
American hegemony, 11, 238, 244, 255, 256
American Revolution, 9, 147, 153
Amtsgewalt, 30
Anarchy, 240–242, 278, 301
Ancestor, 4, 19, 41, 42
Ancient Near East, 8, 123–131
Anglican Church, 152
Anglican tax system, 151
Anglophone journals, 11
Anglophonic hegemony, 11, 247
Anonymization, 51
Anti-landscape, 111
Apocalyptic view of history, 148, 149, 152–158
Arab, 136, 137, 286, 289
Argentina, 188, 287
Argument, 9, 45, 46, 48, 49, 55, 58–60, 85, 116, 135, 136, 143, 152, 154, 164, 170–178, 189, 191, 194, 216, 222, 225, 229, 237, 253, 256, 265, 272, 278, 301, 315, 316, 319, 321, 324
Armageddon, 9, 14, 149, 154, 170
Artillery, 7, 91, 93, 94, 96–99, 101, 104, 106, 107, 109, 111, 113–116
The Art of War, 38
Asia, 20, 55, 136, 287, 289
Assyria, 8, 20, 127
Astral signs, 8, 128
Astrology/astrologer, 19, 128, 130–131
Astronomical diaries, 128
Asymmetry, 33, 81
Attractiveness, 4, 33
Augur, 19
Australia, 257, 287, 290

© Springer Netherlands 2015
P. Meusburger et al. (eds.), *Geographies of Knowledge and Power*,
Knowledge and Space 7, DOI 10.1007/978-94-017-9960-7